# Awareness and Life

# AWARENESS and LIFE

## IN A NUTSHELL

by Dwight Harris

On one side is unfathomable expanse
The other an indescribable point
All Life is God arising from the center
And looking out in wonder
We are not growing older
Our future is growing shorter
We live inside our very own time
Each moment faster than the one before
The end comes shrinking like a wedge
And God is belief in a benevolent unknown

# AWARENESS and LIFE

The Ontological Distinction of Biological Life

by Dwight Harris

**Walls of Ivy Press** (c)

dwightharris.com/book/html

facebook.com/awarenessandlife/



Cover: Life confronts the blur, turning it into the rocks from which it came

*To Susanne*

*Life has boundaries*

*Love has none*

AWARENESS AND LIFE
The Ontological Distinction of Biological Life

TABLE OF CONTENTS

# PREFACE

## What exactly is organic matter?

When I was about 4 years old, I asked my mother if worms went to heaven. I was told some version of "don't be ridiculous." But why did I ask that question? I didn't wonder whether rocks went to heaven. I remember wondering about the phrase- "Bad dog." Nobody would say, "Bad rock. Don't do it again." Did the rock do anything? A worm, on the other hand, could take food from another worm. A tree could block the sun allowing small seedlings to die. What do we have in common with the rest of the living world? We are all alive. Life is alive. Is algae any less alive than a bat? Or a human?

A few years ago, I took a course given by F. Samuel Brainard, on the meaning of awareness. He divided "consciousness" into "conscious awareness" and what struck me, "unconscious awareness." Awareness. Of course. Living beings are aware. The rock is not.

Some say only humans are conscious. If a gorilla is conscious, goes the logic, or a dog, is a hamster conscious? If so, how far do we go? Why stop there? Is a rock conscious? I'd say, "Yes, stop there." A rock is not conscious. Why is it not evident to stop at life?

What does it feel like to be conscious? What does it feel like to be alive? Isn't it the same feeling?

When we look out at the night sky what do we see? Galaxies? No. Stars? No. We see shining lights. Do we see a moon? We see something with lines in a shiny object, and a shape that changes from a sliver to a full circle every 30 days or so. Do we see a sun? No. We see a circular fire, or maybe just a bright light, crossing from horizon to horizon followed by about an equal time span of darkness repeating once every 24 hours. Do we see an hour? No. People have divided the time that the sun crosses and disappears and then reappears by 24 time sections. What did ancient man see? A god? Dead people in the stars? Stories in the stars written by gods?

Only in the 20th century did humans begin to understand how the universe had a beginning. Before that it was assumed it went back forever. Only in the 19th century did people begin to see evolution. Before that it was assumed man, animals, plants and the world were created pretty much at once. This was the thought, at least, in much of the Western world. It was assumed we went someplace else to live after death.

We supposedly now see galaxies and atoms. Except that I don't see either. I'm told they exist. Is the real world the universe described by current science, or is the world what we see and hear? Before Hubble's discovery of an expanding universe, Einstein thought the universe existed forever in its current state. Was he wrong? Or was that reality at that time? What will the universe look like to scientists, and all humans for that matter, in 500 years? In 10,000? Will they laugh at the ancient, ignorant belief in stars? In molecules? Are we learning more about the Universe, or are we just filling in details?

What do cats see? What do cows see? What does a tree see? What do bacteria see? What is the true world? A dog smells about as many things as we see. A dolphin uses hearing like we use sight. A bat can hear in the dark maybe even more than we can see or hear in the light. Are humans more "intelligent" than other animals?

What is the world that all life sees?

What does the world really look like? If you could put all the living beings in the world into one mind, what would the world look like? Imagine being able to sense the world from every species' point of view. If you could be a bacterium for one day, would you have any idea what you were doing? Could you survive in a bacterial world? Is the human world better or more accurate or true than the bacterial world? Or is it merely more imaginative. Is there something true for all living beings? Undefined by any single living being or species but confronted by every one? What is there without words? What is there before interest, before need? What is essential that cannot be ignored?

I asked my mother when I was about 5, "Why is the school bus taking so long?" My mother said, "Don't stare at the clock. The time will go faster." I thought, "Wow. We can control time."

After college in the late sixties, I studied philosophy at the Graduate Faculty of the New School in New York City for a few degree-less years. One course that particularly interested me was the "Phenomenology of Biology," taught by Hans Jonas. Jonas' basic tenant is that life as such is defined as the constancy of form through a change of matter. A rock, or any material, that is, any non-living thing, has a shape that is at one, that is, consistent in time, with its material. On the other hand, we all have heard that every cell in our body is replaced over a seven year span. Yet we are the same.

Of course, anybody who studies philosophy wonders what consciousness is. Phenomenology, which was then, and I think still is, the reigning school of thought at the Graduate Faculty, describes consciousness, but didn't to my satisfaction, explain it.

I also took an aesthetics course in which I wrote a paper on improvisational music. Musicians are actually "playing" one another. I was, and still am, fascinated by the concept of communication. How is it and what is it we communicate? Is consciousness communicated?

I felt that there was something about life, that is, biological life, that is not explained by DNA and chemistry. DNA is necessary, perhaps, but not sufficient.

When I took Dr. Brainard's course I thought, as I mentioned above, this is it, life is aware. Consciousness and life can be explained in terms of one another. If life evolved, why not consciousness?

Of the many books and articles I have researched on the subject, three are most significant: Robert Rosen, *"Life Itself,"* Hans Jonas, "*The Phenomenon of Life*" and Peter Ulric Tse, *The Neural Basis of Free Will: Criterial Causation.* These three authors, together with Aristotle's Four Causes, form the basis of my explanation of what Life and Consciousness are. They are both aware. They are both alive.

What is the audience for this book? Anybody who is interested in life and consciousness. This book is my attempt to explain them in the best way I can. In a way, it is my doctoral dissertation, fifty years too late, but with a lifetime of thinking about it.

# INTRODUCTION

*We have to remember that what we observe is not nature in itself, but nature exposed to our method of questioning*
Werner Heisenberg

Scientists and scientific thinkers have often concluded that there are two areas beyond physical description that science cannot explain: Life and Consciousness. A physical brain does not explain consciousness. A DNA molecule does not explain a living cell.

Is everything that is alive conscious, or at least, aware? Is everything that is aware alive? If awareness and life were the same thing, perhaps that one thing could be explained.

I explain life and consciousness without resorting to or invoking spiritual intervention. God is not an explanation. What I say in the poem about God being Life, is my belief, but as a belief it is not defensible and I have no interest in defending it. I will say, however, that it is not inconsistent with what I have to say about life and consciousness. Since God is not an explanation, maybe life and awareness are. Maybe they explain themselves.

The second part of the poem, about living inside time, actually living time, is essential to my overall thesis. But first, what exactly is it that all life confronts? What does all life have in common? The basic structure of existence is the only something that is given to us. The given is what we can't go beyond. It is a paradox. Everything else is how our organism, indeed all organisms, organize and come to life with this paradox.

I describe it as the smallest, distinct part and the largest, empty whole and the necessary but impossible choice as to which comes first. Existence presents us with the fundamental paradox of infinity: The infinite (as far as we can know or imagine) totality of distinct entities and the infinite (as far as we can know or imagine), endless continuum. In other words, individual things and the in between. Which is more fundamental? It is the necessary choice that cannot be made yet must be made. It is unavoidable. It is choice itself.

The two infinities reach out to one another. The smallest imaginable point leads to infinite mutual externality, point after point, never reaching the largest imaginable whole, which in turn leads infinitely into itself toward the point. Large and small mutually imply one another.

**"Both the largest and the smallest structures of the universe are astonishingly simple. It is here that we find the two 'standard models,' of particle physics and cosmology."**- Robbert Dijkgraff[1]

Whether it is motion, point and distance, the need for limit in calculus or current measurement vs. observation disputes, from Zeno to Quantum/General Relativity unification, this paradox is central. The history of Western human thought from the ancient Greek philosophers to current quantum field theory has been essentially dealing with the same dichotomy. From Zeno's paradox of time and distance to Heisenberg's Uncertainty Principle, how one gets from here to there through infinitely divisible here-and-now's remains the question. Quantum electromagnetic particle forces rise up to, but not quite reaching, Einstein's spacetime gravitational continuum.

Though Descartes divided the world into internal consciousness and external matter, Science chose the road of mathematically predicted matter. What has been unveiled is a remarkably coherent and vastly detailed universe, but so far culminating in the mathematical deadlock between Quantum Mechanics and General Relativity, thereby preventing a complete Theory of Everything. Science has come close in different ways to unify the two theories. Stephen Hawking is noted for his theory that black holes, which are predicted by General Relativity, dissipate at the edges via quantum randomness, rather than pull themselves down to nothing with infinite gravity. The Bose particle is also theorized to unite quantum mechanics and gravity, but quantum gravity, if proven, merely equates mathematically, the smallest known and the largest known. The small and the large are still there. And besides that, exactly where in this Everything equation are the life and consciousness variables?

Consciousness, which is what we generally call our waking experience, although technically we are "conscious" when asleep, and not in a coma, is a specific kind of awareness and must be explained by first explaining awareness.

Some ascribe to what we might call "The Creationist Theory of Consciousness." Consciousness simply showed up at some point. But if our bodies evolved from one-celled organisms, why not consciousness? If human consciousness evolved from a one-celled living being, then the world that is real to e.g., a bacterium, has evolved to a world that is real to humans. What is transcendently real to Life itself? Is biology the way humans organize the world, or are humans the way biology organizes the world?

Time is usually thought of as the 4th dimension of four dimensional spacetime. The smallest measurable length allowed by quantum mechanics is called the Planck Length ($1.6\times10^{-35}$ meters). The Planck Time, beneath which time is said to be meaningless, is the time it takes for light to travel across the Planck length. Physicists claim there is no such thing as now, because what we confront is a series of photon hits, each occurring at and for a separate miniscule undeterminable time.

Yet between the photons, beneath Planck Time, somehow in some way we are living through it. Imagine a 3D sphere, a 2D plane, a 1D line and a point. That point has no spatial dimension. If we are living at that point, it is in pure time. Time in itself has no space. It is the 0 dimension. Pure time is an inside-out.

How is the paradoxical separateness in reality understood by a Cartesian consciousness separate from matter? How is the blur of mutual implication resolved?

Georg W. F. Hegel (1770-1831), begins his *Encyclopedia of the Philosophical Sciences,* not with consciousness but with what consciousness does, that is, it begins with what he saw as the most general, fundamental idea that can be conceived, Being. Which of course, is Nothing, no Thing. Being and Nothing are one and the same. Yet at the same time they are not the same. For Hegel, the paradox is expressed by the back and forth of Becoming. This "dialectic" is the inherent logic of reality.

Existential reaction to the Hegelian system, sees the problem of consciousness as one of common experience. Jean-Paul Sartre (1905-1980), if I understand him, argues that Nothing is not one side in the relation of Being and Nothing, but is the nothingness at the center of current experience. Nothing comes before distinction.

Alfred North Whitehead (1861-1947), on the other hand, sees philosophy from Descartes to the early Twentieth Century, as attempts to apply a subject/predicate division on what is truly organic existence. The relation between subject and predicate is itself, for Whitehead, primary. Hans Jonas (1903-1993), in his important philosophical work, *The Phenomenon of Life: Toward a Philosophical Biology*, takes Whitehead's insight and sees it manifested in the difference between life and non-life. Life is a constancy of form through changing matter.

Food and air passes through us every day, exchanging new matter for old matter, until it is all replaced, yet at the end of a year, we pretty much look the same. A rock, however, is a constancy of matter and form.

Matter causation reduces either to a before pushing the after, or an uncaused randomness. But Aristotle, long ago, insisted on another type of causation, Final Cause, which is purpose. But purpose invokes a causation from ahead, from the future. Action caused by a goal. Scientifically off limits. Not allowed by deterministic material reductionism.

How might life cause or be caused by its own future? Life, the constancy of form, is not just matter flowing through form, but form flowing through matter. Life is at the nexus. Matter alone would have either a future without a past (randomness) or a past without a future (determinism), or both. Life is directed forward toward that which is not given.

If life did not arise from matter, perhaps matter arose from life. We are aware, and only aware, through and within our living bodies. All living beings are aware. Anything that is aware is alive. There is no escape. This logical, philosophical description, however, does have its own confinement. If God did not create life or the universe, what did a Creator create? What we are left with is wonder. Wonder at the universe unveiled by our awareness of it. Perhaps Creation is philosophical.

## PHILOSOPHICAL CREATION

1. All that is, is what is transcendentally given to all life.
2. What is given is small and large.
3. The mutual implication of small and large is the paradoxical choice that must be made but cannot be made.
4. This is choice itself.
5. Without choice, there is only infinite back and forth, directionless change.
6. Choice is the beginning of Life. Life chose Life, the choice of direction.
7. Time is change with direction. The direction is forward.
8. The choice of direction pulls time out of change by maintaining whole form through changing parts of matter.
9. Time is at the point where matter changing through form and form maintaining through matter intersect.
10. As time, the small and large manifest as the infinite dynamic limitations of mutually implying part and whole.
11. The temporal point between form and matter opens forward toward the non-given future, and therefore, outward, through the in-out dimension of time into the past, thereby spatializing time.
12. Awareness is looking at the past as if it were the future.

Though Life may not have a purpose beyond changing forward, life, as lived through individual living beings, must survive to live. Life lived is purposeful.

The book is divided into Parts, Chapters and Sections.

Part I, "Matter," traces the philosophical and scientific progression that mirrors the evolving separation of form from matter.

In Part II, "Form," this progression culminates in the concept of information when form frees itself from matter, as thermodynamics looks back on itself, becoming alive in the awareness of information from the inside/outside of its own boundary.

With Part III, "Awareness in Form," the forms of awareness as such, including consciousness and communication, become applied to the Matter from which Form has separated.

With Part IV, "Awareness in Matter," the forms of living matter such as the brain, animals, plants and fungi, are the temporal structures through which awareness operates.

Part V, "Living Awareness" is the philosophical basis of awareness that starts from consciousness itself. Here, I tackle time, agency, self and awareness directly.

A thorough perusal of science writing is used to underscore and explain my thesis. To that end, I quote liberally from many sources, both scientific and speculative. This book is meant to be philosophical.

To aid the reader in understanding what I'm saying, I've placed in bold, key phrases, and summarized the point following each section. Epigraphs at the beginning of each chapter are also informative.

# I. MATTER

## A. DUALISM

*Behind anything that can be experienced there is a something that our mind cannot grasp and whose beauty and sublimity reaches us only indirectly and as a feeble reflection*
Albert Einstein[1]

### COMMUNICATION

Socrates never wrote anything. For him, philosophy was a living, continuous argument, an active communication. The highest purpose and joy of life is to argue with others about the meaning of ideas. Instead of each participant remaining in defense of a position, ideas are mutually built. Truth is the arriving together in agreement- only to begin again another day with another opponent or another starting point or question. It is said that Socrates' idea of heaven was a place where he could just go on arguing.

He considered the written word to be a dead word. Conclusions previously reached and retrieved again are out of context, and no longer in sync with the current discussion. He also believed, as some do today about internet searches, that writing is deleterious to the active mind; it weakens thinking and memory. The word "philosophy," of course, means "love of wisdom," not "the study of wisdom." **The love of truth is a living truth.**

Plato (428-348 B.C.), Socrates' student, attempted compromise by writing dialogues. His protagonist was often Socrates himself, which is how we mostly know him. But Plato's dialogues, though written in the form of Socratic argument, and not what he would negatively term "rhetoric," arrived, sometimes quite tortuously, at true ideas.

The truth could be reached not simply within argument but through arguing to conclusions. The True and the Good dwell at the top of a realm of Ideas, or Forms, reached by reason, mathematics and geometry alone. The circular table, for example, has its source in the idea of the perfect circle, which can only be approximated here on earth, cut off by the limitations of ever changing sensory observations. Man lives in something analogous to a cave, facing the wall able to only see shadows lit by fire, of the true forms behind their backs. In the "Timaeus," understanding through mathematical argument reveals the true structure of the cosmos. It is the means for determining the reliable vs. the unreliable.

Socrates (469-399 B.C.) was not the first Greek philosopher. There were a number of "pre-Socratics," and of relevant mention here- Heraclitus

(535- 475 B.C.) and Parmenides (515-485? B.C.). Both of them believed the world could be reduced to one thing. For Heraclitus it is change itself, for Parmenides it is the unchangeable itself. For Plato, the latter is Truth, the former is not, but each has its place.

**The Point:** The blur of pure change begins to reveal itself to human thinking as change and not-change, reflected and understood in the immediacy of self-other communication.

## CAUSE

Aristotle (384-322 B.C.), Plato's student, wondered why things did change. Within a world of appearances, he sought causal explanations not in abstract ideas but in sense perception. For things to change he reasoned, there must be a cause- a "why" a thing came to be as it is. Aristotle concluded that there are four types of cause:

1. Material Cause. Things change the way they do because of what they are made of, the wood of a log or a chair, for example.
2. Formal Cause. Things change because of their shape, arrangement, appearance or plan like a design for a chair. Interestingly, number as such can be such a cause. Things change by being part of a collection.
3. Efficient Cause. Things can change because of their interaction with another thing, like a chair bumping into a table.
4. Final Cause. Things can change because of their purpose. The chair is the way it is because it is made to be sat on. The purpose of a seed is a plant.

Change is fully explained by all four causes.

The realm of Platonic Ideas was largely replaced by Aristotle's concept of the First Cause, that is, the First Mover. Each thing being caused by a preceding thing necessitated an eventual beginning. Top down is instead a start point. **Aristotle combined Plato's Forms with time**. To be fair, Aristotle did see a problem with Final Cause existing in the future, and so thought of it as inherent in that which is caused.

**The Point:** Change occurs in time. There is a before. What changes are specific things. Things change for a reason. Asking "why" things change reveals their cause. In order for them to be, they must have matter, form, another thing (efficiency) and purpose. What is before a specific, distinct thing is an infinite series of distinct things. Something must cause the series at the beginning.

## BEGINNING OF MODERN SCIENCE

2000 years of pure speculation more or less ends, and modern science more or less begins, with Galileo Galilei (1564-1642), who first combined observation, hypothesis, mathematical deduction and confirmatory experiment. By measuring the time a pendulum took to sway from side to side, and observing that the width of the swing was not dependent of or correlated with the time span of the swing, **he separated motion from distance.**[2]

Another of Galileo's fundamental insights is that motion only makes sense relative to the motion of something else. In his most famous experiment (the Tower of Pisa story apparently is not literally true), he dropped two objects of different weight (not mass- not yet discovered) and found they fell at a uniformly accelerating motion proportional to time, and contrary to Aristotle, that **weight and speed are not mutually implied**.

**The Point:** Quantitative measurement is added to Aristotle's qualitative observations. Even without purpose, distinct things still have an after. The same thing can move at varying rates over a distance. These things can be observed, anticipated and quantitatively measured to demonstrate the validity of the anticipation. What is anticipated about the observed is generalized as a quality of the thing.

## DESCARTES

But things moving and their causes (with God often inserted as First Cause), remained the basic construction in Western thought until Rene Descartes (1596-1650), who saw a need for a source of knowledge of things, a touchstone of rational certainty. He discovered this in his own thinking, and thus brought forth the concept of consciousness, the Cogito, the "I think therefore I am."

For Descartes, the world therefore, consists of three separate substances. Beside God, there is Consciousness and Extension. **There is an immaterial inner, and a material outer.** Thus was born the "Cartesian Problematic," the general platform upon which Western thought has dealt with ever since. The fundamental question becomes, "What's the relation between the internal reality of our mind and the external reality?"[3] In general the answers are: 1. We and the world are immaterial ideas. 2. There is nothing but matter. 3. We and/or everything are both in some combination.

From his starting point of conscious existence, Descartes believed in deducing knowledge mathematically from the simplest surety towards complexity via the way of geometric proof, with given axioms leading to theorems. He saw extension visually, inventing the geometric three-dimensional coordinate system as a way to visualize algebra. Using the intersection of perpendicular x and y axis' (e.g., in a two dimensional plane), the "ordered" pair (x, y) can represent any point, assuming equally sized units.

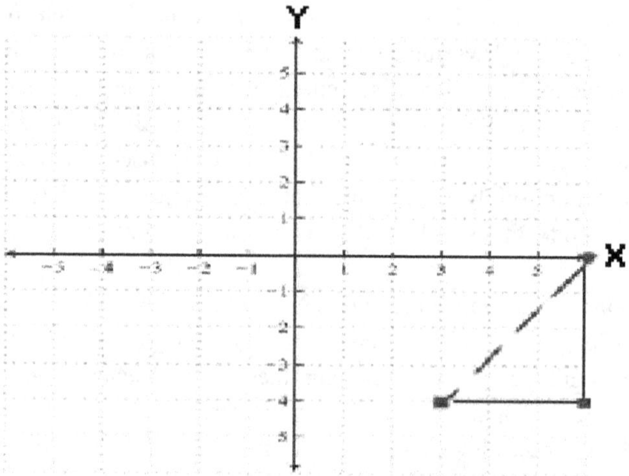


Fig. 1. Cartesian Coordinates.

To use a very simple example, one point can be the pair (6,0), which would be located on the graph as 6 units to the right (positive) side of the y axis and 0 units from the x axis, that is, directly on the x axis. Another point might be (3,-4) located 3 units to the right of the y axis and 4 units below (negative) the x axis. The distance or straight line between the two points can then be easily calculated by imagining it being the hypotenuse of a right triangle whose other two sides would intersect at (6,-4). The vertical (y) line would be a 4 unit change from 0 and the horizontal (x) line would be a 3 unit change (subtracting 3 from 6). The original line is then calculated using the Pythagorean theorem for the length of the hypotenuse, the equation $d^2 = x^2 + y^2$, that is, $5^2 = 3^2 + 4^2$ or $25 = 9 + 16$.

Note that the origin (0,0) at the intersection of x and y axis' is located at an arbitrary extensive point and not at the point of consciousness. For Descartes, such is the split, that even the body, including the eye and brain, is part of externality. Animals are not conscious at all. They cannot feel pain. They are machines. And with the body being completely matter, the body is actually more rationally structured than the mind.[4] This can lead, for example, for the mind to be seen as nothing but an infinite regression of the "ghost in the machine," a body inside a body, etc. In fact, with matter as pure extension, "structure determines and explains function"[5] or behavior. Movement is change within coordinates. Euclidean lengths are not just relations but relationships between numbers.

**The Point:** The observing is simple, internal consciousness. The observed are complex, external things whose different changed positions in relation to one another can be measured geometrically.

## QUANTITY

Mathematics is the science of quantity. And it is through the visualization of quantity, that starting with Descartes, the world came to be constructed. According to a widely accepted scientific view, "nature is the realization of the simplest conceivable mathematical ideas."[6] In place of Aristotle's qualitative categories of changing things, modern thinking ascribes quantitative descriptions to matter in motion.

Quantity is in turn divided into two types: Magnitude and Multitude. Magnitude deals with abstract mathematical concepts like ratios and equality, while multitude deals with the relationships and structure among different quantities. Magnitude refers to size or "how much," while multitude means "how many."[7]

Continuing the definition from Wikipedia, quantity is "the cornerstone of modern physical science.... **Physics renders the abstract qualities of material entities into physical quantities.**"[8] Equations are the mathematical tool for describing relations between entities, physical as well as the purely mathematical. Dimensionless constants in the universe are the ratios between measurements. As such, they are fundamental.[9] **The unimaginable vastness between the parts contains the unknown origin of relations.**

Unlike the geometrically described change from point to point, objects in space themselves move. The motion of objects are either mathematically "scalars," that is, described completely by magnitude, or "vectors," consisting of both a magnitude and a direction.[10]

Magnitude and multitude "imply the principle distinction between continuity and discontinuity,"[11] that is, between the continuum and the distinctly separate. When things move, they move in and through space. Yet "most of classical mechanics deals with things that change smoothly-continuously...as time changes continuously...."[12]

Do objects move in and through a space and time that is ultimately discrete, that is, a series of infinitesimally small points, or a continuum, that is, an infinite smoothness? Again, this is the fundamental paradox. Is everything one or the other? It can't be both. It can't just be either.

**The Point:** Distinct things move in continuous time and space. Things have measurable relations within their parts and with other things. Relations between measurements are apart from the physical universe.

## PARADOX

There is another pre-Socratic, Zeno (490-430 BC), who, unlike Aristotle's "change", concentrated more on the problem of motion itself, that is, things moving within the change. (Aristotle, in his "Physics," does deal with Zeno, as does Plato). But Zeno is famous for his "Paradoxes of Motion." There are nine that are known. I will describe three.

The first is the "Dichotomy Paradox." Before one can cross the room, say, one must first go half-way. Before one can go half-way, one must go quarter-way, or in other words, half-way to half-way. Before one can go quarter-way, one must go half-way to that point. At that rate one can never get anywhere. **Motion is impossible in that distance is impossible, because distance itself is infinitely small.**

The second is "Achilles and the Tortoise." Here the slower tortoise is given a ten meter head start, because the faster Achilles has no problem in reaching that point faster than the tortoise would have. Unfortunately, when Achilles reaches that point, the turtle has moved on, though slowly. Achilles can then reach that second point, again faster than the turtle had, but again, the turtle has moved on. Achilles, therefore, can never catch up. Although the second object has greater speed or velocity than the first object, no matter how much the distance between the two, the total distance can never be crossed. **The distance between two moving objects has "an infinite number of distances, each greater than zero..[and ] would take an infinite amount of time."**[13]

The third is the "Arrow Paradox." Every instant the arrow is in a particular place or position. Does it move within the instant? Inside of it?

Around it? If "time is discrete [then] time leaps from one separated point to another....[If] motion happens in that instantaneous leap...[then] **time is continuous, [and] therefore movement is impossible. But since movement is possible, time is discrete.**"[14] Or motion is discrete. Later on, Alfred North Whitehead (1861–1947) concluded that Zeno's paradoxes mean that "motion is atomistic." Henri Bergson (1859–1941) concluded that they prove motion is continuous.[15]

**The Point:** The fundamental argument rooted in the paradoxical split in infinity between the continuum and the infinite discrete series is applied to discrete positions and continual distance, discrete moments and continual time, and discrete movements and continual movement.

## B. NEWTON

*There is something essential about the **Now***
*which is just outside the realm of science*
Albert Einstein

### CALCULUS

Isaac Newton (1642-1727), realized that Cartesian Coordinate Geometry was not up to the task of calculating motion. To accomplish this, he invented the Calculus (as did the philosopher, Gottfried Leibniz (1646-1716), who took the subject in an entirely different direction).

The difference between rest and motion can be thought of as positions at different moments. In a continuum there is no next point and no next moment, there is always another in between any two. Each unreachable moment has a coinciding unreachable spatial point.[1]

The infinitesimal calculus, "the method of indivisibles, [is] founded on the paradoxical doctrine of the infinitely small...."[2] Motion would be "infinities of points per second."[3] But since the rate of motion over any distance can be different at any one point, rates themselves can be compared. Newton's solution is to begin not with distance or motion, but with velocity.

Calculus divides into the derivative and the integral. The concept of the limit, the approximation of the infinitesimal, is fundamental to both. I will try to present this conceptually. To find out how to actually do calculus, you'll have to look elsewhere.

A derivative measures how much one quantity changes in response to changes in some other quantity.[4] Fixed or standard velocity is an average distance divided by time, e.g. miles per hour. This assumes one measured distance of space per measured length of time. But velocity is not constant. Things slow down and things speed up. At a slower speed, the object (or moving point) is positioned at some of the spatial infinitesimals for at least one instantaneous moment. At a faster speed, the object jumps through more spatial infinitesimals. As the speed changes, it must be calculated at each instant. The derivative of the position of a moving object with respect to time is the object's instantaneous velocity. In the continuum of space there is a velocity at every instant.[5]

**Calculus enables us to measure a rate of change which is infinitesimally small.** This is accomplished with the concept of the limit (although in Newton's time, it was not formalized as such). The limit is the method of approaching the infinitesimal by taking smaller distances traveled in shorter intervals and dividing by those shorter times, or taking the limit as the time interval approaches zero. "The limit means that there is some number (the limit) such that the terms of the sequence become and remain arbitrarily close to that value...,"[6] in other words, an extremely close approximation, the closer the more precise. "The rate of change with respect to time [is] the derivative defined as the limit of the average velocity during decreasing non-zero intervals of time."[7]

Integral calculus is the opposite. It is the sum of many infinitesimal quantities.[8] Integration sees "time as a series of instants, each with its own speed."[9] Add up the distance moved on an instant by instant basis. Distance is an infinite sum, for example, .99, .999, .9999 added together approaches the limit of 1. In general, the integral defines the sum of an infinite series. Integrals allow the adding or multiplying of changing numbers. With integrals we usually need to know where to start, the constant of integration, that is, the initial condition.

The calculus is not confined to velocity and distance. A general concept is the function, which is not just an equation relating two numbers, but a relating of two or more sets of numbers or two or more sets of equations. A derivative measures the change in the function as its input changes. It is "the ratio of the infinitesimal change of the output over the infinitesimal change of the input producing the change of the output."[10] Velocity is an example of a vector (magnitude and direction). When velocity changes in either magnitude or direction it is called acceleration.[11] Acceleration is the derivative of velocity.

In addition to being an arithmetic rate of change, the derivative can be thought in geometrical terms as a slope of a curve. The calculus can be applied visually. But the physical phenomena presented is still an idealization, that is, describing an infinitesimal point particle moving in a continuous trajectory. The assumption of calculus is that reality is infinitely divisible.[12] The assumption, therefore, that mathematics is a perfect reflection of the physical world or that numbers correspond to physical objects breaks down at the infinitely small.[13]

If we look at a normal size object, differentiation (the process of finding a derivative) is breaking the object into pieces and integration is weighing and counting the pieces,[14] but there is no way of calculating the original shape. I will talk more about this in the chapter on Thermodynamics. But at this point we can see the elimination of Aristotle's Formal Cause. "Morphogenesis," as it is also called,[15] is not explained by reductionist physics.

**The Point:** Motion and distance are replaced by velocity and position as the touchstones of measurement. The thing's velocity and position are both infinitesimal and can only be approximated. Differences in velocities is the rate of changing velocity, which is acceleration. The original shape of the object cannot be calculated. Formal cause is eliminated.

## LAWS OF MOTION

Newton, using calculus, not only found a way to describe motion, he found a way to predict it.[16] Starting with one velocity and its change, the velocity and position can be derived at the end of a length of time.[17]Along a straight line or a curved trajectory, the motion between two points of time are minimized, each instant determining the next. The future is determined by a differential equation at each instant.[18] Since acceleration is the derivative of immediate velocity, **prediction derives not from the rate in time, but the rate of change in time.**

Following Descartes' method of understanding phenomena, that is, beginning as simply as possible and proceeding methodically from there,[19] Newton first considered a body at rest, one with no motion. Perhaps in recalling Aristotle, and not finding it in Descartes, he also realized that in order for an object to move, that movement must have a cause, by which he conceived a "force." **He essentially "mathematized Aristotle's 'efficient cause.'"** [20] From there, he produced his three Laws of Motion.

1. "An object at rest will remain at rest unless acted on by an unbalanced force. An object in motion continues in motion with the same speed and in the same direction unless acted upon by an unbalanced force."

In other words, an object either is at rest or moves at a constant velocity, unless acted upon by an external force. Velocity, of course, is the derivative of movement in time and time is a given. So, causal process in time is described by, if not replaced, with timeless logic. Nature consists of "nothing but particles with timeless properties, whose motions and interactions are determined by timeless laws...." [21]

An object then is nothing but <-Inertia Velocity Momentum->. An object is completely defined by the difficulty in starting it to move (inertia) and the difficulty in stopping/changing its movement (momentum) and its velocity at any infinitesimal moment.

2. "Acceleration is produced when a force acts on a mass. The greater the mass (of the object being accelerated) the greater the amount of force needed (to accelerate the object)."

A body's velocity will change (accelerate) in proportion to the strength of the action exerted on it.[22] The change in the rate of velocity can be faster or slower but either way, "acceleration is proportional to the force."[23] Mass is the measure of inertia, or how difficult it is to accelerate an object. Momentum, which is mass times velocity, is a measure of how difficult it is to stop an object in motion. Force and acceleration both have magnitude and direction. Mass has magnitude but in relation to something else.

Force -> <-Inertia Changed Velocity Direction/Acceleration Momentum ->

Objects, of course, do not normally exist alone or in pairs but in multiple numbers. The total force on any one object, as well as the rate of change of the momentum, is the sum of the forces exerted on it by all the other objects.[24]

Perhaps a more accurate statement of the Second Law would be: "The acceleration of a body is directly proportional to, and in the same direction as, the net force acting on the body, and inversely proportional to its mass. Thus, F = ma, where F is the net force acting on the object, m is the mass of the object and a is the acceleration of the object."[25]

Direct proportion is represented in the relation between two sides of the equation (F = ma). Inverse proportion is indicated by the relation of quantities on the same side of the equation (m & a). An increase in force means how much force is required to change the momentum.

3. "For every action there is an equal and opposite re-action."

When an object is forced by a second object, the first object causes the second object to not only accelerate equally but do so in the opposite direction. This means that in any interaction, momentum is conserved. Since the acceleration of any object is the sum of the forces from all other objects, including its own, in a theoretically isolated system, this means the total momentum never changes.[26] Forces, however, are determined in part by the location of the different objects.[27] Various positions mean various distances. Therefore, in the same time period, velocities may vary. **Mass, therefore, must somehow be separate from velocity**.

For Newton, the system is not isolated. In fact, the touchstone of the Cartesian system of arbitrarily positioned 3-dimensional perpendicularly crossing coordinates, is replaced by reference to absolute space and time. The meaning of a point position's location, instantaneous velocity and the implied infinite circularity, is stopped by the assumed reference of absolute space and time.[28] Matter changes in an unchanging Space.[29] Newton's laws, therefore, allow all changes and positions as a whole to be simultaneously shifted.[30] All locations are equivalent as there is no preferred direction in empty space. "The universe is symmetrical."[31]

Newton universalized force in terms of mathematical prediction, for earthly objects like billiard balls hitting billiard balls, as well as the movement of planets and falling bodies or gravity. The same force is action at physical contact and at a distance. Forces change the state of motion of an object. But "a force is always a force on something, there are no free-floating forces."[32] What a force is in itself is not explained. Newton himself, admitted that force at a distance is an absurdity, although mathematically, "Efficient Cause" becomes disembodied force on the "Material Cause" of disembodied inertia.

If understanding nature is to understand how things are caused, causes are now seen as requiring laws of nature to logically entail an effect.[33] Cause, the reason for change, is now explained by making correct predictions which through mathematical deduction entail, that is, necessarily determine, future motions or trajectories.[34] The prediction of the derivative is then deduced by the integral. The cause and effect are deterministic. There is no room for spontaneous change, as cause and effect are separated by a past-to-future flow of time, albeit limited by its infinitesimal nature. Aristotle's "Final Causation," that is, purpose, to the extent that it exists in the future, a cause from ahead, is therefore prohibited. The job of Newtonian physics is to predict the future. "F=ma predicts [future] events completely. If the position and velocity are measured at a given time (t), the rules of calculus compute its position for a subsequent time, t + dt." [35] Gravitational force is the same as the interaction of billiard balls, but how this occurs at a distance is not explained. In Newton's own words, the "great ocean of truth roars unseen" before us.

**The Point:** Things not only have but ARE positional mass (inertia) and momentum (mass x instantaneous yet changing velocity) which require a force to change. This force comes from another thing. Therefore, force and accelerating mass are determined by each other. This relation can be calculated and their future can be predicted in an absolute empty time and space.

## C. CLASSICAL PHYSICS

*Laws of nature are assumed to exist independently from the physicist while equations of physics are the mathematical expression of our own attempts at reaching them*
Albert Einstein[1]

### SCIENTIFIC METHOD

The foundation of modern science is known as the Newtonian Paradigm, "the explanatory practice that explores law-governed phenomena within a configuration space bound by initial conditions, to the universe as a whole." [2] What Galileo, Descartes and Newton constructed (but not completely formalized until the 19th and 20th centuries) is the scientific method, that is, hypothesis, experiment, prediction, observation and eventually, theory- a general framework within which further hypotheses may be tested.

Hypothesis is a tentative proposition that uses inductive or "bottom up" reasoning or logic, consisting of observations and measurements of experimental results that, through intuitive leaps, show a possible pattern to be further investigated. Deductive or "top down" reasoning or logic, starts with the general idea or theory and works down to the more specific examples, that is, experimental measurements. "The assumption at the philosophical heart of science is given knowledge of a system's initial conditions, and understanding of natural law, [you] can calculate the approximate behavior of the system."[3] "If you know everything about a system at some instant, and know the equations that govern how the system changes, then you can predict the future."[4]

The term "useful" in science means "the idea has observable consequences."[5] Science progresses when a "good theory is superseded by a better theory, when new data don't fit the old one."[6] The basis of scientific fact is that it is falsifiable, but so far no one has been able to prove the proposition wrong. Finally, by combining Descartes and Newton, we see that science searches for equations that exist outside the consciousness of the scientist.[7] This assumes "material reductionism," a world explained by divisible material parts whose relations are caused by natural forces. **Observations are explained not by the existence of the object, but the models that are formed based on the mathematical relations of the object's constituents (parts and properties).**

**The point:** The world consists of separate external things. Observations of things and their relationships to one another enable predictions of where these things will be in the future. These relationships are known in a mathematically reduced model of the world.

## EXPERIENCE

For the philosophical school known as Empiricism, what is observed is not what the mind calculates, but what the eye sees (or at least what the mind thinks it sees). **Knowledge must be based on experience.** John Locke (1632-1704) postulated substances, not as consciousness and pure extension, but things in the world with properties or qualities. David Hume (1711-1776) looked further into the mind to realize that what we see are not objects, but ever changing scenes as we look about. He developed the theory of association to explain why we attribute similarly appearing qualities to substances. Immanuel Kant (1724-1804) compromised by claiming that consciousness filters the world through categories like quantity, quality and relations like cause and effect. His mathematical "Antinomies" of space and time (beginning or no beginning); atomism (infinite series of small, indivisible parts or infinitely divisible parts) and incompatibility of cause vs. freedom, are examples showing how the split in infinity presents irresolvable contradictions in consciousness.

As Descartes noted, the fact of subjective experience is the most fundamental fact of all; the reality of empirical results can only be known through subjective experience in the first place. But science, with its productive method and assumptions of a world understood by mathematics and quantitative measurement, won out. Even today, the mutual externality of material reductionism has held sway in biology and neurology, while a disciplined qualitative analysis of internal experience has been largely confined to the relatively obscure philosophy called Phenomenology.

**The Point:** Consciousness experiences the world through its own built-in organization. The categories of this organization derive from the original paradox.

## MEASUREMENT

Science is mainly divided into physics, chemistry and biology, with many sub-divisions in each. At the time of Newton, we are basically talking about physics. According to Wikepedia, "Physics is fundamentally a quantitative science. Its progress is chiefly achieved due to rendering the abstract qualities of material entities into physical quantities, by postulating that all material bodies marked by quantitative properties or physical dimensions are subject to some measurements and observations."[8] In fact, such is the **mathematization of science** that "an equation allows you to predict the results of an experiment without actually having to conduct it."[9]

**The Point:** A world reduced to matter is understood as quantities that can be measured.

## DETERMINISM

Prediction implies that future events are determined by results of the present state of any observed system.[10] This not only means that future events cannot cause present events, but determinism precludes spontaneity, the creation of anything new.[11] In other words, freedom of choice does not exist, as **future behavior is completely determined by an initial state**. The irony is that time, as the absolute variable t, can simply be changed to -t and the same equations will describe the past. Mathematics is reversible. The movement of billiard balls can be "predicted" in reverse.

**The Point:** The quantitative relations of things in space and time, including the movement of things, now and in the future are completely determined by the past. Time, therefore, can be seen as backwards or forwards.

## MODEL SYSTEMS

A system is actually a sub-system, containing objects and forces (and energies, which we will get to shortly), which acts as if nothing else exists. These model systems are defined by laws and initial conditions. The scientific method progresses "through the simplification of models, contingent rules and continually corrected observations." [12]

We might notice here that the accuracy of predictions depends on the size of the system being measured and the corresponding resolution of the measurement. The smallest system would be a single particle. It consists of the three dimensions of its position and the corresponding instantaneous velocities at that position. (Mass and force require another particle).The largest model would be a duplication of the entire universe.

**We understand the model, not the underlying nature.** That nature works like the model is demonstrated only by repeatable results. The initial conditions, no matter how precise, will on each measure occur at a new moment. Most physicists agree that theories are approximations of the world, even though physics itself is supposed to be about certainties.[13] Mathematics only provides precision in relations that describe the simplest events and whose effects can be measured.[14]

Scientific theories are also limited in that they describe what and will happen but not why.[15] Physics itself doesn't interpret physical theories,[16] and remains ignorant of what the ultimate causes are in themselves.[17] The properties of matter are only known by their interactions.[18] In fact, the word "matter" itself, has no scientific meaning.[19]

The mathematical description of the universe, besides relations, contains "constants," changeless numbers that are required in order for the equations to "work." They are not part of the theory, but "part of what we don't yet know."[20] The most famous is the speed of light. Some others are the gravitational constant, Planck's constant, and the strength of the electromagnetic force.

> **The Point:** Science understands the world by making models of it and testing those models by making predictions. The smaller the model, the more precise are the predictions.

## TIME

The most significant assumption of physics is the concept of time. Time has a two-way direction, it runs uniformly and can be compared at different locations defined by three spatial coordinates and one time coordinate. **Laws are considered timeless, yet in a universe with a beginning, there is no explanation of where the laws were before**. And the laws cannot be tested before the events that act in accordance with those laws. In the words of David Bohm (1917-1992), "timeless laws can't be, because time itself is part of the necessity."[21] Scientists and their measuring tools, including the clocks that measure change in time, are outside the system. A clock is not included within any temporal relation.[22]

In relativity theory time is spacetime, which make it a mystery as to how it flows. It is a four-dimensional coordinate system. In quantum physics, time exists forwards and backwards in an "infinite-dimensional space."[23] If mathematical laws and equations are atemporal, there is no real distinction between past, future and present, since mathematically the laws and equations are logical rearrangements.[24]

The subjective feeling of time passing, or the fact that, as Hume noted, our impressions have a temporal order, differs from objective time where the change between two events must be recorded by a clock.[25] The configuration space (all possible configurations of any subsystem) seems timeless when studied by much of physics.

**The Point:** In a fully determined, changeless universe, time has two directions. Life experiences one directional change.

## SIZE

**Though the universe may or may not be structured at many sizes,[26] physics basically splits the universe into three**. General relativity, a geometric spacetime continuum, describes the very large scales of planets, stars, and galaxies, and is accurate at the speed, or near the speed of light. Quantum physics describes the very small ($10^{-14}$cm and less). Newtonian physics (classical physics) describes the length and timescales of the objects in between. Since Newtonian physics is an "approximation" of general relativity, there are only two main "real" scales. The word "particle" is usually used in reference to things in the quantum world, and "object" is used in reference to things in the macro sized, or generally visible world. In this strict vernacular an object, whether big as a star, or small as a hydrogen molecule, would ultimately consist of particles.

The five levels of "matter" that make up the universe are "molecular [the objects of chemistry], atomic [including electrons], nuclear, hadronic [protons and neutrons] and quarks."[27] Below the molecular, the quantum world is invisible to the naked eye. Quantum physics is probabilistic as opposed to deterministic.

Force is reduced to four kinds: Electro-magnetic, strong and weak nuclear forces, and gravity. The first three reduce to variations of the same force described within the quantum infinitesimal, but gravity stands by itself in the largest scale, the geometrical continuum.

**The Point:** The objects that we experience are approximations of a real world that is made of particles too small to be directly experienced and a space-time continuum too large to be directly experienced. Three forces at the small size and one at the big size are all determined by the mutual relations of their respective objects.

## D. RELATIVITY

*To us believing physicists, the distinction between past, present and future has only the significance of a stubborn illusion*
Albert Einstein[1]

### ENERGY

Energy explains force in terms of what happens within the actual interaction of objects. Force, which does not exist by itself, does have a direction in relation to something forced. Energy has a magnitude, it is scalar, but has no direction, it is not a vector. **Energy is what is transferred from one object to another by a force.** In terms of motion and position (or rest) of objects, there are two complimentary types of energy: kinetic and potential.[2] Kinetic is the energy in the movement of an object, potential is the energy in an object at rest. Energy is conserved in the totality of the two, but each can change in proportion to the other. What is known is the difference between the two, not what either is in itself.[3] Force is always directed in pushing the object toward lower potential energy, that is, increasing kinetic energy or faster movement.

Mass is a measure of how difficult it is to accelerate an object. For example, an object rolls down hill toward lower potential energy while picking up speed, and therefore, gaining kinetic energy. Momentum (mass times velocity) is a measure of resistance to stopping. So momentum is slowed going up the hill, which means the kinetic energy is reduced as the potential energy is increased. The total energy is conserved.

Of course, since we are usually talking about sub-systems, that is, small models of the universe, energy will not be conserved if parts vary or the system is affected by another (the hill, for instance). "Translation" is the term used for the static, timeless symmetry in simultaneously shifting the Cartesian coordinates of all the objects or particles in a system by the same amount. But the symmetry connected with energy conservation involves a shift of time. If energy has no direction, but its totality in the sub-system is shifted, then there has to be a shift in time. If energy has no velocity, then force requires space. The force within momentum requires time.[4]

Furthermore, the sum of the potential energies of all the objects in the system at a moment in time, that is, the configuration (the distances among them) of all the objects, is greater than the total potential energy of the system as a whole. This is because again, contrary to Newton, force takes time.[5] If the total energy is conserved, forces are time delayed.[6] Potential energy is the energy of position and increases with the distance from any potential location. So energy is a property of a particle that is different than its mass, force (or charge) and momentum.

Electric forces are related to the distance between electric charges[7] but what a charge (negative or positive) is in itself is unknown. So if matter is anything, it is simply energy moving about in lots of configurations[8] defined by mass, charge, and momentum, which are nothing in themselves but the relations among them described by equations. There doesn't seem to be any direct representation of physical reality in either space or time.[9]

**The Point:** Force is the external explanation of how one thing moves (or stops) another thing. But force needs time as well as space to cause an effect. Energy is the internal (within the object) factor that includes time.

## MASS

But objects are not only understood to be relational, but relative. In other words, a universe of related moving points is also a universe of points of view. Motion is not only change of position, but it is viewed from another changing or moving position. Therefore, we can't tell which object is moving and which is at rest. Now, as we have seen previously, a body's mass is it's inertia, whether as positional mass (with potential energy or minimum kinetic energy) or moving mass (with kinetic energy). Since energy varies with time only (it is scalar), and momentum is a vector and therefore involves distance, mass does not vary in total energy and momentum. But energy and momentum differ as varying ratios of an object's time and distance relative to the view from other objects. So, relativity differentiates mass from phenomenal matter,[10] that is, things with mass, volume (size) and density (mass per unit volume) that we can "see." In fact, "**there is only a single property, the body's mass, which in different frames appears as different combinations of its energy and momentum. In this sense, relativity unifies energy, momentum and mass.**"- Marc Lange[11]

**The Point**: A thing's mass does not change. Its energy and momentum are at different ratios depending on the position from which it is viewed. It is only in that respect that energy, mass and momentum are equivalent.

## LIGHT

Another invariant is the order of cause and effect in time, that is, the interval between the start and stop of an object's directional movement. But the combination of time and distance will vary from different perspectives. What will not vary is the speed of light.

Albert Einstein (1879-1955), of course, put time and distance together as a single space-time in his Special Theory of Relativity. If someone on a moving train walks forward a certain distance at a certain speed, that distance and speed will appear differently to someone standing at the station. If speed is the ratio of distance to time, then the walker's speed measured by the clock on the train will be slower than the walker's speed measured by the clock at the station. Einstein argued, "that because this is a greater distance for the light to travel than if the clock were standing still, time will pass more slowly because the tick takes longer."[12] From the perspective of the man on the platform, time on the train has slowed and distance has shrunk.

Instead of time and space being the reference for position and movement, the speed of light (which had already been calculated at 299,792,458 meters per second, or 670 million miles per hour) is now the reference for time and space. "Space and time are defined by the propagation of light,"[13] which traveling at the universal speed limit, is the same in all frames of reference. Although the speed is constant everywhere it has a limit because it must cross a given space in a given time. But since that given space and time are defined by the maximum speed, space at that speed is an infinitesimal point and time an infinitesimal instant. In other words, "in the reference frame of light, there is no space and time."[14] Looked at a different way, mass would be infinite and energy would be zero.

Light itself, which Einstein discovered is packaged as elemental photons, has no mass. The energy is not a function of its momentum, but the frequency of the waves that make up the photon. Particle/wave duality will be discussed when we get to quantum physics.

Since the speed of light is the maximum speed, nothing can travel faster than the speed of light. Therefore, the causal relation between two events is limited by the speed from one to the other. But if two events are far apart enough in space and correspondingly close enough in time, the action of one cannot reach the other. In that condition, one event cannot be the cause of the other.[15] **Causal structure is built into space and time by light**. What we observe is the *ratio* in which space and time manifest in our frame of reference. For every 186,282 miles of space, there always is one second of time. It is this ratio that is constant for all observers, no matter how fast they are moving.[16]

**The Point:** Cause and effect are separated by time and distance. Time and distance are relative to any point, but unified at the speed of light which is absolute at any point. Time and distance at any point at the speed of light are infinitesimal. Mass, which is changeless, is infinitely large at the speed of light. Light has no mass. Cause and effect are limited within the confines of the speed of light.

In other words, light, mass and energy are matter without form.

## SIMULTANEITY

Another implication of the relative motion and position of any two objects is called the relativity of simultaneity. There are no such things as simultaneous events. Observers in relative motion to one another will each see two events as either simultaneous or not if the two events are separated by distance.[17] "Thus, there can be nothing objectively real about simultaneity, nothing real about 'now.'"[18] The present moment depends on the observer's observation.[19] So, the lack of any direct representation of physical reality is further confined to a moment in which the representation is received.[20] In "Einstein's theory, you cannot divorce perception - the information you gather from the environment - from reality."[21] Neither moments of time nor their flow, are objectively real. **Each reference system has its own special time and place.** The separation of perspective is complete.

Now, science is supposed to be about invariant quantities and invariant relations. We know that this invariance is beyond perception, and the world consists of perhaps, infinite viewpoints, each appearing to one another as a different combination of space and time. Relativity combines these spaces and times into one invariant space-time. "The fundamental equations between objects take place in space-time, not objects in space or time separately."[22]

**The Point:** In a world in which time and distance are relative, the now, that is, the present moment, is an individual perspective. Objective or invariant reality occurs in a unified space-time.

## GRAVITY

Still, there remains a force at a distance, called gravity, which appears to be immediate or at least effective at distances seemingly traveled faster than the speed of light. Light has no mass. That leaves mass and space-time. They must interact directly. Time slows with the increase in velocity until the speed of light, at which time stops. Since mass increases with velocity, time would also slow with the increase in mass.

Special relativity describes objects from what is called an "inertial frame," in two parallel running trains (or physics labs), "moving at a constant speed and direction with respect to one another."[23] Laws of physics would be the same in both train cars. If one suddenly sped up (or down), people in that car would feel a force that people in the other would not ("things" too of course, as some things would get knocked forward or backward). Einstein saw this as the "equivalence principle," that is, acceleration of any inertial mass is the same force as gravity.

Newton's derivative understands the rate of change of motion of an object as its instantaneous velocity. The derivative of velocity is the rate of change of velocity which is instantaneous acceleration. Since velocity point-moments must change to reach the next point-moment, where is this change located? And if it deviates in velocity, where is this deviation? It must be in a direction, a vector. If it is change, no matter how close to instantaneous, it must have movement, which Einstein conceived as the minimum curvature of a trajectory.[24] The structure of space-time is dynamically curving, with the larger the mass, the more apparent the curve. Circling bodies, which includes all bodies, are continually changing direction and therefore always accelerating. **Mass doesn't force other mass, mass distorts space-time which then affects other masses**. Massive objects encircling other objects are actually accelerating around a curve, which in a curved space-time is the shortest path.

With objects on Earth we don't notice the largest effect, because at our scale it is far too small to perceive. On the largest scale, gravity even bends light when it travels through space-time close to a giant mass (but only "traveling" from the space and time point of view). With the increase in mass, time slows. Gravity, as the relation between mass and space-time, is a negative pulling force, an inward curvature. Distance and time theoretically stop at the singularity of infinite gravitation.

Gravity is thus not really a force at all, but a dynamic geometrical continuum of mutually defined mass and space-time, described by some[25] as four dimensional Cartesian coordinates in constant fluctuation based on the moving center of the mass of what we see as multiple giant planets and stars.

The universe is a place where time doesn't even flow because time is spatialized.[26] In other words, "Einstein merely substituted a 4D absolute external entity for a 3D absolute external entity."[27] In 4D space-time, the "flow of time is an optical illusion."[28] Descartes' absolute geometry becomes relative geometry. Space-time is a "block universe," with both the directional flow of time and the instantaneous now set aside. Though gravity is immediate in different ways for both Newton and Einstein, the electromagnetic force at the tiny end of things is somehow mediated.[29]

**The Point:** Gravitational force at a distance is explained as a mutual dynamic relation between mass and a curved space-time continuum. It is equivalent to the acceleration of mass.

## E. QUANTUM

*The many become one, and are increased by one*
Alfred North Whitehead[1]

### SMALLEST

Space-time, light, mass and gravitational force make up the principle relations behind the large continuum. At the small end, we have parts represented mathematically as an infinitesimal collection of points and moments that can be measured only approximately compared to objects moving nearer to the speed of light. From the large perspective, the smallest is the minimum acceleration, that is, deviation from velocity, i.e., the "curvacle."

So between the infinitesimal series of point/moments and the continuum is there a substantial basis for the phenomenal - an "emergence of the discrete from the continuous?"[2]

Quantum physics posits a realm of atomic scale between $10^{-14}$ cm and $1.61619926 \times 10^{-33}$cm (the Planck length, the shortest measurable distance. Planck time, remember, is the time it takes light to travel the Planck length). Anything below this number is considered by physics to be unmeasurable and therefore, meaningless. **It is the closest physics gets to the infinitesimal.** But as such, quantum mechanics is a mathematical theory, "developed to describe the interactions of elementary particles."[3]

**The Point:** The smallest physical parts are mathematically described interactions between particles as close to the infinitesimal point that can be measured.

### DISCRETE PARTICLES AND DISCRETE JUMPS

In this realm we find photon particles and atoms made up of particles. Atoms consist of electrons and a nucleus which is in turn made up of neutrons and protons held together by quarks, (and many other particles making up the "Standard Model,").These are the particles of mass and energy that are "fundamentally discrete."[4]

The standard definition of a quantum "is **the minimum amount of any physical entity involved in an interaction...[with] magnitude that can take on only certain discrete values.**"[5]

The photon is a quantum of light. Quanta of light mediate every electromagnetic interaction. Exchanges of energy require exchanges of photons.[6] Electrons are particles of negative charge that, considered as particles, travel about the positively charged atomic nucleus in orbits separated by energy levels equal to the energy of a photon. A photon can raise the level of an electron's orbit or carry an electron in the outermost orbit to another atom. The photon "from its own point of view leaps from here to there, distance doesn't exist for a photon."[7]A discrete entity, as we have seen, would have to leap over a position to move.

**The Point:** Discrete particles of mass/energy leap through the continuum by "trading places" with mass-less photons whose energy is equal to the distance leapt.

## PARTICE-WAVE DUALITY

Light was once thought to travel in waves. Einstein discovered light travelled as particles, later named photons. Quantum physics discovered that it traveled as both. When a beam of light passes through a two slit filter, the interference pattern imprinted on the back wall will look the same as when photons are observed passing one at a time through a single slit. Each photon has the same energy as the beam. This is what is known as particle-wave duality. **The particle-wave will be either a particle or a wave depending on what the experimenter is measuring.** This is true for all quantum entities.

**The Point:** Depending on how it is observed, a particle is either a discrete particle or a continuous wave.

## POSITION-VELOCITY UNCERTAINTY

Furthermore, if we try to measure a particle's position and velocity simultaneously, we run again into Zeno. But this time it's in the form of Werner Heisenberg's (1901-1976) Uncertainty Principle. Zeno's arrow paradox showed that any object is both at rest and moving in a space the same size as the object. The Uncertainty Principle shows that position (rest) and motion are complimentary properties, whose measurement can only be described by probabilities that are more or less certain depending on which property is being measured. The more precise the position is measured (the smaller the region measured), the more uncertain the velocity is known. The more precise the velocity is measured, the less precise the position is known. So the search for the instantaneous velocity via the infinitesimal calculus ends up in quantum size not with position and velocity but with either-or. The paradox of the split infinity, discrete and/or continuous, remains in the form of simultaneity of two complimentary qualities of one entity.[8] **In Newtonian physics, things are are almost there. In Quantum physics, things might be there.**

> **The Point:** Zeno's paradoxes of position and motion, which reflect the ultimate paradox of discrete and continuum, manifest in the quantum realm as an inverse relationship between the certainty of an object's position and motion, that is, the more certain is an object's position, the less certain is its motion, and vice versa.

## WAVE PROPERTIES

To better understand this, we have to look at **three properties of a wave.**

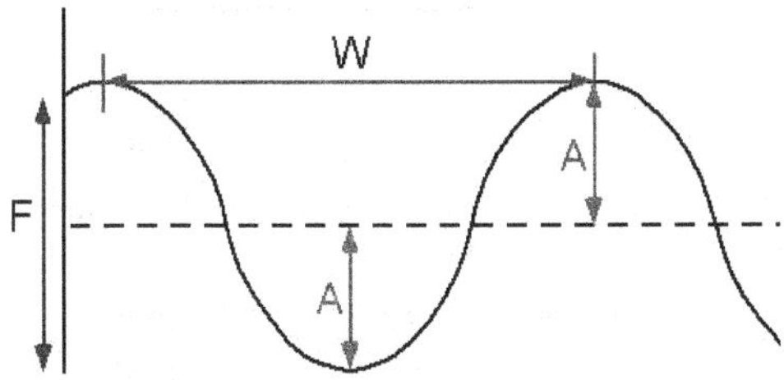


Fig. 2.

For a wave, "the maximum vertical displacement in one direction (crest to zero) is called the **Amplitude**....The number of crests passing a given point in a time period is called the **Frequency**. The crest-to-crest distance is the **Wavelength**."[9]

The amplitude of the wave associated with a particle corresponds to its position, and the wavelength corresponds inversely to the velocity. The shorter the wavelength, the more certain the position. Very short wavelength corresponds to higher velocity, and velocity (which is a vector or vectors) means **all directions, and therefore a large uncertainty**. The odds are greater that a particle is located where the amplitude is greatest. But that is where the wavelength is more ill-defined, which in turn determines the velocity of the particle. So a localized wave has an indeterminate wavelength. Its associated particle, while having a definite position, has no certain velocity. A well-defined wavelength is spread out, so the associated particle has a more predictably certain velocity, but its position is uncertain.

> **The Point:** Wave properties correspond to particle properties. Large vertical amplitude (which has shorter wavelength) = more certain particle position and greater and therefore more uncertain velocity. Longer wavelength = more certain velocity, less certain position.

## TIME/ENERGY UNCERTAINTY

**The Uncertainty Principle also applies to time and energy**. The shorter and, therefore, more precise the time interval the higher and more imprecise the energy. The more precise the amount of the energy, the more uncertain it is how large a period of time the particle (short wavelength) has that energy.[10] In reverse, the more energy, the more uncertain that energy is, the shorter the time period, that is, the more certain the time period is.[11]

> **The Point:** The shorter, more precise the time interval of the wave (position), the greater, more uncertain, is the energy. The more precise, or certain, the energy, the more uncertain the time that the particle or short wavelength has that energy.
>
> Just like Position and Motion, Energy and Time are complimentary, co-existent properties, in other words, two sides of the same infinitely distinct point.

## WAVE FREQUENCY

To better understand this, we need to consider the wave frequency which, as previously stated, is **the number of wave crests passing a given point in a period of time.** This rate (also called "oscillation") is proportional to the energy. And the higher the wave's frequency, the shorter is the wavelength- which is where the particle is most likely to be. And "potential [positional] energy varies almost infinitely small for a change of one wavelength...."[12] But the shorter wavelength is inversely proportional to the total energy. "To go to shorter distances, we need higher-energy particles....So the lower limit on length is based on the upper limit on energies."[13] Energy is directly proportional to its frequency and inversely proportional to its wavelength. Higher frequency and shorter wavelength are intertwined. The more certainty in an energy measurement the more uncertainty in the time during which the measurement is made, how long the particle has that energy. Both time and position are more imprecise the more precise is the velocity. Time and energy are inversely paired at the position.

**The Point:**

| | |
|---|---|
| high frequency | low frequency |
| short localized undefined wavelength | long defined wavelength |
| uncertain energy | more certain energy |
| more certain time | uncertain time |
| more certain position | uncertain position |
| uncertain velocity | more certain velocity |

The paradoxical point/continuum manifest at the smallest point, so far imaginable, as co-existent, not separate, properties.

## MOMENTUM

Since momentum = mass x velocity, there is an inverse proportion between particle mass and its wavelength.[14] A particle-wave is a composite of many wave (or "path") probabilities. It therefore, has many momenta, or a single momentum of large, maybe infinite, probability paths. Once a single momentum is measured, the uncertainty principle states that position, energy and time are probabilities. So, the quantum state is not simply positions of particles, but "is a dynamic relationship between position and momentum that changes in time."[15] **It is a dynamic pattern where space manifests as mass, and time manifests as equivalent energy**.[16] Unlike at the macro or Newtonian scale, "...force determines the probability of scattering, not the magnitude of deflection."[17]

**The Point:** In the particle-wave, momentum, mass, energy and time are all in a dynamic relationship of mutual uncertainty where force determines probability of outcome rather than macro space-time location.

## LEAPS AND CLOUDS

Instead of the picture of a nucleus with planet-like orbiting electrons, we have **a probability cloud of indeterminate size and shape**. (Sub-nuclear particles are also probabilistic). Instead of a defined location, an electron is spread out according to a probable percentage at every location. This "wave packet" itself, changes energy in whole number "quantum leaps."

The explanation can be seen in the nature of a wave-length. Imagine a vibrating string held taut at two ends or nodes. There would be one wave. Three nodes (one placed in the middle) would produce two waves, "with a wavelength half as great as the one with two nodes....The number can only be an integer and can change only by jumps."[18] The leap size depends on the whole number[19] as the amplitude in the wavelength crests depending on energy difference between the atom nucleus and the orbit. Atomic size of an atom is not a macro-measurable size, but is called the atomic radius, which is the energetic difference between the nucleus and the outer or valence electrons.[20]

**The Point:** Particle-waves or wave packets are formless probability clouds without definite location or velocity. Wavelengths are whole numbers determining quantum energy leaps.

## WAVE PACKET AND PARTICLE/WAVE WHOLE

The wave packet, though indefinite, can be thought of as a seamless whole. Erin Schrodinger (1887-1961) calculated that if waves have mass, then a "moving object is a packet of waves."[21]A wave is significant ontologically, whether energy or mass, because it is in some way a whole with parts, albeit known only mathematically as a wave function. **"Belonging has no meaning for Newton billiard balls, but it does in a wave-based universe"** - Jonathan Edwards.[22] But the particle aspect and the wave aspect are in a direct relation via a precise probability calculation of their relative motions and positions. The either/or nature of the wave packet is called a superposition. "The quantum wave function [is] a mathematical description of the superpositions associated with a quantum entity at any particular moment."[23]

**The Point:** The wave packet is a part-whole relationship in which the part is the whole and the whole is the part, both particle and wave are in "superposition." In some sense, the wave packet "belongs" to itself.

## ENTANGLEMENT

The wave packet itself, as a single entity, connects to other wave packets through a phenomenon called entanglement, where pairs of waves interfere with each other such that the uncertainty is not only whether it is wave or particle but which particle "belongs" to which wave. The interference pattern results from the peaks and valleys of different waves of possible magnitudes adding and subtracting as they cross. And of course, the particles can also be considered entangled. Entangled quantum systems can be described as a single wave function. "The particles' identities merge, their collective state is precisely determined, but **the individual identity of each of the particles has disappeared**."[24] But the increased energy and mass in the entangled whole is detectable.

> **The Point:** Wave packets connect via entanglement, in which the identities of each are mutually uncertain. They may "share" a particle, in which case the collective state is a precise whole. The parts lose their former identity.

## NON-LOCALITY

As waves and the corresponding particles exhibit complimentary coordination, so too do waves and particles between and among quantum systems in entanglement. But particles can interact in ways that are permanently correlated to the point where, **though separated, they behave as a single entity**. This is a phenomenon known as nonlocality, where entangled objects display qualities, such as opposite spin states, instantaneously over large distances (faster even, than the speed of light). If first one type of spin is measured in one particle, the opposite will always be measured in the second. If something else is measured in the first, a corresponding value of the quality will always be measured in the second. In some way, they are not actually separated.

> **The Point:** Entangled particles can demonstrate non-locality when the measurement of a particle quality will always be reflected in its entangled "partner" no matter the distance. The whole determines the parts without regard to distance.

## SUMMARY OF INDIVIDUAL PARTICLE/WAVES

The discrete position and motion of a point is resolved statistically as relative probable certainty and uncertainty of a quantum particle.

Particles are wave properties measured as particles.

The formless particle/wave probability cloud is a whole with parts so intimately entwined that the wave packet can be said to belong to itself.

The properties of multiple waves can be entangled so that it is unknown which wave a particle "belongs" to.

Particles can be distinct in space and time, without evidence of location, but remain identified, communicating through corresponding qualities within a greater whole.

## CHARGE-FIELD RELATION BETWEEN TWO PARTICLES

There is one more quality of a particle that we have merely mentioned-charge. Negative charged electrons relate to positive charged protons within a field. As we said earlier, it is not known what a force is in itself. Like other properties of a particle, charge is only known by its interactions.[23] An electric potential is relative, there is no single magnitude. "The electric field determines the potential difference between two locations."[26] **The field, itself, is the force**.[27]Although acting on particles, fields themselves interact by the entanglement of waves. So electrical force (along with the two nuclear forces), unlike classical Newtonian force, does not depend on spatial discreteness. In fact, force becomes a function of a continuous field and positional points with potential energy.

Mathematically, the standard quantum model is accurate only if particles have an absolute minimum value. Calculations of the uncertainty principle require non-zero momentum or kinetic energy. A particle can never be completely at rest. And the electromagnetic field (the magnetic field represents the movement of electrons) propagates outward in time at the speed of light unless slowed by phenomenal sized objects.[28] The energy of the wave is stored as a property of the electro-magnetic field.[29] Electric charges generate the electric and magnetic fields which interact, forcing the electric charges to move.

**The Point:** Electrons and protons have negative and positive charges respectively, and are measured relative to one another. Together they can form an electro-magnetic field which is its own force ever expanding outward at the speed of light unless it is slowed or directed by macro-sized, phenomenal objects.

## OBSERVATION OR MEASUREMENT?

De-coherence of the wave function, or wave collapse, is how the measurement of the super-positional quantum state is described. In other words, it has no definite state until it is measured. Before measurement "all outcomes are present in combination."[30] Whether the measurement break occurs because of the interaction between the micro-quantum level and the macro measuring device or the interaction between the quantum state and the experimenter doing the measuring is a source of controversy. De-coherence can be said to cause the collapse or only serve as an explanation for the observation. The scientific community standard is the "Copenhagen" interpretation, which is that the observer is defined as the measuring device.

Either way, the collapse occurs when the micro-level meets the macro, that is, "the measurement is any interaction between classical [or macroscopic] and quantum [atomic-scale] objects."[31] **Probabilities are reduced to the single actuality as seen by an observer.** Observation or measurement forces the quantum system to a precise value, as it requires light to interact with the object. "Light that bounced off the electron to hit your camera forced the electron to appear."[32] It is impossible to measure a quantum sized particle smaller than the wavelength of the measuring device's light, as photons will cause the particle to move. **The act of measurement affects that which is measured**. In the words of John Wheeler (1911-2008), "No phenomenon is a real phenomenon until it is observed."[33] In other words, "The behavior of all particles and objects is linked to the presence of an observer."[34]

> **The Point:** De-coherence of the wave function or wave collapse is the interface between the quantum and the phenomenal. The particle does not "appear" until the encounter.

## MEASUREMENT PROBLEM

Beyond the act of measurement is **the real space-time lag between the measuring and the reading (and understanding) of the measurement**. Is the only reality the measurement? Some would say that the experimenter both disturbs and produces the object,[35] and in doing so, creates an observable historical reality,[36] "defining a before and after relative to that which is measured."[37] "Your probe is altered by the interaction [as] you look at the alteration to understand your measurement."[38] Observation in one sense cannot be done without a scientist measuring. Perhaps the scientist and the observed are what is entangled.[39] The moment of each measurement is a phenomenal approach to the instantaneous, the "classic limit," but that which is measured is a world of probabilities.

The quantum-phenomenal boundary is itself not a measurable point but an "overlap" of a phenomenal measuring act, or series of acts, and a range of quantum probability.[40] "If you have sufficiently strong decoherence, you get classicality independent of your measurements. And if you have coarse-grained measurement, you get classicality independent of the interaction with the environment."[41] The speed of the wave function collapse can only be expressed in relativistic terms, and the collapse itself can only be expressed in quantum terms- two incompatible understandings of reality. In another way of looking at it, the Uncertainty Principle states that position and motion are inherently indefinite with regard to one another, though each accurate in probability with enough measurements, yet classical measurement is approximate. **The two realities are compatible only in our understanding of them.**

**The Point:** The wave collapse is a discontinuous point in an otherwise space-time continuum. The human observer observes the measurement in space-time. Superposition of a particle/wave (before the measurement) is the logical solution to the paradox of discrete things moving through the continuum to the next discrete position.

## SEPARATION OF QUANTUM FROM CONSCIOUSNESS

The quantum state becomes definite in de-coherence, as the wave-collapse is measured while collapsing in a phenomenal mixed state, world of time and space, perceived by observers. Scientists choose when, where, what, how and how often and by whom, to measure. The question arises as to whether or not their choice affects the quantum state.[42] If the particular measuring event that actually occurs is dependent on the selection process of the experimenter,[43] is the collapse or "reduction" of the wave-function only a change in the experimenter's knowledge?[44] "A probability wave isn't an event if nothing happens until the event is observed."[45] In the largest sense, the quantum world doesn't exist without observing scientists.

The wave collapse could be imagined to be "a property of the observer [which is the] uncertainty about the thing observed."[46] In the words of Henry Stapp (1928-), "...the observed system must be isolated in order to be defined, yet interacting to be observed."[47] If the chain of events is an infinite series of interacting wave collapse/observations then we might "end the infinite regress by saying it is the consciousness of the observer that gives actuality to the event."[48] Measurement is not completed until immanent awareness is included.[49] A timeless immediacy is itself the discontinuity of measurement. Since special relativity tells us that the shorter the period of observation, the larger the velocity,[50] a timeless immediacy measures infinite velocity. "Someplace on the scale between big molecules and humans there is this mysterious process of observation and collapse. Conceivably, it's indeed at the last step, at awareness," Bruce Rosenblum.[51]

The Copenhagen interpretation of the wave collapse is the measurement, not the observed measurement. Yet the collapse is the very point at which measurement changes from probable outcomes based on a series of temporal measurements to the approximate measurement of spatial resolution.

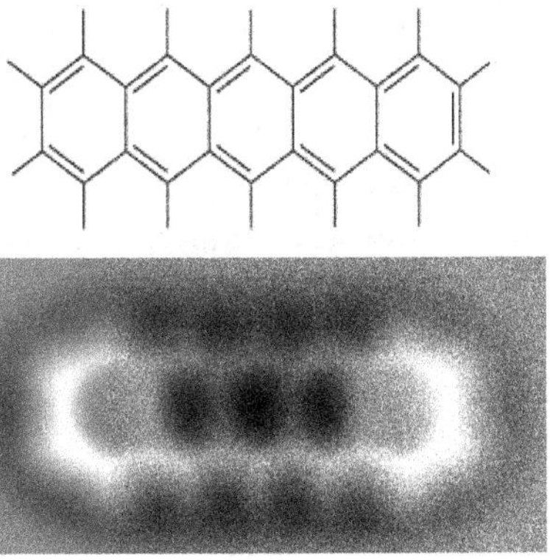

Fig. 3.
An photo image of pentacene shown underneath a diagram of the same molecule

"Location of the electrons will always be blurred. There's an inherent uncertainty in the quantum mechanics of these systems, the images are blurry because of this uncertainty – not because of the lack of resolution.'"[52] In the above photo, I see five benzene molecules. But it is all by reference. In other words, I don't see atoms and I don't see electron bonds, **I just see the relationships**.

**The Point:** Approximate measurement becomes more and more accurate until it reaches uncertainty. Finer and finer resolution is the sliding scale of awareness. Within immediacy lies uncertainty, along with an objectivity that is separate from the aware observer, yet is consistent with his measurements.

## QUANTUM IS MATH

Even if it is not consciousness or awareness that arises out of or causes the quantum world, that world is certainly known only through the "abstract relationships among mathematical structures,"[53] said Paul Dirac (1902-1984). There are no phenomenal forms at the quantum level. Max Born (1882-1970) goes so far as to say that quantum waves are purely probability predictions and have no material basis as Schrodinger suggests. Some would say that quantum equations were invented to explain the "smoothness" of observations. The scientific ideal that reality is simpler than direct representation, means also that simplicity is more general.[54]

Quantum equations are linear, meaning summations are equivalent on both sides of the relation. A "wave packet is the sum of an infinite number of terms."[55] Effects are directly proportional to their cause, e.g., $x = y$. The effects of the nonlinear equations that describe phenomena are "proportional to a higher power of the cause,"[56] for example, $x = y^2$. **In time, initial conditions always change according to previous results.**

> **The Point:** The quantum realm is known not as observable forms, but only by mathematical equations. Observable forms are mathematically nonlinear, that is, they change in time.

## QUANTUM IS NOT PHENOMENAL

The quantum level is not perceived by the senses, that is, directly. So the assumption that quantum statistical correlations, rationalized as small phenomenal particles, are continuous with the phenomenal is problematic.[57] Phenomenal objects that we can possibly sense, are somehow more than calculations.

The past of a classical particle is a deterministic single trajectory. A quantum particle has infinite possible paths before detection.[58] The simplest classical system is a particle with position and momentum. The basic quantum system is a quantum state, an abstraction called a "Hilbert space," a state of all possible positions and momenta that is infinitely dimensional. The probability of a wave collapse over a particular period of measurements is divorced from the phenomenal world over that same time period.[59] Special relativity shows that the mass of any object is absolute, but the ratios of time and distance of any object are relative to any other object. **Every point of view is really a dimensionless ratio**. A point of view assumes an object's relation to a hypothetical observer. **Quantum uncertainty is the point without the view.**

**The Point:** Space-time consists of points which have speed and distance only relative to other points viewing them (or that could view them). The points in themselves are beyond viewing.

## CAUSATION

Prediction (the exact knowledge of the past and present enables the complete calculation of the future) at the macro level, is replaced by extremely precise calculation of a probability distribution at the quantum level. The wave function is potentially infinite, so therefore, the future is not determined.[60] But macro causation is random in terms of any single encounter with a quantum system. Each encounter is chosen at random, and each measurement alone has a neutral probability, that is, random. Only when the same experiment is run thousands or millions of times does the distribution become statistically accurate. So for quantum reality, statistics replaces direct causation. **Probability itself is a measure of randomness.**[61] Inside the quantum world, quantum fluctuations occurring with decohering wave functions, as waves go in and out of phase, is also a source of random variation.[62]

Since it is correlating incompatible qualities, superposition of position and momentum uncertainty is another fundamentally random aspect. And if a position is in some way everywhere, then uncertainty is more fundamental than determinism.[63] So, causality, as such, may be statistical.[64]

**The Point:** Causation defined as one thing causing another is countered by randomness found in both quantum and phenomenal realms. Random motions of things in large enough quantities can be understood as statistical cause. (Aristotle, by the way, included collections in formal cause).

## QUANTUM/MACRO INTERACTION IN NATURE

The measurement in which entangled particles de-cohere is part of the phenomenal environment. Interaction among entities is a universal phenomenon independent of any single observer/measurement. In what way does the observable, measurable world meet the realm of quantum physics? "Somewhere between molecules and pears lies a boundary where the strangeness of quantum behavior ends and the familiarity of classical physics begins."[65] The quantum system is theorized either as an electromagnetic wave spreading outwards[66] or as a huge number of particle/photons merging into a wave that "pulls a macroscopic effect out of the quantum world."[67] If photons are involved in every energy exchange at the molecular level, then in some way wave collapse touches the macro world whether observed or not.[68] "The world looks classical because the complex interactions that an object has with its surroundings conspire to conceal quantum effects from our view....[69]

"The entanglements are primary. They interconnect quantum systems without reference to space and time....We must explain **space and time as somehow emerging from fundamentally space-less and timeless physics.**"- Vlatco Vedral[70] Perhaps the dividing line is the magnitude of the space-time distortion resulting from the mass of the particle. Space-time gravity itself is hypothesized by Roger Penrose (1931-) to cause quantum wave collapse.[71] The effect of gravity is itself a measurement that causes decoherence.[72]

All parts vibrate synchronously in superposition, yet one theory says that if the superposition takes place in a large enough object, the time-dilation of gravitational pull at different ends should affect the frequencies differently. Relativity posits time dilation, that is, time moves slower closer to a massive object. If the object is large enough, the difference collapses the superposition. It is estimated that the limit of superposition of Earth should be approximately one millimeter.[73]

A caveat to be contemplated is that what is entangled is a measured property of entangled states. The unmeasured properties remain indeterminate. A measurement of up or down spin is indeterminate as to any entanglement of left or right spin.

**The point:** Entanglement of wave packets and superposition of wave packet states somehow emerges or creates the phenomenal world. If the mass totaled in multiple entanglements is great enough, or if the wave that collapses is somehow large enough, then phenomena might result.

## TIME, CLOCKS AND PHENOMENA OUTSIDE QUANTUM THEORY

Quantum theory itself does not explain or contain within it, the wave collapse that becomes a definite state when seen at the phenomenal level. The quantum time factor does not carry over, as experimenters in general make separate interactions using their own clocks. And any clock does not necessarily mark the same time as others.[74] Quantum explanation is a method for predicting measurements,[75] not objects in time. But **the collapse of the wave function is irreversible, and as such, constitutes a one way ar-**

**row-of-time phenomenon at the point of that which distinguishes a quantum-state from a macro-state.** The universal points of view of relativity theory, concretizes to a singular moment of micro-macro interaction. The time reversible equations of quantum physics and the "block universe" of general relativity are both inside a boundary of time but divided by a timeless demarcation.

Of course, according to Schrodinger, quantum waves are "matter waves" that do evolve in time. The field is a function of space and time and therefore, physical.[76] But if mass is defined as a resistance to acceleration, and particles are actually "energy-storing waves and resonances,"[77] it is hard to see where the phenomenal real world begins. All we can say is that anytime a quantum system interacts with its environment in a thermodynamically irreversible way, the wave function will collapse.

**The Point:** Time reversible quantum mechanics and the "block universe" of space-time meet at the timeless point of quantum to phenomenon interaction. This timeless point is where the one-way phenomenal "arrow of time" begins.

## ESTABLISHMENT OF FORM

Quantum particles/waves, either as mathematical constructs or as real entities known mathematically, have no dimension. They are formless. A nucleus of an atom "looks" the same from all directions. It is spherically symmetric surrounded by electron clouds. In addition to being formless, electrons are all identical; and they never age or break.[78] Structure requires something else. David Bohm sees the beginning of the form effect in the varying energy potential charges within the field. The "quantum potential is a formative field on the movement of electrons."[79] If Aristotle's Formal Cause is to be found at the quantum level then "matter waves are the formative cause...the structure of the particles forming a system is the actuality of that formative field."[80]

An electron is a point particle, but it is located via a probability wave, a formless cloud. Somehow when we look at it, the wave collapses (whether or not our looking has "caused" the collapse), or billions and billions of waves collapse, and **what we see is an average of random locations spread out over time.**[81] And we see our near surroundings about a billionth of a second after the collapse.[82]

**The Point:** The difference between the quantum realm and the phenomenal realm is form. What we are aware of, as opposed to what we calculate, is shaped material. From what comes form?

## F. THERMODYNAMICS

*With large numbers, the ideal and the real converged*
Dorian Sagan[1]

### MICRO TO MACRO

There are a number of essential differences between the micro-quantum level of reality and the macro-phenomenal. Objects the size of an atom and smaller are formless probability clouds. They have no physical structure.[2] **Waves and fields are both calculated to interact at a point.**

Macro or large complex systems, are unique. They exist in a spatial continuum that is infinitely differentiable. Quantum properties and particles are limited in kind, but with a multitude of identical copies.[3] As formless particles whose waves interact at points, they are inherently non-separable. Within both the large and the small, the split of the continuum and the discrete remains, but in the latter the split is uncertain, and the former the split is approximate.

But when decoherence of a particle/wave interacts with the phenomenal world we live in, the phenomenal state is time irreversible. Once you're out, as it were, you can't go back to the same quantum state.

**The Point:** The discrete and the continuum manifest as fundamental, uncertain particle/waves and uncertain energy/time at the smallest scale and at the phenomenal scale as objects moving through a continuum separated by infinite edges, but experienced and measured as a one-way past-to-future flow, that in turn is bounded at the largest scale by a curved space-time continuum containing space-time points with locations determined only by other locations.

### DIRECTION

The nature of probability changes from random but accurately probable positions to a dynamic expanding volume where the highest probability is the maximum. The directionality of time is from small to large volume,[4] as light spreads out towards the future.[5] In other words, **the future is the point of maximum probability**. Quantum time is symmetrical, phenomenal time is asymmetrically directional.

From the macro perspective, the lowering of temperature, which is a slowing of the molecular average velocity, causes the breaking of symmetry. This manifests itself in e.g., "phase" transitions like freezing water as molecules in random symmetrical possibility, crystallize in specific spatial directions. Though energy is directionless, momentum has direction.

The higher kinetic energy of greater momentum does not alone account for temperature. The density of multiple molecules in relation to the total volume of any system must be included. With volume comes curvature, acceleration, and irreversible divergence and convergence of parts, both calculable by nonlinear equations.

**The Point:** Quantum randomness measured as probability of discrete states is symmetrically directional (toward the maximum probability). Time in the phenomenal world is asymmetrically directional in a three-dimensionally structured space. Macro-states are statistical totalities of micro configurations. The maximum probability is toward the future.

## PHYSICS VS. CHEMISTRY

Physics covers the quantum realm and the macro realm, which includes astronomical General Relativity and the Newtonian approximation or idealization at the middle, phenomenal realm. Physics covers matter as such, in other words, "properties of matter in themselves."[6] **It is chemistry that specializes in the immense kinds of matter with structural differences, not formless points of limited kinds**. With direction and dimension the atomic elements are spatially arranged[7] in molecular chemical bonds. Although the wave collapse is sharp, the classical picture of chemical structure with electrons circling the atomic nucleus and multiple directional lines between atoms indicating molecules, ignores the role of wave mechanics and electromagnetic force in forming the bonds of molecular shape. (See the pentacene photo in the previous chapter). Quantum physics (another term is "particle physics") and chemistry are somewhat superimposed.[8] Pure structure by itself does not account for the randomness beneath nor the directionality of time.

Quantum linear equations in which summations are reversible are transcended in the chemical realm by nonlinear equations. The wave packet is, as we have said, an infinite summation.[9] But the myriad yet organized complexity of chemistry factors in another level of dimensional direction.[10] The straight line that represents a linear equation is replaced by a nonlinear dynamics to powers other than one. Non-proportional relationships represented best by geometric visualization, and uncountable collections of molecular entities are understood as statistical aggregates.[11]

**The Point:** Chemistry is the science of complex, unlimited molecular structures. Physics is the science of the underlying physical laws, in both cosmic vastness and simplest particle description.

## MEASURE/RESOLUTION and SIZE/SHAPE

The macro level is not just larger in size, it is size. Size is continuously infinite the closer you get to anything. Measuring size requires resolution. Resolution varies with distance. Particular size, therefore, is an illusion,[12] that is, it depends on point of view. With size, points are relative to one another. In three dimensional space, shape is the relationship among at least three positional points (in reality, an infinite number of points). The shape is the surface of the three dimensional volume. **Shape (or form) and volume together make up the visible object**. Shape can differ independently of the size or total volume. In a three-dimensional universe, volume changes as a factor of three, surface by a factor of two. For any given volume a sphere has the smallest surface. If the shape is irregular for the same volume, the size of the area of the shape is larger. If the volume is reduced and the shape remains the same, the surface size will be larger in proportion to the volume.

**The Point:** At the phenomenal level, size or volume comes into being. Size has a form. Size and shape/form reflect a distinction between quantity and quality.

## FRACTALS/SCALE

Form and size are more clearly differentiated when understood visually in the resolution of fractals. At any size a fractal will appear to have the same shape. **A fractal is a way of seeing infinity[13] in the split between the continuum of infinite resolution of size and an infinite series of discrete shapes.** A fractal is an "infinitely long line surrounding a finite area."[14]

With any scale model, scale means the same shape at a different size. As the observer gets closer to a fractal, the scale gets smaller and a never-ending series of new but self-similar shapes appear. No matter the resolution, the shape is "locally identical to itself."[15] The irregularity of shape remains the same throughout all changes in scale.[16] The "curvature of a fractal space increases to infinity as resolution approaches zero."[17]

Fractal self-similarity, it is thought by some,[18] is the visualization of the cosmic unification of quantum non-locatable discrete spatial relations and the undifferentiated expanse of general relativity. Perhaps a quantum wave that exists as temporal symmetry in a curved space-time can't be imagined, but is sensed phenomenally as three-dimensional position. Self-similarity is a partial view, as close as we can get to the two ends of existence.

But whatever may be, the observer senses the passing of time. However it happens, the quantum world meets the phenomenal in asymmetrical time.

Fig. 4. "Fractal Tree No. 2", Robert Fathauer © 2007.
[To view more of Dr. Fathauer's work, please visit www.robertfathauer.com]

**The Point:** Scale is the same shape at different sizes. Fractal self-similarity combines infinite resolution of size with infinitely distinct shapes at all sizes. Fractals are the phenomenal representation between the largest continuum and the smallest part.

## THERMODYNAMICS

The unidirectional, irreversible arrow of time, itself, has to be considered. The only scientific law that does so is the Second Law of Thermodynamics. The first law, the Law of Conservation of Energy, says that energy cannot be created or destroyed inside an isolated system. The second law declares that if an isolated (closed) system is not in thermal or chemical equilibrium, the "entropy," almost always increases toward the state of equilibrium, a state of lowest energy. In an open system, energy is lost to the universe outside in the form of heat at one opening, but can gain energy at another. The third law states that the system's entropy approaches a changeless or maximum state of no return.

The second law describes the flow of energy ("free" energy) from order to disorder, or from more to less useful. Unencumbered by any limited volume, this is a flow of pure energy directed toward total dispersion. Looked at universally, the probability of outward direction is total, and total probability is really necessity. If direction is not instantaneous and there is a difference between beginning and end, then it happens with time. The amount of entropy is at one with the amount of time passed. In and of itself, **entropy is the form of time**. Both are essentially irreversible direction and the measure of the difference of the distribution."Without time, entropy cannot happen."[19] "Without entropy, time need not exist."[20]

A solid form like a vase may break into a thousand pieces, but except in a reverse video, we never see the pieces reconstruct the vase themselves. The vase will not reassemble. Logically, it's not just the number of pieces that are problematic. The order and variety of shapes are virtually incalculable and irretrievable.

Although thermodynamic processes occur in and between solids, the second law is more commonly observed in the flow in gaseous and/or liquid states. The parts on this level refer to atoms or small molecules.

Any single configuration of molecules, the microstate, of the macroform contains a precise number of single molecular energy states at any moment in the totality of molecules. In an extreme microstate, the molecules are all in random motion with one another and therefore, equally probable,[21] because each gas molecule by itself moves independently. Each step is statistically independent. The configuration is unknowable. This is the state of diffusion or "Brownian" motion. Just this side of quantum symmetry, says Charles Lineweaver (*Complexity and the Arrow of Time*), "dynamic instability with many degrees of freedom induces the arrow of time."[22] A state of energetic equilibrium is potential direction.

Variables at the phenomenal level are fewer, and although they have a statistical underpinning, the manipulations of temperature, volume, and pressure on a system can more clearly demonstrate relative free energy and entropy. The density or closeness of energy increases with the increase in the number of molecules in a large volume,[23] but the freedom of motion and direction is also restricted by the number of molecules. Temperature is the average of all molecular motion within an enclosed system or an isolated part of the system. Isolated subunits, that is, single atoms and molecules, have no temperature.

Microstates, the myriad configurations of subunits, can be thought of as different possible ways the system can achieve a particular macro-state. The probability of any microstate at a given change of temperature or volume,[24] differing from entropic equilibrium, at any given moment,[25] was first formulated by Ludwig Boltzmann (1844-1906).

Thermodynamic free energy is defined as the ability to do work. Entropy is the measure of useless energy, the degree of disorder, the measure of a system's energy per unit temperature that is unavailable for work. The more random all the movements are, the less the odds are that they will assume some sense of order. Boltzman entropy means that the less possible microstates there are with a given macro condition, the more useful the energy state is.

The force of entropy depends on free energy. If the system is open, the energy can release at one end, thereby increasing the entropy of the outside universe. Reduction of entropy in any local area must increase the entropy somewhere else.[26] But the entropy can also be decreased in the system relative to the macro-state forces of reduced system volume, pressure and/or temperature increase or any other energy source from outside. Entropic heat does not by itself reverse direction.

The second law determines the direction of chemical reactions from higher free energy to higher entropic energy released as heat. Without any outside force, the direction is toward the more stable state of equilibrium, a state of lower energy.[27]An outside force (a catalyst) that helps combine chemicals decreases the entropy by supplying energy.

**The Point:** The Second Law of Thermodynamics represents the asymmetrical arrow of time as a flow of sized shapes (forms) from order to disorder. This flow can be looked at as ordered free energy or low entropy flowing toward total equilibrium or high entropy.

Order can be thought of as "useful." Free energy is energy available to do work. Instead of a flow, energy can be understood as a state, defined as a quality that always increases and never decreases. The quality can be measured as a relation between possible configurations of atoms in a system, i.e., microstates, and a change in outside macro variables.

Entropy can be looked at as a natural flow, or as a state that can be manipulated. Disorder then becomes statistical order.

## GRADIENT

The second law describes energetic direction. The gradient is the rate at which a thermodynamic process changes relative to time, distance and macro form level variables. Equilibrium is the state of a reaction in which the rates of the forward and reverse reactions are the same. The production of entropy would be a gradient reduction.

As a rate of change, the gradient is, of course a derivative. It is not a motion itself, but a direction to move. The gradient points in the direction of the greatest change relative to the macro variables affecting a system. Any work related force, including the source of free energy, can have or affect the gradient.[28] With an electro-magnetic force, electrically charged particles form the gradient of a potential energy.[29] The potential energy itself, is minus its own gradient.[30]

The variation of energy flow in and out and the changeable rate of the gradient, means that the thermodynamics within an individual system can remain in a state of non-equilibrium, that is, slowed for a period of time from the outside point of view, or from inside, as local time itself slowed. If the direction of time is formed by an increase in entropy, then if decreased at any point, theoretically there is a local time-reversal. Micro randomness can spontaneously add energy and thereby slow or reverse the local flow of time. Stephen Hawking (1942-2018), in his work on the entropy of black holes, has claimed a "time reversal where entropy decreases."[31] In fact, he has suggested that **time itself, the difference between past and future, is a gradient**.

**The Point:** A gradient is a measure of the speed and direction of the thermodynamic flow in a system constrained by distance and macro variables like temperature and volume.

## WORK/CONSTRAINT

The gradient is necessary but not sufficient to maintain an isolated system in a state of non-equilibrium. **Free energy is the capacity for doing work**. Work involves movement directed by a force.[32] But the second law is not just about doing something, it's about restrictions on what can be done.[33] Work transfers energy from one system to another by overcoming varying constraints.

| **The Point:** For the gradient to affect constraints, work is required. |
|---|

## UNIVERSE

The asymmetrical flow of time means the universe itself is not in equilibrium. This means that there was a point in which there was total equilibrium and/or total order. The Big Bang is theorized to have begun in a singularity. One idea, proposed by Lineweaver et al,[34] is that at this point, or very soon thereafter, there would be a state of maximum energy in the electro-magnetic potential matched with a gravitational field which would at first be in a smooth, low-entropy state.[35] So mass and radiation (the movement of electro-magnetic waves) would be in thermodynamic equilibrium; **two kinds of lowest entropy in a relation of maximum entropy**. There was a "dynamic compromise between gathering and dispersive tendencies."[36] But to break the equilibrium a source of free energy is required. One version of the inflation hypothesis provides the answer.

The Big Bang was proceeded, or followed, by exponentially expanding spatial energy of indeterminate time, energy "inherent to space itself" that expanded exponentially.[36] This converted the equilibrium into the kinetic energy of mass and radiation. The entropy gap began at the first quantum of time (the Planck time) in a temperature difference between photons and mass. The first discernible diameter of the universe is the Planck length, which, as we said earlier, is the distance light travels in Planck time. In other words, light waves are the size of the universe from the very beginning, which might explain why, from the perspective of light, time is zero and the universe is the size of a point, and/or light is everywhere at once, even as the universe expands.

Small random fluctuations in the equilibrium produced "tiny ripples in its structure"[37] that were then amplified by gravity which caused the temperature of particle mass to cool faster than radiation. As the photons and mass particles diverged, the entropy gap formed.[38] So as the free energy was degraded, the universe both expanded and cooled as particles attracted and the gradient energy spread them out,[39] creating a space-time curved by the difference.[40] The curvature produces direction by breaking the symmetry. The deviation from simple motion or singular velocity is the curvature of acceleration. As the entropy gap widens, "matter" progressively differentiates becoming more and more complex, depending on the availability of free energy. An unlimited variety of forms are created. The entropy gap is a rise in both gravitational and energetic entropy.

The Bose particle, or Higgs Boson, is theorized, and some say proven, to be the source of mass in particles, and therefore the whole universe. For some, this explains the unification of the universe. The "rippling," stated above, creates mass particles by breaking the original symmetry. A special mass producing rippling particle, however, would still leave separate the quantum electro-magnetic forces and the space-time gravitational force and would not contradict the thermodynamic evolution. Even if these forces are proven to be equal mathematically, unification still requires something to be unified.

**The Point:** One theory is that the universe begins in a burst of free energy expanding out from a point and breaking the absolute equilibrium between the two low entropy (high order) gravitational and electromagnetic forces, leading in time to zero energy and infinitely dispersed particles, with complex structures forming in between.

## EMERGENCE

Emergence is the idea that at some point a large group of constituent particles behave as a single entity differently than the collection of particles themselves. **The whole, as it were, is more (or different) than the sum of its parts.** The general electromagnetic and gravitational increases in entropy are reflected in localized forms as dynamic relations of quantitative magnitude. A universe that began with quantum oscillations and ripple size gravitational distortions,[41] manifests locally in forms consisting of an enormous number of micro-parts and a few macro-state parameters, that, together, are reflected as size and shape.

Local conditions range from gaseous states in thermodynamic equilibrium, where the microscopic level is maximally dispersed and the macroscopic state is a simple uniformity,[42] to those varying in changing macro parameters, chemical compositions and gradients. At the other extreme, complex structures form in and after sudden phase transitions from gas to liquid to solid. Here the motion of elements and molecules become slower with declines in temperature, allowing quantum electro-magnetic forces to bond them together and to crystallize with orderly direction.

As the local processes become increasingly complex, they become more stable even as they move away from equilibrium.[43] The individuality of the form arises from within its own dynamism and so can be seen as self-organizing and spontaneous. "Isolated systems differentiate as they individualize in complexity."[44] They are self-maintaining in relation to the surrounding environment. With increasing separation of size and structure, comes more uniqueness and complexity, as long as local gradients remain stable. But overall, the universe's gradient from free energy to entropy decreases.

> **The Point:** Within macro objects, the relation of size and form is increasingly complex. With increasing complexity, individuality of wholes emerges out of constituent parts.

## CHAOS

In a state far from equilibrium, the energy flow of an open system begins to generate spatiotemporal patterns. Local particle flows coordinate into "extended flow patterns."[45] Random fluctuations at the micro-size become orderly at the macro-size. The internal or self-interactions within the contained flow, fed by energy from the environment outside, produces a coherency of repeating structures.[46] The classic example is Benard convection cells. When a container of water is heated from below, the water at the bottom heats and expands upwards. At a certain temperature, a phase transition occurs and the cool, denser water sinks to allow the warm water to rise. **The flow exists independent of its material.**[47] Heat energy converts to a "moving equilibrium" of "rolling cylinders."[48] Molecules move in a coherent fashion creating a "global dynamic order"[49] that gradually dissipates as entropic heat through the top. Change in shape becomes motion of patterns.[50]

Turbulence is the point where the flow's movements are self-similar. As things change in time, simple patterns become complex rhythms. Initial conditions and variables are in constant flux, never the same. The relations are nonlinear, eventually becoming incalculable; "predictable at first, but eventually unpredictable, as the time scale within the system is affected by its own dynamics."[51] In convection cells entropy eventually dominates, but the force of gravity also affects the water molecules. So from disorderly motion within a flowing form, orderly patterns can spontaneously, but temporarily or contingently appear. Chaos is the science that studies both the forms that appear in the middle of gas or liquid gradients as well as the dynamic relations among collections of solid forms.[52]

Chaos is the point at which a smooth flowing river becomes a whirlpool, or a tornado suddenly changes direction, or multiple planets come under each other's gravitational pull.[53] The "three-body problem" (three bodies or more) where each body exerts a gravitational force on the others, most notably in astronomical systems, is incredibly complex for mathematical calculation. The solutions among objects as well as multiple macro-variables like temperature, volume and pressure in weather systems and turbulence require a different approach.[54] The immense non-linearity of the equations reflect the dynamic initial conditions that lead the individual trajectories to diverge exponentially.

With computer technology, previously incalculable data points of movement can be simply traced and graphed. In one example given below, a point is measured at a fixed distance from each of three variables. The three variables' trajectories together would be represented by the corresponding trajectory of the point. The trajectory loops indefinitely, but at some point crossing over to another loop and then back again without ever intersecting itself. A shape begins to appear, something like butterfly wings, or a one-sided folded piece of paper- a moebius-strip. This shape is called a "strange attractor," similar through any change in parameters. Since the outer boundary of the shape remains limited, the trajectories become infinitely close to one another. This is again a fractal visualizing infinity. But here, with infinite change comes a changeless quality. Dynamics becomes geometry,[55] and measurable quantity becomes topological quality.

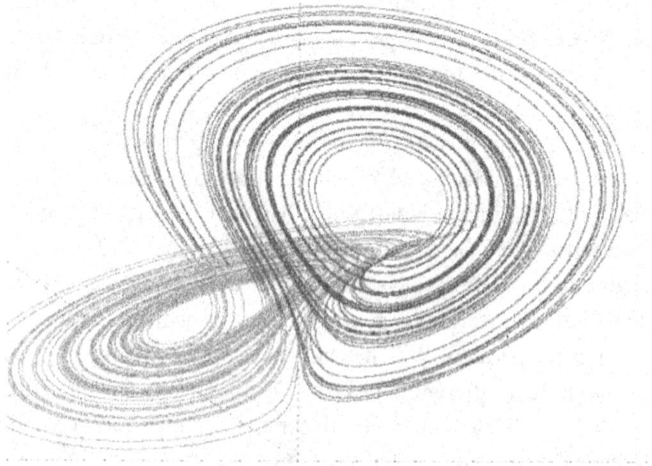

Fig. 4a. Self-Similar Trajectory

Each round of loops is self-similar to the others over continuous scales. They are self-referential in that the behavior of each loop is in accordance with others within. Although fractalization takes place in time, fractal change represents correlations between values of macro or phenomenal level variables,[56] not correlation with a clock.

> **The Point:** Chaos is the point at which flow spontaneously self-references into a form in time, but not measurable by a clock. These forms are fractals, with loops changing sizes while maintaining similar shape. The forms are patterns, separate from the underlying space-time matter.

## FORM WITHOUT SIZE

Every rhythm in a chaotic trajectory would in totality not only produce an infinitely long line in a finite area- a fractal, but an infinity of infinitely long lines. Between energy input and entropy, Leoanard Susskind and George Hrabovsky in *The Theoretical Minimum* ask us to imagine a theoretical state in which all possible starting points and all possible trajectories simultaneously exist in a "fictitious space filling fluid,"[57] where each individual particle is symmetrical (that is, "formless," circular with uniform angular velocity) and would, therefore, retain its energy value on the same energy surface. This state of extreme equilibrium would be neither an entropic "divergence" or an energetic "convergence." The volume would, therefore, be incompressible. The volume would remain throughout any change of shape.[58]

This is a description not of a fractally linked chaotic pattern (same shape, many sizes) but an infinitely layered volume with many shapes. Since that which is incompressible is also irreversible, and irreversibility marks both the general arrow of time and the local time of an energy to entropy flow, then **time, we can imagine, is concomitant with both progressive individuality of form and free diversity of forms**. In other words, time is at the apex of the infinitesimal and the infinitely discrete.

**The Point:** In a thought experiment that takes the chaotic moment to its ideal abstraction, pure size, regardless of shape, links within entropic time the infinite continuum as individual form and the infinite discrete as multiple forms.

## COMPLEXITY

Beyond randomness and orderly chaos lies the province of Complexity Theory, or perhaps, just complexity. Thermodynamics and chaos are necessary but insufficient to describe complexity. There is no 2nd law of complexity. Complexity is random order that emerges and increases with energy and an increase in entropy, far from equilibrium but always in isolated forms. Though the universe as a whole increases in entropy, increases in complexity occur not everywhere, but locally in already complex objects.[59] **Where there is complexity there tends to be more complexity.**

No two hurricanes are the same. The trajectories of three macro parameters, in this case the macro gradients of pressure, temperature and humidity, are not only incalculable, but the values of the parameters themselves are in constant flux and are certainly different in separate storms. Variety in minerals derive from "the condensation of hot fluids driven by a temperature gradient."[60] Most or all of the natural elements formed since the beginning of the universe were created in the varying temperature and pressures within stars. The varying gradients of countless chemical reactions have produced an immense variety of molecular constructions, and without any known limit, albeit tempered by electro-magnetic structural restraints. Uneven entropic dissipation with energy loss, along with gravity and friction, affects the original thermodynamic construction.[61]

In general, simpler objects become more complex, often structures within structures, but not in any consistently measurable rate. They emerge in a multiplicity of sizes. Non-uniformity remains paramount, unexplained by the ideal of mathematical simplicity. The simplicity of symmetries in formless particle-waves breaks apart increasingly in complex directions as the temperature lowers. Out of chaotic patterns comes increasing individuality.

"Complexity itself may not be a science but simply the pattern of nature."[62] And as we shall see, in the "inherent drive of matter to form ever more complex structures, life seems inevitable."[63]

**The Point:** Simple patterns arising in chaos from random structured flow become complex structures with progressive individuality. Each whole is made of unique variations of parts.

## SYNCHRONICITY

We've discussed the emergence of orderly patterns in chaos and the individuality of singular forms at the crossroads of free energy and entropy production. Synchronicity theory deals not just with rhythmic patterns in change, but with simultaneity among objects.[64] "Harmonic oscillation" occurs between particles when random movement of frequencies match, that is, are in harmony. Oscillation is "mutual synchronization"[65] which **occurs in time, not in space**. As opposed to entanglement, energy is absorbed but below the threshold of full transference to another particle.

In observed conditions, similar to the thought experiment in a previous section, in a closed thermodynamic chemical reaction, Steven Strogatz, *How Order Emerges from Chaos*, describes a point at which an area of total equilibrium consists of symmetrical particles, while the total flow moves independently. The individual particles remain in the same energetic place, but the "pattern of excitation" forms a "chaotic wave" rather than a periodic one.[66] As the particles form larger chaotic patterns, "the patterns synchronize when a changing value of a macro variable is shared."[67] The surface of the chemical mix does not move. The oscillations themselves are patterned. The chemical waves become self-sustaining, deriving energy from the oscillating particles upon which they circle. Form is separated from matter, as the patterns are moving in sync without disturbing the material substrate. As a "liquid surface without volume, it is fractal, with less than three dimensions and infinitely thin."[68]

In a laser, energy is funneled into a collection of atoms until they "oscillate together in phase...energy input and dynamic order are dynamically linked."[69] In contrast, a light bulb emits light in a random continuum without resonance. At phenomenal levels, larger and longer lasting examples of self-organization called "oscillons" or "phonons" have been demonstrated.[70] They emerge as a single-frequency oscillation across a crystal lattice structure. There have been quantum experiments in which photons have entangled as a single phonon when split beams were aimed at two closely positioned diamond plates.[71]

**The Point:** Chaotic patterns are rhythmic space-time matter. Synchronization is rhythmic simultaneity of forms in time separated from matter.

## NETWORKS

Network theory concerns the organization of complexity. Thermodynamic systems themselves separate when the reaction rates between the energy source and entropic outlet differentiate. The dynamic patterns of interaction among systems form space-time geometric relations. An "architecture of relationships"[72] defines a "network topology."[73] Phase transitions of all kinds, such as freezing, boiling and magnetizing may vary across scales[74] but all are "saddle points" of limited areas where transitions are most easily accomplished.[75] What is generally observed but not yet fully understood[76] is the "universal pattern of complexity...short chains, high clustering, and scale-free link distributions."[77] Collections of randomly shaped fragments will gather, forming cascades of energy, but only within a certain range of distances. If there are too few parts in a cluster, they will easily dissipate. But the more complex the pattern, the more robust the form.

In liquid crystals, systems of individual forms that, starting with resonant vibrations from photons to phonons over macroscopic distances within the form, spread among multiple forms evolving into a network that is in turn a self-organizing structure.[78] **Different patterns occur in the same circuit**, or flow areas. Parts can be organized in unimaginable combinations in different patterns. The same parts can form different patterns. "Complex networks are the next step in a quest for a science of spontaneous order. If the day should ever come that we understand how life emerges from a dance of lifeless chemicals or how consciousness arises from billions of unconscious neurons, that understanding will surely rest on a deep theory of complex networks,"- Strogatz.[79] What emerges in complex networks is the dynamic structure with which to innovate.[80] The "robustness of patterns emerges from the network itself."[81]

**The Point:** Complex structures organize as networks. Parts and wholes become independent of one another. Patterns themselves become individualized.

## INFORMATION

If the 2nd law of thermodynamics explains where everything is heading, then information is where it's come from. If mass is the embodiment of energy, energy is the embodiment of information.[82] Macro-level properties, like elasticity, conductivity and transparency "arise holistically from interactions of a large number of atoms."[83] They are statistical properties. **Thermodynamic dissipation is statistical embodiment.** What we see is statistical knowledge of what we cannot see. In other words, for every macro-state there is an immense amount of possible microstates, i.e., an indescribable combination of micro-parts. Therefore, the same macro-state can be derived from many different microstates. Measurements like density and temperature are averages of particle motion and distribution,[84] not a detailed description. In a cooling gaseous filled container, the relative amounts of gas and liquid can be known only in terms of statistically based whole quantities.[85] Correlations among the immense amount of parts are unknowable.

Entropy represents unknown information, in other words, the lack of fully described microstates.[86] This is why statistics has been called the "theory of ignorance,"[87] that is, the micro-level causes of macro-events are merely insinuated through patterned regularities both phenomenally observed and found within statistical distributions of individually unknown particle trajectories.

On the one hand, observable forms are broken into smaller and smaller components until finally reaching clouds of probabilities. On the other hand, the degree of macro constraint on a system adds a different kind of knowledge, as entropy is reduced and non-equilibrium is increased. Mechanical or efficient causality, that is, "specific antecedent physical conditions"[88] like force, energy and work are non-geometric[89] and, therefore, also not directly observable.

Thermodynamic entropy is the "arrow of time," that is, it is time directional as order becoming disorder. Energetic motion can be understood as procedural information. The degree of entropy can be understood as a computer algorithm.[90] If quantum states are thought of as bits and changes of state, e.g., excitations of energy in electron orbits thought of as operations, then the number of bits (particle qualities) and the number of operations performed can theoretically be measured. Chemical reactions, as we will talk about later, can perform logical operations. The mathematics of dynamic interaction will be seen as procedural.

In information theory, correlations reduce randomness, which then lowers the amount of information needed to describe any phenomena.[91] In information terms, energy is thought of as a message. Thermodynamic entropy is a statistical measure of the amount of information contained in each message received. For Claude Shannon (1916-2001), the founder of information theory and a formula for information often thought of as an inverse to Boltzman's, "information is counterweight to entropic disorder."[92] **Entropy is a measure of disorder. Information is a measure of order.** It "quantifies instructions" required to construct an organization,[93] which e.g., is the minimum number of bits needed to decode (or encode) the message. The more possible combinations, the more uncertainty, the more information is needed. Redundancy, or identical sections, reduces the information needed by increasing predictability. Total certainty means no more information needed.

**The Second Law means matter flows from order to disorder, past to present. Information flows from disorder to order, past to present.** According to the First Law, the total energy of the universe is constant. The 2nd Law says that entropy always increases. Yet energy is not lost. The increase in entropy is pure direction. The direction is forward into the past, turning thermodynamics into information.

There is a certain ambiguity with talking about mass/energy as information. Is information a separate entity from the knower, from knowledge of the information? Does the observer or the object decide? There are two schools of thought in quantum theory: the "epistemic" and the "ontic." In the epistemic view, the wave function collapse represents a block to knowledge. All that can be known is that there is a discontinuous change in knowledge.[94] Multiple wave functions could represent the same state, leading to uncertainty as to the true nature. In the ontic view, reality lies in the entanglement of wave functions. Each wave is entangled slightly with others and with different parameters in each entanglement. As the entanglements increase, observed interactions are less likely to encounter superpositional realities. Particle entanglement arose from a "shared cosmic causal past" in the big bang.[95] The present interaction in this view is the most certain, as it is the most individual. The farther away something is in space and/or time, the less we know (is present to us) and the more knowledge or information there is to be acquired. **Knowledge of information flows from present to past.**

**The Point:** Micro-states, the various organization of parts, are known statistically, not directly. Entropy is the measure of disorder. Information is the measure of order. Entropy is unknown information. The thermodynamic arrow of time flowing from order to disorder can be thought of as information becoming more certain from the point of view of the present. The totality of information is information to be known. Knowledge flows from present to past.

## KNOWLEDGE

Quantum Information theory defines a "pure" state as a state in which "nothing else can be correlated." It is "maximal knowledge," that is, there is no unknown information and as such, is self-contained.[96] A mixed state contains only partial knowledge, that is, some information is hidden or uncertain. Isolated models in Newtonian mechanics are artificially pure. Thermodynamics is mixed since detailed information of single particles are not known, only their statistical distribution. In quantum mechanics, at the individual particle level, randomness and information is uncertain and not complete, so we initially have a mixed state. But in quantum information theory the Purification Principle "states that if one takes into consideration a sufficient number of systems, it is always possible to find a level of description where all systems are in a pure state."[97]

In other words, **we can have maximal knowledge about a composite system but remain ignorant about its parts**. An un-entangled particle A is in a mixed state but an entangled AB is pure in its correlation with itself. The information is pure because it is self-sufficient. If information is contained within the knowledge of two observers, or two events, but not either alone, then that knowledge, and only that knowledge, is pure."The system as a whole is in a pure state, but the state of each individual particle is 'mixed' with that of its acquaintance."[98] Entanglement means that information is encoded so that knowledge of one part comes from knowledge of another part.[99] Informational purity, an unobstructed message, becomes knowledge when received through entanglement with another message. A higher level of information is achieved. The pure individual is self-knowledge.

Information defined as message, unlike the spread of free energy/entropy, is dimensionless.[100] The question still remains, how do we get from formless yet discrete particles to a space-time continuum? Quantum equations are themselves time-symmetrical but temporally irreversible and irretrievable at the point of wave function collapse. If quantum or particle information is known in a reversal of entropic disorder, how does quantum mechanics, that is, particles themselves, account for the "arrow of time?"

Seth Lloyd (1960-), the founder of Quantum Information Theory, finds that as particles become entangled, a single particle state (e.g., "1" or "0" for a type of spin) becomes "mixed." A "pure" state can only be applied to the entanglement, as uniqueness of that state occurs within the sharing. **The purist individuality is a relation**. With increasing entanglement of entangled states with one another, the collective state becomes less uncertain. When the totality of correlations contains all the information and the individual particles are completely unknowable, even as probabilities, the total correlation is in a state of informational equilibrium, that is, changeless. Locally, the information about a cup of coffee, for example, "becomes smeared over the entire environment," and therefore, does not spontaneously heat up. "The arrow of time is an arrow of increasing correlations."[101]

De-coherence can be viewed as the loss of information into the surrounding environment, in effect sharing or transferring information into the universe.[102] Quantum information theory looks at the universe as accumulating entanglements approaching macro certainty. Local totalities of entanglement are the observed phenomena in the present. We can't remember the future because the present is "the process of becoming correlated with our surroundings" when the photons reach our eyes. "The universe as a whole is in a pure state," Lloyd says, "but individual pieces of it, because they are entangled with the rest of the universe, are in mixtures....Whether the entanglement with an observer is objective information or subjective ignorance, either way, it would seem that **time is local**."[103]The more certain the information the more local.[104]

Though the quantum realm is not geometrical, "entanglement is the fabric of space-time," or at least "we must think geometrically about entanglement."[105] Space-time is explained as emerging from knowable levels of collective properties of interlinked nodes in a complex network. A entangles B. AB entangles CD. ABCD entangles EFGH, etc. "Smooth, continuous space-time emerges from discrete bits of quantum information."[106]

**The Point:** The purest state of certainty is the state of correlation itself. Total certainty is total self-sufficient individuality.This is not knowledge. Pure knowledge is shared informational certainty between two less than certain entangled partners. Entanglement is the individualization of relation. Greatest certainty is a state of greatest entanglement just before the present moment of de-coherence - a dynamic interaction with the universe. Total certainty is lack of information.

> Life is lived in the state of present knowing, looking forward, with certainty always, immediately behind.

## SPECULATION

A two-dimensional system rises from the formless. "Space-time is a geometrical representation of quantum information."[107] A hologram is a three-dimensional representation of a photograph. If the universe is a three-dimensional holographic projection of two-dimensional information, expanding networks of entanglement, creating two-dimensional images, that in turn project a three-dimensional world, may be plausible.

We could also imagine this concept without using information. Since a massless particle at the speed of light doesn't have dimension in the direction of motion, it is flat, or two-dimensional. Mass, therefore, only manifests in slower three-dimensions. Perhaps the Higgs Field causes three dimensions along with particle mass. The probability cloud, however, remains formless.

But we might also ask, does a universe beginning in a point and expanding in every (infinite) direction necessarily create curvature along with volume? Equidistant to the Big Bang, narrow information or energy points would, barring all other factors, seem to necessarily connect in an expanding sphere, or in some way at least a curved space-time continuum. I can imagine that space-time distance between "information streams" is curved. Einstein apparently speculated that the smallest space-time point is a "curvacle."

These most general boundaries, small and large, and all subsequent boundaries or limits, are necessary for there to be anything for knowledge to work with. Particle/wave, entanglement, the holographic principle that three dimensional space-time is encoded in two-dimensions, space-time/mass and the connection, mathematically or intuitively, between entanglement and the holographic principle may be the best way so far that man understands how to put the given together.

But facing the limits and their connectivity, we must add the Second Law, the concomitant Information Theory and the knowing of that information,[108] the knowing of what has been.

In the irreversible arrow of time, past states, once they have reached the present, must still be quantitatively inferred, and what is calculated is quantities of information. But if the Second Law points in one direction, then the state of maximum information is all the way back, the state of maximum free energy order at the beginning. In information terms, the beginning is the total uncertainty of all information. From there, entropy spread outwards creating complex forms out of streams of information. The energy gradient, together with macro variables like temperature, provide the information.[109] Is each stream also equal to a "line of time,"[110] each its own arrow?

In some way, **information is an entity itself, or a quality of matter like mass.**[111] The pure state of entanglement is contained information."The monogamy of entanglement guarantees that no third party can be correlated with their random bits and decrypt the message,"- Guilio Chiribella.[112] The final correlation in time is the present correlation with a knower. Information is not information without an observing awareness. Information is everywhere. Meaning is observed and ascribed to everywhere. The future however, with dissipating energy, is unreachable.

But information is restricted at the lowest end of size by the uncertainty principle. The fundamental attributes of position/motion, and energy/time are simultaneously unknowable.[113] Some theorize that quantum non-locality/second-law-of-thermodynamics, that is, the information/energy duality we have been discussing, is itself a block to information.[114] But the combination of atemporal distance and asymmetrical time might not be a block. Assuming information is not information unless it is known, the combination may be the window.

And what of the universe, that is, our perception or conscious picture of the universe? Let's combine particles with a thermodynamic large picture. At the beginning, momentum and energy are infinitely large and therefore, totally uncertain while position and time are singular points and therefore, completely certain. In the end, momentum and energy are zero and therefore, completely certain, and position and time are infinitely large and therefore, totally uncertain. The extremes of the large seem to somehow mirror the extremes of the small. There are incompatibilities, of course. Universe values take place in a time span. Particle values, including time, are probabilities.

It is sometimes argued that entropy as disorder applies only to macrostates, like shards of a broken vase. Total entropy in a volume, means micro-parts randomly spread out with a sameness of temperature. But disorder can also mean there is no way back. Within shards, however, molecules and particles retain structure. Standard measures of entropy assume a system with a volume, but systems lose heat to the surroundings. Conscious imagination necessarily defaults to maximum surroundings, an infinite volume. The "heat death" of the universe, supposed in the 19th century, that all systems eventually lose all their energy, though incorrect according to current physics, reflects the development of awareness through the way consciousness necessarily organizes the given.

At the smallest, we are talking about quantum particles (the beginning of the universe according to many theories was actually pure plasma- the source of free energy). At the largest, we are talking about parts in general. Positions of all parts (not quantum particles) would not simply be evenly spread out, but spread out beyond measure, and therefore, uncertain in the sense of unknowable. Energy and motion between parts would intuitively be zero, and as such, certain. It should also be noted that particles have velocity. Parts have relative motion. The Uncertainty Principle applies to all particles at any time. Here we are comparing a particle to a theoretical infinite distance between parts, whether or not this applies to probability clouds.

SMALLEST (particle)

| High Frequency<br>short localized undefined wavelength | | Low Frequency<br>long defined wavelength | |
|---|---|---|---|
| energy | (high, uncertain) | energy | (more certain) |
| position | (more certain) | position | (uncertain) |
| velocity | (uncertain) | velocity | (more certain) |
| time | (more certain) | time | (uncertain) |

UNIVERSE (parts)

| Beginning: point<br>order, free energy | | End: infinitely spread out<br>disorder, entropy | |
|---|---|---|---|
| energy | (high, uncertain) | energy = 0 | (certain) |
| position | (small, certain) | positions | (infinite, uncertain) |
| velocity | (high, uncertain) | motion 0 | (certain) |
| time | (0 certain) | time | (infinite, uncertain) |

Seth Lloyd says, "The universe as a whole is in a pure state...of zero entropy"[115] Yes, but the pure state is pure if entangled with something. The entanglement is the something. The informational beginning is a single mixed particle. If we could look at the universe from outside, it would be one giant uncertainty principle. Inside are gradients of information driving complex networked organizations. But what is information if it is not recognized? And what time is it when we look? When is now? Is free energy driving entropy from behind? Or is informed entropy driving into the past? It could be argued that the universe is not like a quantum particle, because we cannot look at the universe from outside. Of course. But the thermodynamic picture is the universe from inside. And the universe from the inside can be imagined to look like a particle from the outside.

The universe itself is both certain and uncertain, both point and continuum. If time is a continuum of past to future, is the present an infinitely small point? Planck time supposedly is the smallest time that can be understood. Yet we are living through it.

**The Point:** The universe is coming towards us as we go out towards it. In time at the chaotic center, life knows what it can. It has to.

II. FORM

A. EMERGENCE

*Being is becoming*
Werner Loewenstein[1]

## CHAOS

In one perspective, life emerges from inorganic matter. Quantum mechanics, General Relativity and the 2nd Law of Thermodynamics set the stage, but specific patterning requires Chaos, Complexity, Synchronicity, and Network explanations. We have seen how order arises spontaneously out of chaos. "Patterned structures"[2] emerge that are both increasingly complex, individual, highly organized and "stabilize further and further from thermodynamic equilibrium."[3] Turbulence arises in a combination of many micro-parts and a few varying macro-level influences. Randomness and the macro-determinateness of "statistical mechanics and continuum physics"[4] merge. In solid structures like crystals, free energy and entropy mix as high order and low energy. In the example of a heating liquid, structure self-creates as rising and falling columns separate. What emerges is a dynamic relationship between parts and whole.[5] And the whole is distinct from the sum of its parts.

Although the advent of biological complexity can be seen as an emergent phase transition like solids-to-liquids-to-gas, life is qualitatively, ontologically distinct from inanimate complexity. **At some point complexity becomes self-similar**, that is, symmetrical through differences in size and time, a recursively repeated "pattern inside of pattern."[6] "Beyond [mere] accumulation"[7] periodicity becomes rhythmic with fractal shapes originating from within. The individuality of organized energetic flow endures while still open to its surroundings. It is a self-organizing leap above chaos. The identity of local particulars is subsumed to a circle of causation, a form independent of external causation.

Chaos also creates information. "The unpredictability in the dynamic relation of parts and wholes, chaotically changing between macro-regularity and micro-randomness, yet persistent through that change, is the embodiment of information," Gleick, *Chaos: Making a New Science*.[8] Life is an uncertainty becoming certain. **In life, information directs its own material, and as such, reveals itself.**

**The Point:** Life emerges out of chaos as self-sustaining patterns separate from the material. Information becomes self-aware. Complexity becomes self-similar.

## RANDOM

"Quantum fluctuations...form the source of random information in the universe."[9] Random motion, that is, "stepping to the right or left millions of times per microsecond,"[10] may be known only statistically, but is fundamental as thermal energy in orderly living processes. **A living being is a dynamic relationship between micro-randomness and macro-determinism.**

**The Point:** Random quantum fluctuations and thermal direction interact in the living moment of free energy becoming entropy and information becoming certain.

## CONSTRAINT

Reversed entropic processes, such as living beings as well as inanimate turbulence, are constrained by macroscopic parameters like temperature, pressure and composition. The values of these parameters are independent of entropic time.[11] Variations in these values as well as higher level loops constrain biological chemical reactions[12] at a local thermodynamic time. Each organism operates at an optimum temperature which constrains all levels. The least constrained process is at equilibrium.

**A biological system's own history of environmental interactions serves as an informational constraint.** Higher temperatures of a living body increase the motions of its atoms. But probabilities are increased with the knowledge that these constraints bring[13] and the temperature is stabilized. In life, the "thermodynamics of organized complexity" is a heterogeneous mixture of micro-states and macro constraints.[14]

**The Point:** Organisms differentiate and are differentiated by the interaction with the macro level environment, which includes both thermodynamic and informational constraints or limitations.

## NON-EQULIBRIUM

Living beings are "open dissipative systems." Constantly interacting with the environment, they use low-entropy energy, "degrade the energy, and expel high-entropy energy."[15]

Matter seeks equilibrium in dissipation. Living beings gather the dissipative gradient by moving towards it, store the energy and selectively release it producing their own gradients. Unlike inanimate things, **life is a process of "durable patterns of ephemeral entities"[16] organized to "resist thermodynamic equilibrium."[17]** This durable resistance is accomplished by the complex variety of "interconnected cycles"[18] flowing in and out of nested compartments, each with its own stored energy optimally timed for the persistence of the whole organism. The Second law is resisted within each loop by the "activation energy" required to break the chemical bonds of storage.[19] In life, stability is dynamic, with a balance of reaction rates. Each reaction is a variance of near equilibrium, the dynamics of which culminate in an organism far from equilibrium.

> **The Point:** The thermodynamic flow of energy is interrupted by a living being, kept above equilibrium, temporarily and selectively released in the nested buckets of its parts as required by the whole organism.

## NONLINEAR

Mathematically, emergence is nonlinear. If the whole does not equal the sum of its parts, the process is not represented by an equation, but a dynamic procedure. The equations are iterated, looped with ever changing starting points or inputs. Instead of cause and effect being represented as points in time, reality is represented as an ever changing continuum.[20] Separate events are represented as a single occurrence. The structures described are not static but fractal. **Living systems are complex, interactive and dialectical**.

> **The Point:** Mathematically, living beings are more like a computer program than a static equation. The relations are nonlinear with dynamic starting points in ever moving loops and sub-routines.

## GRADIENT

The direction of matter (mass/energy) is explained by the Second Law of Thermodynamics, that is, increasing disorder or entropy. The gradient from free energy to entropy, itself in concert with gravitational space-time, creates increasingly complex patterns that give rise to complex individual forms. **When the gradient turns in on itself, form is freed from permanent attachment to its material substrate,** and in doing so, reverses temporarily the arrow of time in individual loops informed of the direction whence they came. The directional change of energy to entropy is locally reversed. In matter, ordered energy flows toward disordered entropy. But a self-sustaining relatively closed system temporarily reverses its own entropic flow.

The initial free energy source for life on Earth is the solar gradient which is far more concentrated than the temperature and pressure differences that e.g., drive hurricanes. The entire ecosystem is itself one large gradient as matter flows in and out, up and down the myriad of local gradients, producing on the whole more entropy than an equivalent inanimate mass.[21] The energy is degraded as it moves up the food chain. Energy originating in light is stored along the way by living beings as chemical and electric energy. In general, a gradient is "simply a difference across a distance."[22] Life reduces the universal thermodynamic gradient by distinguishing differences in its own movement through particularized gradients.

Living beings make use of the high energy, low entropy energy sources by reducing the corresponding gradients while increasing their own order and releasing greater entropy in their waste product of heat and chemicals.[23] **With circulating metabolism the organism "frees itself from the entropy it produces."**[24] The closed energy loop is maintained with an increasing complexity of loops within loops that can store the energy until needed. Internal time awareness consists of thermodynamic loops, each of which has its own degree of thermodynamic non-equilibrium. All the loops, that is, chains of interactions performing a function, work together in timed interactions that in total are the cell or the organism. **The energy is stored from past encounters, leading to a gain in time distinct from its environment**. The resulting autonomy is necessary for survival. Indeed, it's the autonomy which survives.

Living organized complexity originates not only in evolutionary species selection but "top down via gradient breakdown."[25]**The detection of gradients introduces selection**. At first, movement is either directed toward nutrients or away from toxins.[26] This direction arises in life as an orientation toward the future. The origin of purpose, starting with directed movement, eventually culminates in conscious planning, as a coherent internal time is extended relative to environmental disruption. With stored energy living beings have the ability to choose when and where to act. Need becomes will to summon energy. "Thermodynamically based intelligence is the link between sensation of a food gradient and the expenditure of energy to arrive at its source."[27] Stored energy is used in "second-by-second" behavior manifested as both thinking and moving and is independent of direct genetic control.[28]

**The Point:** The energy gradient of the environment is captured, reversed and stored by a living being by selecting, i.e., choosing, needed sources ahead of itself, and using more energy to stay ahead of the environment and thereby releasing more entropy. The reversal of the gradient frees form from both the outside thermodynamic flow and its own entropy. Stored energy from its past, selectively released, creates its own time. Within that time, awareness of need and choice of sources arises.

## WORK

The First Law of Thermodynamics states that the quantity of energy remains unchanged (in a closed system). **The Second Law concerns the quality of energy**.[29] Free energy is the "part of energy still concentrated and useful."[30] It is the capacity to do work. To do work, free energy must be organized toward a goal for the work to do. Unlike the First Law, free energy changes as it does work. Its quality (exergy) decreases[31] as the heat increases. The random motion of heat is low quality.

**Unlike purely physical movement, work is movement with meaning**. It is intentional movement, toward a goal of choice. It is behavior. It requires effort both thermodynamically and logically as information. The conveyance of meaning is a two-way street.

Though living beings, i.e., organisms, are constrained by macro-variants, living cells are generally without internal temperature and pressure gradients (though chemical and electronic gradients are pervasive). Unlike a machine, external temperatures and pressures do not specifically drive work. Indeed, at the molecular level of biochemical reactions, these forces are absent. **In metabolism, work is created in the exchange of information and energy**. New knowledge is stored up to a point and then erased as released energy, causing information not yet known to increase, and the corresponding entropy to increase.[32] An increase in more order than the environment, produces an increase in disorder. **Information turns work into agency**. To a living being, thermodynamics is information.

**The Point:** Energy quantity is turned into energy quality by meaningful categories in work. Information and energy are exchanged, turning work into agency.

## B. FORM

*The ability of form to be active is the main characteristic of mind*
Michael Talbot[1]

### LIFE/FORM

In thermodynamic chaos, patterns become increasingly individualized forms or shapes. In life, form itself is liberated. Energy flows through matter. It transitions from e.g., water pressure to electricity, or as heat from metal to liquid or vice versa. Life is form flowing through matter (information through energy) as matter flows through form (energy through information). The ontological relationship between matter and form is reversed. **In the inanimate, form is a quality of matter. In life, matter is a quality of form.** What in matter is an "abstract distinction, becomes concrete reality....Form is the essence, matter the accident."[2] The form persists through a change in matter. Most of a living body's material is continually replaced. All of it is replaced during a normal lifetime. Every last atom in a living being is different after a period of time, while the body, more so in an adult, remains relatively the same. "Over a period of weeks you are materially a totally different person."[3]

In matter, the forms change along with change in matter. Knock a rock into pieces. The forms change, but the matter changes right with them. Now instead, imagine sand (little sand matter-forms) constantly pouring into one side of the rock-form, filling the rock-form (including the surface) and then that sand pouring out of the other side, all the time the big rock maintaining the same form.

A living being, a living form, is not a static thing, it is an organized process, "dynamically stable." It is a process of "being an organizing whole."[4] Here, we finally do reach Aristotle's formal cause. Form is no longer caused by material laws. It is formal cause in that the form, "organized with respect to its own persistence,"[5] causes itself. Formal cause does not mean form causes or gives rise to life. It means living form, as an organizing principle, is its own cause. Causation is from within. Life is a self-sustaining flow in basic metabolism all the way up to neuronal patterning.

**The Point:** In life, form individualizes to where it is freed from matter. This does not mean separate from matter. It means form remains through a change in matter. As such, individual form flows through differentiated matter as differentiated matter flows through individual form. Freed form is a self-organizing process, maintaining itself as life within the flow.

## LIFE/MATTER

Living form is not independent or totally separate from matter like Cartesian consciousness is separate from extension. In fact, life must have matter readily available.[6] At any given moment (usually pico-seconds) form and material are one in structure. Structures within a living being exist temporarily, assemble to perform a particular function and then disappear.[7] The form returns again and again, with different material constituting the same structural organization.

Though inanimate matter "obeys precise mathematical laws,"[8] living matter acts according to rules. "Laws are forever, rules can come into being."[9] Biological rules are unknown laws of both physics and chemistry.[10] But the laws of matter are still unavoidably present. Life makes it own use of them. Living order is not a mechanism. It is an organism. And as such it is teleological. In biology there is "doing" and "meaning" where in physics and chemistry there are only "happenings."[11] Organisms also adapt to their surroundings, as no mere "organization of atoms"[12] at the level of physics does.

From the point of view of life itself, what is matter itself? Each species of living being is organized to survive in its own environment. Elements in its own environment are selected, that is, made meaningful by the organism. What is food to bacteria is something else entirely or probably nothing at all to elephants. **Life in general makes use of what it has been given.**

Forms flowing through matter interact with a changing something. The totality of changing somethings has been described by humans. If it works, and is consistent with overall theory, it is true. Human awareness up to and including scientific awareness has explained the world with progressively useful categories and definitions, that up to now have only taken us back to the starting point, expressed currently by quantum particles and a space-time continuum. Even mass and energy as well as the categories called "life" and "matter" are not originally given. Although life itself cannot by definition experience that which came before, the given cannot be simply non-life, or we are back to Descartes' insurmountable division, with life inventing reality out of nothing. It needs something to work with and to be in. **Awareness arises in life with the conversion of the changing-something into information.**

Though living forms may be temporary, they do endure for awhile. The energy of change flowing through the form is temporarily resisted by the form,[13] but eventually the form is worn down. Just as the internal structures are constantly turned over during an organism's lifetime, so does the individual organism in the context of all Life. In death the form and the material become one. Imagine a live body lying on the ground. Imagine that same dead body lying next to it. What is the difference? What was in a dynamic relation now becomes static, as form dissipates with matter. In death, the animate become inanimate. Form is no longer animate. If a living being could live forever would it remember an infinite past? Life requires time.

**The Point:** Life is not matter, it is living matter. It lives inside the fundamental choice within the given that arises in form that is free from matter. The choice to survive necessitates making what is given meaningful. Matter is given meaning.

Death is essential, for life is lived in its own time keeping one step ahead of everything else by looking forward at everything else. Living forever would be timeless.

## TIME/FUTURE

Death implies life. Death implies life time. Life is "time-dependent"[14] "with a built-in history."[15] Living processes die. Non-living things don't. Instead of static consciousness or awareness observing change in matter, change is incorporated in life itself. Life is both form flowing through changing matter and matter flowing and circulating through form. The flow is a continuum between the separate distinctions of form and matter.

**Time is change with direction**. But which direction is future and which is past? Future and past are themselves defined by the direction in which life flows. The relation between form and matter is directed in life. The environment to the living being exists as flow from the past. Within the living being an extended but temporary present continually opens out to a not-yet, creating a future flowing into its past. From the infinitesimal current moment of pure time in living awareness, past and future are directed.[16] Life looks out at the past in anticipation as if it were the future, though the future is ever non-existent. In life, form frees itself from matter in the choice to continue. And in doing so invents the future directed from an eternal now. It does so in any way it can and as long as it can.

As I will show later, even the infinite regress of consciousness, when directed solely at itself, is a loop of new time. Each loop of the regress takes and/or makes another infinitesimal moment of time. It takes its own time. Between subject and object, distance is temporal. The immediate future is an "abstract possibility space explored by complexifying life," called by Stuart Kauffman (1939-), "the adjacent possible."[17] **Life/awareness is a feedback loop moving into the absolute nothingness of an anticipated future.**

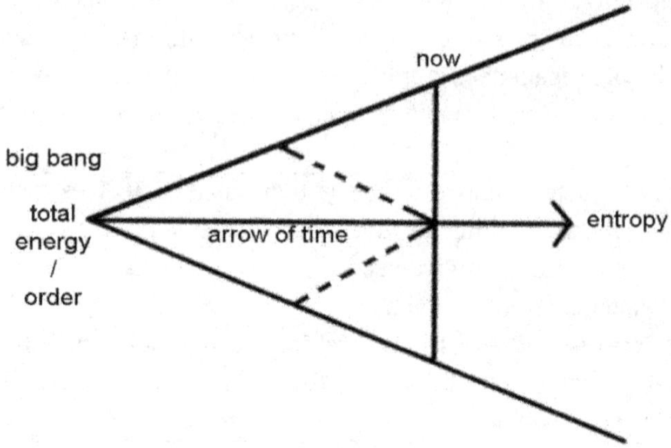


Fig. 5.

The dimension of time, represented by an infinitesimal point, has in life an inside-out through which the clock-time universe passes. Matter is outwardly extensive. Life has an "inwardness,"[19] lived in the present. But our sense of outside time passing is based on Thermodynamics' 2nd Law.[20] Clock time in the matter universe is relative. Cause before effect is concomitant with universal clock time. In living systems, causation is circular, a blending of cause and effect. This creates an "extended present."[21] Life is always in the present. Within that extended present, extended in loops of interconnected processes, the future is opened by a forward rolling anticipation. It is within this dynamic window that life looks forward at the past, and out from itself.

Looking at the above graphic, I ask, "If there were no life anywhere in the universe, how old is the universe?"

**The Point:** The opening at the nexus of form and matter is pure time. Life is a loop of rolling time. Pure time is the result of directed change necessitated by the necessary but impossible choice which changes infinitely between the small discrete and the large continuum. Living time is directed change. External time is the measure by life of the matter that passes through form. External time is spatial with future determined by the past. Internal living time points in the direction of the future, creating its past.

## SELF

**Self-identity emerges with life.** The living being retains its self "by not remaining the same matter."[22] The retaining of its own identity is fundamentally what it does. The form that maintains itself is the whole organism. The parts are ephemeral, with maintained functionality sustaining the whole. Self-reference emerges with self-causation.[23] Awareness is the "selfing" in its interacting direction of discrete-continuum change.

In self-identity self is self-othering. Life is self in a changing environment. Every living being has a sense of self as difference between its self and what is not itself. In the words of Hans Jonas, "Life is essentially relationship, implying transcendence."[24]

**The Point:** The progressive individualization of formed matter reaches its apex in the freed form of life. As such, being free is what life is. The discrete-continuum contradiction frees form from matter as an individual self over against the non-self in its own time, looking forward from the in to the out.

## BOUNDARY

Life is not lived on one side of the interaction with the environment. It is the interaction. As such, it creates its own boundary. The boundary is fully integrated within the living process. The self that is the organism does not exist outside its own process. **Between the living being and the environment there is "no material continuity...only the continuity of constraints."**[25] Beginning with the cell wall, the living boundary does not separate self and other. It contains self and other.

The rolling form in dynamic interchange with the environment maintains an organization that includes a permeable membrane "self-imposed and actively maintained."[26] The closed thermodynamic circle describes an inside/outside that is its own inside/outside.

**The Point:** In order for living form (living matter) to be separate from its non-self, the edge of itself must be part of itself. Beginning with the cell wall, the boundary contains self and other within it. A boundary is the dynamic limit of awareness.

## PURPOSE

With living awareness not only Aristotle's Formal Cause but Final Cause arises. Direction toward the future is end-direction. The one-way "downhill" thermodynamic direction[27] has no discernible purpose. Though a reduction of gradients occur on the path to entropy, physical laws explain results in terms of past occurrences and external forces. But as Aristotle observed, "living beings are prior to their parts."[28] **The living being in its outer and future direction is prior to itself.** The inwardness of formal cause is necessarily teleological.

Biological explanation "connects events and things with their future goals."[29] For example, thymine in DNA has the purpose of reducing mutation. Skin has the purpose of protecting the body. Life has the purpose of survival. "The whole vocabulary of biology is teleological."[30] Self and purpose are interdependent.

**The Point:** With the opening of the future, life's necessity of freedom is also its purpose. Its reason to live is to live. Purpose is at one with its freedom.

## TASKS

In a living being parts are not things. Parts are tasks. Structures are assembled to perform a specific function, a role with a purpose. Structure is the material instantiation of any living form at an infinitesimal moment. After performing the function, the structure either dissolves immediately or stays together for a limited number of repeated actions. Either way, the task continues as a repeated identical structure materially replaced. The physical material is transient. Structure and function at all levels of the organism are "intertwined."[31] Though the function outlasts the structure, each structure is formed specifically for a given task. Formal cause is not material, it is "geometric,"[32] that is, structure as such.

**Life itself does not exist at the molecular level**. The cell is the smallest unit of life. Molecules alone do not reproduce or redistribute their atomic structure.[33] The functional structures within don't repair themselves.

The cell repairs its parts and in doing so repairs itself. Their function is to receive and pass on specific messages to and from specific other functional structures. Each single function is operationally both cause and effect. All molecules in a cell are linked functionally. The function only takes place in the relation between temporary structures. Functions occur because the organism causes them, and the organism continues because of the maintenance of the functions. Resolution of the function is effected between molecules but understood by the organism in networks of timed interacting loops upon loops. The process is cause-effect-cause rolling through the instances.

Immanuel Kant summed up the part-whole relationship in living beings with what is known as the Kantian Whole: "In an organized being, the parts exist for and by means of the whole, and the whole exists for and by means of the parts."[34] The organization causes the parts to act differently than they would by themselves.[35] We will later see how even genes are not understood by their sequences but by the roles they play. We can see here how responding to situational change is best survived by evanescent functional structures. Conditional responses demand flexibility in the assembly of structures.

**The Point:** Living parts are not matter. They are pure functions or tasks made up of passing matter ephemerally formed into structures serving the purpose of the organism.

## PATTERNS

The closed loop of flowing energy/information expands in a nonlinear procedure of loops nested within loops, forming "durable patterns on ephemeral entities,"[36] "resilient patterns in a turbulent flow."[37] These patterns of loops remain through the fluctuations of the part/tasks. Each loop, including the grand loop of the organism itself, is persistent while incrementally malleable. Microscopic processes loop into macroscopic patterns, all at a distance from equilibrium maintained by constraints, including the pattern itself.[38]

Because patterns are abstractions and not "physical instantiations"[39] they can retain a sameness at different sizes and times. Positions of parts and rates of movement are not quantitatively measured absolutes but topologically, qualitatively, relative.[40] Organizational complexity actually simplifies in biological systems by retaining patterns across all life. Basic patterns are "building-blocks" that are reused in different combinations.[41]

The basic loop of self-metabolism is the most durable form extant through all types of organisms. The pattern of amino acids in proteins, nucleic acids in DNA and membrane lipids are universal re-combinations of recurring structures.

Various loops become relatively concretized over evolutionary time. Patterns are a hierarchy of not only timed loops but speed of turnover, relatively faster in smaller loops. They range from metabolic molecules, working up to the intercellular matrix, networks in organelles, organelles in the cell, cells in the organ, cellular turnover within the organs and finally the organism turnover in replication of the species. Tasks/structures may be determined at a local pattern level conditionally, but **nowhere in the network is there an area of "privileged" causation**. Everything is working for the sustenance of the organism.

**The Point:** The looping time of life is progressively complex as the looping tasks become patterns of loops, yet progressively simple as the same patterns are repeated throughout levels of life. All patterns work for the sustenance of the individual pattern, the organism.

## C. INFORMATION

*Information is the currency of life*
Christoph Adami[1]

### INFORMATION

Awareness is the conversion of the transcendental reverberating continuum/discrete of the given into information. From the material point of view, there simply is a process of converting Second Law energy into information. Information by itself is quantitatively measurable. This includes the rate of information. But Life creates qualitative meaning. A chemical by itself doesn't mean anything. To a bacterium it means food. To a human chemist it means a particular chemical. Looked at from outside, the "essential function of living beings is consumption, processing, preservation and duplication of information."[2] But to life itself, information is meaning.

The total random motion of energetic micro-parts is total information, that is, total information yet to be known with any certainty. According to information theory, "a message has more information the more random it is."[3] But the less random the more specific the message. In a living being a part/task's function is to pass a specific message. The **specificity is accomplished with coding, the recognition of similarity in patterns.** The coding is accomplished by structural shapes. Shapes are recognized by other shapes in what is called "structural complementarity."[4] In a living cell, structures moving randomly on the motion of atoms, "bump" into one another. When appropriately recognized by the right receiving structure, they fit like a "lock and key."

At the smallest scales, the movement of an individual particle from a high energy end of a system to high entropy end does not reveal any cause. There is no "Maxwell's Demon" making a decision about each particle's direction. There is no purpose or external force.[5] But if energetic particles are delayed and accumulated locally, the pace and amount of accumulation and release is accompanied by a local gain in time. The thermodynamic, straight arrow becomes circular in life.[6] Each of the many loops turns the arrow around locally. Environmental information changes into local meaning by messaging in a reversal of time. Memory is the accumulation of meaningful information. Stored energy is memory of its own past. As stored energy is released for purposeful work, local task memory is erased, making room for new information. Meaning is circulated in an ever adapting organization that as a whole can store as much information as it can retain ahead of its own entropy.

The organization is its own constraint. Pattern recognition is a top-down constraint of entropy.[7] At the macro-level, uncertain information is reduced by "contextual information processing"[8] in a self-communicating system. Meaning is contextualized information. In shape arises semantics. Shape holds more information than a quantified series. Different series of identical length can all be coded differently e.g., 13649821...13-64-13-64. Organizational meaning is wider than the informational source and therefore, is in a position to direct cellular computation.[9] It is context that is concomitant with independence from incoming information. Quality drives quantity. "Imposes" is from within. According to Shannon, if there is no uncertainty there is no information. Because of selection, meaning brings its own uncertainty, the uncertainty of that which is not selected. Between random total uncertainty and deterministic total certainty, the present moves toward the future, ahead of total certainty. There is wiggle room for something different.

At all levels, in all loops, the organism brings memory forward. The distance between the present and the environment is temporal. **Time gained is future borrowed**. Out of nothing appears possibility. Life's own ever establishing future is its force. Knowledge at a distance is turned into expectation. Purpose becomes the object of action. Primal urge for "attainment," as Hans Jonas says, is "self-concern from want."[10] "Striving for what helps and withdrawing from what hinders is action born of foreknowledge of both possible satisfaction and pain."[11] Effort is work. "In freeing itself from laws of physics...the organism becomes sentient... A coherent free being, moment to moment is free to explore and create possible futures, Shapiro, *Evolution: A View from the 21st Century*."[12]

Stored information, though adaptive to the environment is dynamically independent. Independence is further individualized as a whole, consisting of structures within structures reflecting differences in entropies,[13] that is, stored informational time spans. Living development is iteration of forms remaining through these collections of local spans. Self-similar structures arise from "efficient redundancies in coded meaning.... Compression can generate resolution independent of target images."[14] External parts become the living beings' own ephemeral parts.

At the same time the information, the meaning, is not an isolated abstraction, it is directed at an object.[15] The shape of sounds and colors are ephemeral. But the shapes provide the information in a way necessary to extract meaning. Tasks are exactingly specific in their encounters. Meaning is relatively constant. Further combinations are pliable.

Possible objects are chosen objects. At bottom is the choice to survive. Choice within the individual manifests at first as a distinction of sustenance and harm. Conceptual choice in stored meanings deepen with distinction of gradation and self-other. One-celled bacteria will climb a gradient toward food and/or sense information about others of its kind.

**The Point:** Awareness turns information into meaning by selectively reducing information and storing it as meaning with a gain of time. Time gained enables the freedom to choose and energy for work to obtain the choice.

## SELECTION

Information enables selection.[16] Living beings retain their own reduction of the flow of environmental information. This reduction is a picture or a model conformed to, and formed within, its own organization. What a one-celled organism like a bacterium experiences as knowledge is at a miniscule time and environmental scale. The structure of knowing what to do in that environment and what to do with recognized and unrecognized phenomena is part of being a bacterium.

Selection of qualities from the blur of environmental change, produces the knowledge of what a living being can act upon. The living being selects from the given and makes it mean something. At bottom, it is whatever meaning it requires to keep going. The "changing blur" is in itself an absolute necessity to make something out of it when confronted. **Choice is the first cause.**

The environmental future is nothing. It does not yet exist. The organism is choosing from future possibilities that it selects from the models formed in its memory. The environment that the organism confronts is the external world of its own past. What has happened in the environment, that is, the environmental changes and the organism's retaining of its reactions or adaptations to what has happened, provides the information. The environment that the organism lives in is the environment that it has organized in a way that best suits its survival. What a living being can do is a combination of its maintained structure and knowledge of what is possible for that living being. The ongoing moment is ongoing anticipation. The future is unknown except by prediction.

**The complementarity of task structure, the lock and key, demonstrates that message selection is built into the very fabric of life**. Looping evanescent patterns of established meaning are constantly interacting with new information in the environment and propelling the organism forward and outward from its internal time structure, its rolling present.

The selection becomes present to the living being as something to be acted upon. Any mistakes in selection or messaging will elicit a mistaken response. Life either learns from mistakes or perishes.

**The Point:** Meaning selected from past information and applied to its own future is reflected in the living structure of each individual organism. Material discreteness flows through continual form. Life selects to its own specifications of what is useful to continue the flow of its form.

## COMMUNICATION

Though information may be thought of as an isolated entity, meaning demands a receiver. The ambiguity of quantum entanglement as either pure information or pure knowledge reveals the necessity at some point of an end user. Is information really information if it is not known or received, that is, communicated? What does it mean if it doesn't mean something?

There are five stages in Communication Theory: "Information source, encoder, communication channel, decoder (or receiver) and a user."[17] The information source is a small-large blur redirected within the living being's own time frame. Interaction with the environment is dynamic motion. Meaningful information is transmitted two ways in a forward flowing circle. Signals from the environment are received, interpreted by the organism's own past receptions and reactions, which in turn affect its ever new actions, affecting its new interpretations. **Life is not merely reaction or merely action. Life is over-reaction.** Using means adjusting.[18] Decoder and encoder are interchangeable. Correct anticipation is rewarded. An open-ended communication cycle that implies a future necessitates a future object. Life's own future is receiving itself.

We might ask that if each living being is acting within its own time, how does a coherent collection of individuals arise? From an objective point of view, we know the answer...space-time is relative. From a subjective point of view the receiver is another subject. Enough distinguishing information is passed from individuals to mean self and not others.[19]

**The Point:** For information to be meaningful, there must be communication between information source and receiver. Organisms are both.

## COGNITION

Cognition derives from choice. Once set free, life must choose to survive by recognizing nourishment and rejecting toxins. Life chooses itself in recognition of itself in others of its kind**. The first internal choice is prioritizing self-other and self-sustenance.**

There is a distinction between the ever-changing informational environment and the meaningful environment. The informational cycle becomes a cognitive cycle when expectation is altered. Meaning is both reduction and enhancement within. A priority of the moment must be determined. Tasks weigh alternative routes in if-then-not-then processing. Structures come and go conditionally in an otherwise stable organization that is an ongoing reflection of its environment.[20] Cognition reorganizes and converts sensory input into meaning. With cognition, new meaning is built.

> **The Point:** Choosing among possible meanings requires cognition of importance and what is new.

## D. FREEDOM

*Liberty is its own necessity*
Hans Jonas[1]

### FREEDOM

Form's immediate identity with matter becomes in life mediately separate. Life is free from the start, but its freedom is necessary to its very independence. It is free as long as it can keep one step ahead of its own material. Biochemically, the organism's internal gradients produce in complex timing its own coherent time apart from the environment. "Living beings function in their structural present."[2] The present is felt presence. Free will is a feeling that is real in its dependence on its own continuance. Creativity is born within the organism's self definition. In the rolling now, there is no spatial or temporal separation,[3] only awareness of possibility. **In a circle of causality, freedom is its own cause.**

> **The Point:** Life is free in its flow through matter. Its future is its own creation, which in turn causes it to continue.

### ACTIVITY

Living beings are not defined by their physical body but by their behavior.[4] A rock does not do anything. Freedom means freedom to act. Biological parts are not completely constrained by laws of physics.[5] Physical reactions run down to equilibrium. Living non-equilibrium dynamically stays ahead of equilibrium with an organization consisting of interlocking feedback loops.

With feedback loops an organism stabilizes by regulating its own production.[6] Part of the output is redirected backward in a circle of causality.[7] Specific tasks function to monitor conditions and decide whether to promote, that is, continue production (positive feedback) or inhibit (negative feedback) production of required selections. Negative regulation can buffer fluctuations in the rate of production.[8] With feedback the organism responds correctively to change. When a needed selection consists first of only sub-selections, feed-forward loops wait in anticipation.[9]

Feedback systems in continuing communication circles of need, action and fulfillment extend the process, effectively creating time. Temporal change contributes to spatial-temporal complexity which loops back as downward causation, interacting with upward and downward loops of causation at all levels.[10] **Timing is required for goal-directed behavior.** The goal is survival. Living action is over-reaction. The life force that counters determinism is not magical. A living being's capture of its own time lets the future in and sets it apart from the world. Awareness, agency and knowledge all arise in the dynamics of timed loops.[11]

**The Point:** Feedback loops select for the purpose of forward timed resolution culminating in the organism's survival.

## AUTOCATALYSIS

The fundamental structure in a living being is a loop. The foundational loop is the autocatalytic reaction, the most basic Kantian whole.[12] Catalysis in the inanimate world is a chemical reaction in which a substance called a catalyst speeds the reaction but with less energy. Because the catalytic agent is not itself dissolved in the reaction, it can be used again.[13] **Autocatalysis is a reaction that catalyzes itself**. It is the most basic self boundary of life. Though a catalytic reaction generally takes billions of years, autocatalysis takes pico-seconds. This severe distinction in time and rate alone distinguishes life from non-life.[14] Unlike linear inanimate catalysis, the speed/energy ratio of nonlinear autocatalysis is exponential.[15]

An inanimate part does not catalyze itself. It requires at least two molecules. What autocatalyzes is a loop. An experiment performed by Israeli biologist, Gonen Ashkenasy, demonstrated how starting with a series of proteins, a protein joined two parts of the next protein into a second copy of itself. This procedure continued around a circle of nine.[16] But this describes a self-catalyzing chain, not a loop. No protein catalyzed itself. Continuing the chain requires a supply of new substrates (parts to catalyze). Otherwise, the same parts would continually be split and put together. This process would stop at the end of the chain or quickly run out of energy or out of substrates.

Since this loop is really a chain in a circle, there is no message, merely growth without reason. Life requires a receiver. The most basic autocatalytic loop would be a structure containing a message coder, a message decoder and a communication link. For the message to be meaningful, an autocatalytic loop is needed containing senders and receivers that make use of the messages. Meaning is what is retained through structural recurrence in a flow of transitory parts. **The unit of living awareness is an information loop.**[17] As the diversity of shapes increases, catalytic circles spontaneously increase as well.[18] Life did not spontaneously emerge one time.

Life is what is spontaneously emerging in its own continuance. Loops beget loops.

> **The Point:** By catalyzing themselves, living loops retain their form beyond the loop's matter. Self-similar processes outlast their material, but retain their form.

## CYCLES

**Loops connect by communicating with one another in timed cycles.** A single cell has many loops crossing over and through one another, each at its own pace and spatial direction.[19] Interdependence is based on expectations of continuance and detection of change in coupled "intervals of awareness."[20] Nested cycles upon cycles maintain mutually dependent circles of cause and effect. A myriad of cycles build up a complex rhythm, **reflecting in timing the self-similarity of loops**. All processes within the total organism are correlating fractal structures,[21] as micro-randomness and macro-determinism combine. Every cycle is regulated by the cell (and by the organism) to do the "right job at the right time in the right location,"[22] in a myriad of interlocking cycles.

Intertwined at all levels are the two main cyclic processes- the input and output, timed to keep the whole ahead of entropy, and ahead of the outside.

> **The Point:** Form maintained in repeated instances of loops is further internalized in the timed cycles of multiple loops.

## LEVELS

The living process is a multileveled[23] "communication web."[24] Most biological entities, from basic tasks, to loops, to patterns of loops, to cycles, to patterns of cycles, are both senders and receivers of information.[25] Although patterns of causation are both downward and bottom up,[26] there is a hierarchy of complexity. Higher level, more complex patterns within a nested dynamic structure constrain, regulate and organize those below.

Looked at thermodynamically, spatial restraints speed the time of entropy production. Turnover at the lower level is faster (though interlocked with cyclic timing) than the more persistent pattern above.[27] The gradient potential of stored energy interacts with the kinetic energy below. At that lower level, many resultant reactions may reach equilibrium. In feedback, feed-forward communication loops,[28] the larger function determines the future actions of the smaller, that in turn "feeds" the larger.[29]

Controlling levels constantly affect the quantities and structures of forms (patterns and tasks) below.[30] But dependence goes both ways. Form and function of both are mutually affecting quantities and structures continuously, but **each level of organization is the observer of the one below**.[31] With more complexity, the user has more information.

The "continuous and nonlinear causal spiral factors of interdependence at multiple levels of the life cycle,"[32] operate in a continuous, not discrete but not uniform, and therefore, fractal space-time. The initial and boundary conditions are both steadily changing, not in sequence, but infinitesimally timed to one another. The standard calculus does not quite catch up to the continual recursive flows as functions are integrated at a level.[33] Levels are moving resolutions. The organism as a whole consists of many levels acting as a "unified agent."[34]

**The Point:** Looping forms and timed cycles communicate in levels of complexity, with the receiver of knowledge and director of action at increasingly broader levels.

## NETWORK

While up to this point we have been concentrating on complexity, network analysis looks at the living process as a simplification by topological repetition. The network is generally and similarly described at all levels and in all organisms. The structures of different living beings may differ from one another and from themselves in time, but the organization as such is the same.[35] Similar or identical networks can be observed appearing in separate evolutionary periods. **With the network, "organizational durability emerges."**[36]

A living network is a pattern defined by nodes and edges. A node represents the coder/decoder/informational gate/decision-point. Edges are the message path/interaction links between nodes. A feedback loop would be a two or three-node pattern, with the edge a halting or delaying interaction. A feed-forward loop is a three-node pattern, with "activation edges."[37] Self-regulation would be one node. Different combinations are known as network "motifs."[38] Communication widens through the connected accumulation of nodes and edges. Multi-level domains within can be open or closed, further controlling tasks and loops and speed of reactions.

Growing networks have their own dynamics as concentration of nodes affects the pathways of links. Network locations are not determined by where a node is but the temporal distance between or among nodes.

**The Point:** Network topology is what is universal and simplest about all living processes throughout any level of complexity. All processes are combinations of nodes and links called network motifs.

## AUTOPOIESIS

Thermodynamically speaking, self-organization results from non-equilibrium and higher level constraints.[39] In life, however, self-organization is also unassisted self-maintenance, self-assembly and self-fabrication.[40] Unlike a machine, or a material system, it is its own source. If the material were its source, the same functioning organization would have to be built repeatedly. If not end-directed, the output would be arbitrary.[41]

Autopoiesis is the description of the living process as further entangled directly and dynamically with its environment. The living system is fundamentally an organization "structurally coupled"[42] with its surroundings. The organism and its environment are each the organizer and the organized.[43]

Organizational complexity as such, is not material, but informational. Since the organizational components are dissociated from the material,[44] its reality is its abstraction. In this way, **the top-down complexity of living is knowing**. The organism's independence is conceptual.[45] Interaction is not determined by molecular properties but by how the complexity knows and actively adapts.[46] Living rhythms are a "hierarchy of coherence times,"[47] that is, hierarchical not in terms of causative points, but in terms of complexity, which is the greater good of the organism. Abstracting is an action. The system is causation.[48]

**The living system remains more complex than the totality of its environmental, interactive history**. Structural coupling is a "sensory-motor coupling."[49] Parts of the environment are selected and obtained to fuel tasks to continue selecting. The organization is not determined by an incoming flow of material parts, but maintains itself by producing and defining its components with informational complexity.

All living beings are in some way attentive to and, therefore, aware and causative in their environment.

**The Point:** Autopoiesis extends the self-communicating organization to the whole organism-environment structure. Just like levels are the users of the communication below them, the organism is the user of the total communication with the environment.

## REPLICATION

The task structures within the cell are ephemeral, lasting pico-seconds and reconstituting as required by the cell. Self-maintenance includes repair and replication of its parts. Loops and cyclic patterns are maintained through passing task structures throughout the one-celled or multi-celled organism.

**The requirement of information to communicate is what drives replication.** Meaningful complexity requires growth. At some point, growth becomes unsustainable. Eventually the meaningful whole must replicate.

The organism does not replace itself, it communicates beyond. Instead of self-replacement, the whole becomes a part. Then the part becomes a whole.[50] Thermodynamics brings about order from disorder, but an organism creates "order from order."[51] It is not duplication of the individual but expansion of the kind.[52] Reproduction is replication of a similarity.

In the material world, stability follows a reaction when energy flows downhill. In life, stability is found in forward action to make more. What makes more is the information communicated as meaning. In doing so, it frees the individual from itself.

**The Point:** A living being constantly maintains its ephemeral parts by repair and replication. Meaning grows in complexity until growth is unsustainable. The whole individual does not replicate. It reproduces. The individual makes a part which becomes a similar whole.

## E. STRUCTURE

*Despite the profound advances in molecular biology...*
*we still do not understand what life is*
Addy Pross[1]

### BIOCHEMISTRY

Living beings are real in their abstraction from the material. Forms outlast their material structures. Standard biology analyzes and explains these structures as causal in two ways: From the reductionist bottom up parts in biochemistry, often understood statistically, and the historical determinism of evolution. Neither of these approaches are wrong. They are the necessary roads that life travels in understanding itself from the outside. Indeed, each in their own way reflects micro-parts and macro-determinism. In this chapter, we'll examine how the object of biology itself consistently demonstrates the true reality of life.

Organic life is not separate from the universe. It uses the universe.[2] The reduction of solar energy is Life's first reduction of external information. Biological organization begins in photosynthesis with the absorption of sunlight photons. The material photon turns virtual in the quanta of energized electron orbits. The excited electron separates from the molecule and its excess energy is utilized as it runs down to its original energetic state.[3] The energy is dispersed throughout the organization. Logical circuitry is manifested in a myriad of chemical reactions.

On the macro-side, not only is the organization kept in check by thermodynamic constraints, but by the ultimate macro-constraint, gravity. For example, the size of single cells is inversely proportional to gravity.[4] Though both inter- and intra-cellular processes are chemical and electronic, physical force affects the whole, initiating a "cascade" of chemical and electric events that in turn alter physical responses in tune with the environment.[5]

On one end of the size scale, the smallest cell is 10 times larger than the largest molecule, a protein (molecule:100 nm diameter, smallest cell:1000 nm). At the other end, the largest life form on Earth is dwarfed considerably by the planet, let alone the cosmos. Living beings are very much in the middle, both in size and in the ratio of magnitude to multitude. For instance, the typical one-celled E. Coli bacterium is $10^{-15}$ the mass of a human but is made of 100,000 billion atoms.[6]

The living being converts inorganic chemicals such as carbon, nitrogen and oxygen into biological molecules in an open, far-from-equilibrium, networked flow of matter and energy powered by its own storage of energy. Tiny bio-molecules are neither confined by gravity nor chained spatially. They flow through one another in every direction, structures often colliding and breaking. Bio-molecules are structurally functional with a head-tail polarity that gives them a sense of direction.[7] Molecules can freely move about yet interact without losing direction. In this way separate molecules are functional in relation with one another. Their structure allows them to pass information, receive messages and contain both.

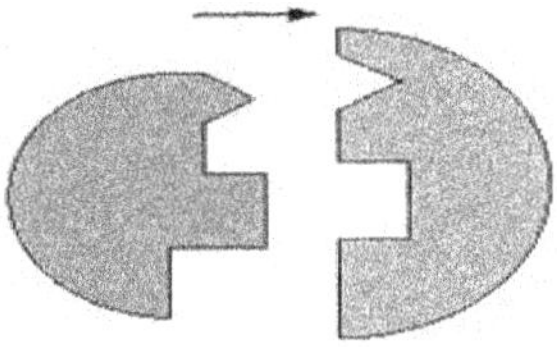

Fig. 6.

As the structures are ephemeral, the interactions between molecules are transient, typically lasting picoseconds (one trillionth of a second). The large energy gradients in the cell are achieved by the totality of many tiny reactions. **The molecular structures determine the directionality of the loops.** With lock and key shaped molecules, the loop is already prepared to receive itself.

Locomotion occurs in two ways: 1. The random motion of atoms in cellular water bumps the molecules around until they find a structural "fit." 2. The molecules change shape powered by stored energy.

Biochemical reactions compliment the directionality by achieving equilibrium not at the lowest energy point but in a circular rolling movement that never quite catches up to itself. Equilibrium lies in the future. Transient reactions are stabilized in replication.

> **The Point:** The flow of matter both selected and transformed by a living being are bio-molecules whose lock and key shape drive and are driven forward by and into the loops that they form.

## MATERIAL

In the most general material sense, a living being is "a carbon based system operating in a water-based medium."[8] Bio-molecules are on average thousands of times larger than non-biological molecules. Small molecules of carbon, hydrogen, nitrogen and oxygen are converted by stored energy into large lipids, nucleotides and amino acids. First, plants using sunlight energy, convert carbon dioxide and water into organic molecules, releasing oxygen in the process of photosynthesis. In the opposite process of respiration, animals convert organic molecules and oxygen into energy and more molecules, releasing carbon dioxide and water. Both processes convert free energy into entropy. Within the organism these processes first produce energy storage and sugar.

**The three essential factors of life- a boundary, memory and sustenance (procedure), are physically manifested as lipids (membrane), nucleic acids (DNA), and polysaccharides (proteins).** These all demonstrate chemical complementarity and hydrophobicity [separation in water], perfect for directional processing.

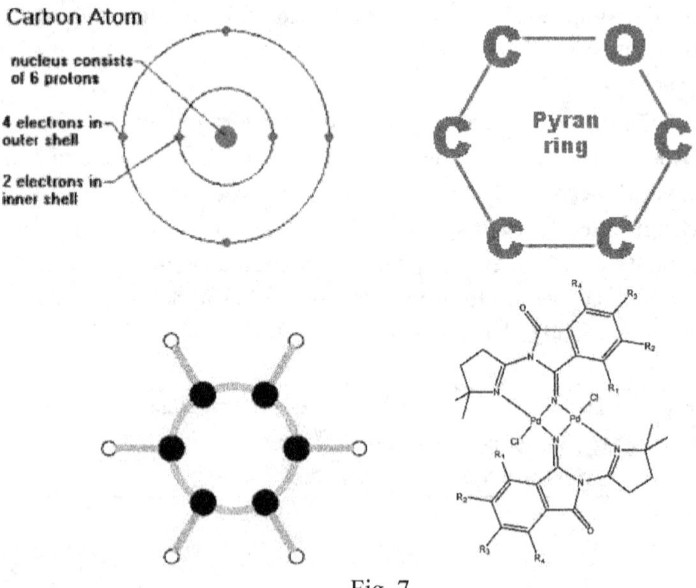


Fig. 7.

The flexibility of carbon's structure makes it a progenitor of organized complexity, as it is amenable to long chains and branches of infinite diversity, especially in three dimensions. For example, the pyran ring is a hexagon consisting of five carbon atoms and one oxygen atom to form a molecular template and link up in a growing chain.

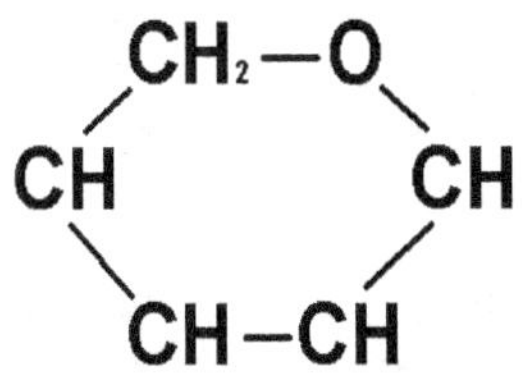

Fig. 8. Glucose

Glucose is a sugar consisting of carbon, hydrogen and oxygen. It is first made during photosynthesis from water and carbon dioxide. Oxygen contains much more energy than carbon dioxide. When it reacts with oxygen, glucose's potential energy is released for work. Oxidation (loss of electrons) and reduction (gain of electrons) are both parts of a central biological redox reaction.

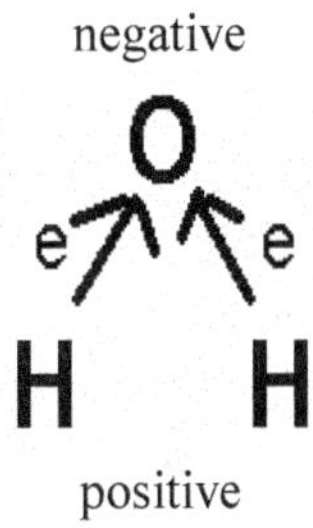


Fig. 9. Water

The structure of water gives it a remarkable organizing ability. Water or $H_2O$, consists of one oxygen atom and two hydrogen atoms. The hydrogen atoms' two electrons are bonded to the oxygen atom, leaving one end of the water molecule positive and the other negative. Water molecules are attracted to one another.

Unlike other liquids, increased pressure increases viscosity- the resistance to deformation.[10] The energy of water is reduced when squeezed and organized in smaller areas, increasing entropy. The hydrophobic force separates water from non-polar molecules like oil, which cannot bond with water. The molecular fluidity of water (the hydrogen bond exchanges take picoseconds) is the thermal motion pushing the large molecules around.

The extremely fast motions randomly make room for the large molecules to connect.[11]

Random thermal energy passes to chemical and electrical energy into deterministic mechanical energy and entropy.[12] Large molecules like proteins have, in turn, "exclusion zones" which merge on contact, creating more space for smaller molecules and more entropy.[13]

Living material- lipids, DNA and proteins- is neither solid nor liquid but in between, similar to a liquid crystal. Because liquid crystal molecules align in one direction, while at the same time freely exchanging places, they exemplify the organizational potential of molecular chemistry for life.[14]

**The Point:** Beginning with energy from sunlight, the structure of non-living matter that life uses, is physically structured for the living direction.

## CELL

The cell is the smallest discrete unit of life. The cell is the primary subject of objects as well as being an object of itself as an observing individual.[14] Life does not exist at the molecular level. Molecules don't have selves. Outside of the cell they just run down. It is the structure of the cell that contains the three basic functional structures required for life:

1. A permeable membrane that separates the organism from the environment, allowing energy to come in and waste (entropy) to go out, and is a boundary of "informational individuality"[16] that enables a one-celled being to compete with others. The cell maintains gradients of potential energy as gradients of molecules trade places across the membrane. Multi-celled organisms and one-celled eurkaryotes, unlike one-celled prokaryotes, also surround the nucleus and some other structures with a membrane

2. DNA is the repository of information about the environment and that which has been retained to survive in the environment. DNA functions as the cell's long-term memory.

3. Proteins made by the DNA carry out the activity of finding, receiving and using energy to keep on going. This includes replacing DNA and membrane lipids, as well as proteins. The cell monitors its parts for damage or change, and maintains itself with the goal of survival, growth and replication.[17]

All living beings are made of cells, each with millions of synchronized interactions. The cell functions as a whole, responding to the environment not randomly, but decisively. Each kind of environmental occurrence or signal demands a particular protein or protein combination for which the cell must not only choose to produce but calculate the correct amount. Proteins and lipids in the membrane are asymmetrically aligned to couple the environment with the internal network. The molecular level performs computations and makes logical decisions with "well-defined functional objectives,"[18] but at the direction of a cell aware of itself and its environment. Any disturbance is immediately communicated throughout. This decision making cannot exist outside the cell. As the cell organizes the correct structures, it dissipates the disorder as entropy.

Although living beings are individuals in a way that inanimate objects are not, life is the same beneath variations in individuals. There are no basic functions in any living being that are not already performed in a single cell.[19] In fact, humans may be said to be the result of "collective amoebic consciousness,"[20] differing in immense complexity as environmental meaning expands in accordance with the organism's functional structure.

Frame of reference is a matter of time and space. Function is to time as structure is to space. The cell as a whole is a network of functional structures that at any given moment point to their own next moment created by the rate of interaction. **The rate at which energy is transferred for work is called power. Power is energy divided by time.** So the more power in work the less time. Time slows relative to the extra energy used in the interactional movement.[21] The energy in the molecular bonds are, therefore, a measure of time. The energy used is a reversed gradient determined by the cell's anticipated requirements. The sense of time for the whole cell is the state of all functional structures at any moment. "Life constantly emerges out of its own time dimension."[22] Meaningful response is seen from the outside as the cell's space-time frame of reference.

The total frame of reference, internally, consists of the expectation of memory, learned by experience and evolution, applied to the environment in anticipation of change. Unexpected change adds to memory, all through loops of recognition. Short-term memory is a continuing comparison of the conditions as better or worse than just before now,[23] giving the cell the ability to predict events. In this sense, a one-celled organism hints of higher life forms.[24] An amoeba can be observed preparing to capture prey by first folding its shape.

Since even in a single pico-second, the entire cellular situation bears on any action, the cell cannot react simultaneously to more than one stimulus.[25] Every self-created moment is a moment of attention, attending to the effort called for by anticipation or surprise. Imagine the cellular symphony of billions of relative space-time moments, each with its own "tempo,"[26] but being whole at any one moment of decisiveness. The changes of internal rhythm, matching and remaining a beat ahead of the one outside, as we will show later, creates internal representation.[27]

> **The Point:** The cell is the basic unit of life. Life does not exist at the molecular level. The totality of loops form a complex rhythm- each loop powered by its local future, the rate of energy release timed in accordance with other and greater loops, with the cell as a whole the furthest from equilibrium.

## MEMORY/GENOME /DNA

**The genome (the totality of genes within the DNA) does not tell the cell what to do. It tells the cell what it can do**. It contains structural, functional, procedural, regulatory and timing information for the running of both the rest of the cell and itself. It's fundamental role then, is to serve as the cell's long-term memory both individually and ancestrally. Memory contains information.

All living beings have the same modular structures in their genome. But the combination of structures are different. Information is stored with a four-letter alphabet: A, G, C & T. DNA (deoxyribonucleic acid) consists of combinations of four base subunits: Adenine (A), Guanine (G), Cytosine (C) and Thymine (T) in a directionally linked double chain.

The double chain is essential to cell division and heredity. A single chain serves in the ongoing operation of the cell. The single chain does not copy itself but serves as a template for its replication. This basic molecular structure consists of carbon, nitrogen, oxygen, phosphorous and hydrogen. Depending on the organism, it can contain over a billion atoms.

Combinations of any three letters constitute words called codons. The codons are strung together in a myriad of combinations to make sentences called genes. Three of those combinations function as a period or stop at the end of a gene. Letters and isolated words are codes. Codes with dispersed repetition contain information. Information must be translated into a contextual activity to be meaningful. Codons correlate with amino acids.

61 3-letter combinations translate to twenty amino acids. Gene sentences are translated into the language of protein sentences which consist of amino acid words. There are approximately 20,000 gene-sentences in the human genome which are organized into 23 chromosomes (23 pairs, or 46,

in a double strand). There are at least 3 billion codon-words, most of which reside in DNA areas outside of the gene-sentences.

> **The Point:** Information about the environment and past encounters with the environment are stored in the words and sentences of the genome within the DNA molecular structure. This information is translated into meaning within the words and sentences of proteins.

## PROTEINS

Proteins in all living beings consist of up to twenty different amino acid words. To better visualize the translation logic, codons and amino acids are listed below according to how they fall in the translation chart. Listing them alphabetically or any other logical way would be irrelevant. There is no given reason why any individual codon or group of codons translate into a particular amino acid. The protein is what is meant.

### Twenty Amino Acids

| | | | |
|---|---|---|---|
| 1. Phenylalanilne | 6. Serine | 11. Histidine | 16. Glutamic acid |
| 2. Leucine | 7. Proline | 12. Glutamine | 17. Cysteine |
| 3. Isoleucine | 8. Threonine | 13. Asparagine | 18. Tryptophan |
| 4. Methionine | 9. Alanine | 14. Lysine | 19. Arginine |
| 5. Valine | 10. Tyrosine | 15. Aspartic acid | 20. Glycine |

### Translation Chart

| **First Pos.** | | | **Second Position** | | | | | | **Third Pos.** |
|---|---|---|---|---|---|---|---|---|---|
| | **U** | | **C** | | **A** | | **G** | | |
| **U** | UUU | 1 | UCU | 6 | UAU | 10 | UGU | 17 | U |
| | UUC | 1 | UCC | 6 | UAC | 10 | UGC | 17 | C |
| | UUA | 2 | UCA | 6 | UAA | STOP | UGA | STOP | A |
| | UUG | 2 | UCG | 6 | UAG | STOP | UGG | 18 | G |
| **C** | CUU | 2 | CCU | 7 | CAU | 11 | CGU | 19 | U |
| | CUC | 2 | CCC | 7 | CAC | 11 | CGC | 19 | C |
| | CUA | 2 | CCA | 7 | CAA | 12 | CGA | 19 | A |
| | CUG | 2 | CCG | 7 | CAG | 12 | CGG | 19 | G |
| **A** | AUU | 3 | ACU | 8 | AAU | 13 | AGU | 6 | U |
| | AUC | 3 | ACC | 8 | AAC | 13 | AGC | 6 | C |
| | AUA | 3 | ACA | 8 | AAA | 14 | AGA | 19 | A |
| | AUG | 4 | ACG | 8 | AAG | 14 | AGG | 19 | G |

| G | GUU | 5 | GCU | 9 | GAU | 15 | GGU | 20 | U |
|---|---|---|---|---|---|---|---|---|---|
| | GUC | 5 | GCC | 9 | GAC | 15 | GGC | 20 | C |
| | GUA | 5 | GCA | 9 | GAA | 16 | GGA | 20 | A |
| | GUG | 5 | GCG | 9 | GAG | 16 | GGG | 20 | G |

Since there are 61 codons correlating with 20 amino acids, most amino acids are represented by multiple codons. For half of the combinations, the first two positions points the combination to a single amino acid. In general, the 2nd position determines the direction in which the amino acid will link, affecting the 3D folding of the protein. The third position is generally thought to be information free but is flexible in determining connectivity.

Chemically speaking, water is again essential as protein folding is partially determined by the lowest energy configuration, with hydrophobic amino acids attracted toward the inside and hydrophilic amino acids attracted toward the outside,[28] the degree of which is determined by the 2nd position of the codon. The final shape is the most stable, which is, of course, the lowest energy state.

Other factors, such as the pH number of the water are instrumental. Acidic water is determined by the amount of negatively charged hydrogen atoms (ions). Bases dissolve as negative hydroxide ions. If the pH number is less than 7 the water is acidic, the most acidic being 0. Above 7 to 14 is the base range. The pH number can be affected by the kind of amino acid itself, in turn affecting the protein functionality. Electromagnetic charges within the protein are open to alteration by multiple factors selected within cellular processing.[29] The protein does not "self"-fold in isolation. It is also built in accordance with other proteins and cellular chemistry. (A more complete description of the chemistry involved would, of course, be found in a biochemistry textbook).

**What is lost in direct translation of words is gained in the ambiguity of contextual meaning.** "Bank" means "savings bank" and "river bank" in English. French has separate words for each. Here, context arises as 3D spatial structure. Protein-sentences, made of amino acid words, are connected together by the directional, complimenting, "lock-and-key" fit with other molecules. Much like human language, a single sentence out of context says little or nothing. Protein loops are self-explanatory instructions, reading molecules and writing molecular products as needed. The loop of many interactions is like a paragraph or a manual. The DNA-protein division is analogous to, and probably the progenitor of, genotype and phenotype.

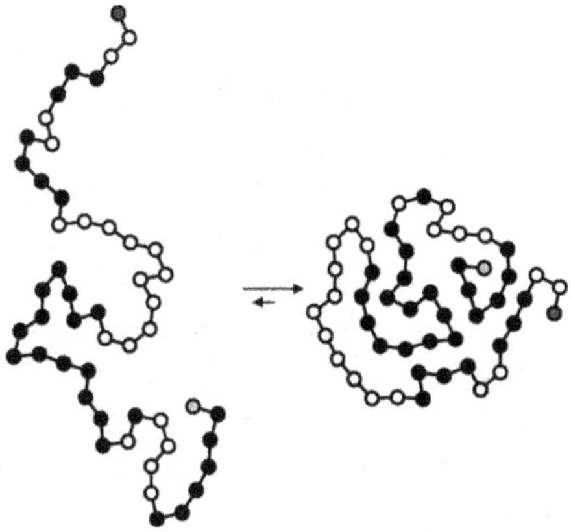

Fig. 10. Protein folding. The hydrophobic amino acids in black.

By folding into three-dimensional shapes, the code becomes both singular in structural function and immense in possible combinations. If there are for example, 200 amino acids in a single protein then $10^{267}$ is the number of possible combinations, more than could have been created in the history of the universe.[30] But is the idea to have an instruction for every single possible event? Information theory teaches us that total certainty is no information. Information to be informative must have some repetition, some uncertainty. Meaning is not infinite information. The DNA code has a structure which allows interpretation. The average human protein contains 483 amino acids. The number of kinds of proteins in the human body is estimated at 100,000. The protein language is highly flexible in meaningful interpretation yet efficient in modularity.

**The Point:** The information stored in the genetic language has a structure that can be translated into a protein language of three-dimensional meaningful, forward pointing shapes.

## THREE-DIMENSIONAL ACTIVITY

Proteins perform virtually all activity in the cell, and each performs a specific function. Proteins control events by e.g., recognizing molecules that mean, or inform, that they should not produce some other molecule. The specificity of the shape allows there to be receptor proteins for every environmental signal that the organism recognizes. All recognition is a reduction from the total information provided by both the DNA and the environment and made meaningful in the context of the total cellular activity. Information is exchanged for entropy. But the "information potential" is the opposite of the energy potential. The receiver already has information to interpret meaning.[31] The message is internally amplified. The specificity allows for many reactions to happen at the same time. The random association of signaling proteins is coordinated in a neural-like network.[32]

One of the principle functions of proteins is to catalyze or help chemical interactions. Catalytic proteins are called enzymes. Protein specificity means each enzyme has its unique purpose in catalyzing a specific reaction.[33] A protein's location is determined by the current demands of the cell and the protein's structure.

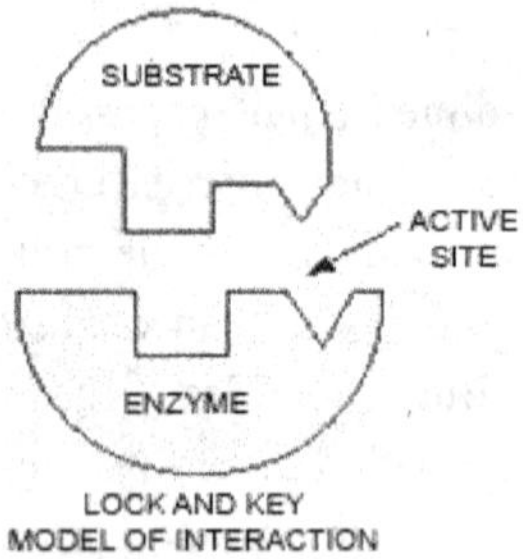


Fig. 11

Enzymes provide the cell with the ability to control itself and the environment. Cellular regulation is performed by enzymes that either activate or repress reactions and thereby affect cellular timing. The enzyme's "active site" recognizes a certain molecule (the substrate) by the fit of its "pocket." The particularity of the pocket's shape assures the chance of a correct match.[34] If there are any errors, side-loops of proteins will usually recognize and correct them.

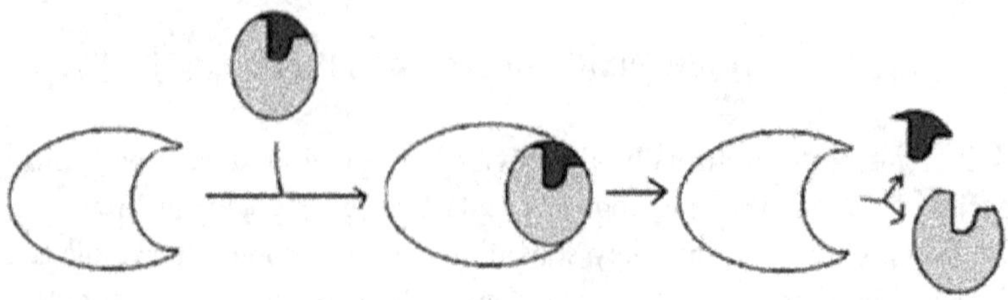

Fig. 12.

The catalytic loop is the most basic and smallest information loop. The enzyme recognizes the substrate, its cavity is "locked" to fit with the help of random water molecule motion (10 billion collisions per millisecond) and the substrate is positioned. Depending on the information contained in the protein, the substrate's bonds are weakened or strengthened (in an allosteric enzyme, two substrates are combined into one). The cavity then snaps back to its original open state, releasing the molecular products. Energy to snap back comes from its original configuration plus energy received from bonding with the principle energy storage molecule, ATP.

The protein or protein loop makes a cognitive "information profit."[35] No energy is transferred to the substrates during the reaction. The perfect fit enables the reactants to overcome the activation threshold and naturally react "downhill" towards reduced energy and increased entropy. By positioning the bonds advantageously, the enzyme influences only the speed of the reaction.[36] The transition takes approximately $10^{-14}$ seconds, which is less time than the bonding. Information passes to cellular meaning while gaining energy for work with the release of structural "translational" entropy,[37] without thermal dissipation.[38] At loop's end the enzyme is once again open and unchanged and has "forgotten" the information that has passed through. No cost has been paid by the enzyme. Higher free energy in the reactant molecules spontaneously converts to lower energy products.[39] **Structure recognition, then, is non-energetic causation.**

Living motion finds its physical origin in molecular motion. The enzymes and molecules move around the cell in two ways: 1.The random buffering by thermal diffusion of water molecules, where the constant motion allows every molecule to contact every other molecule in very short order. This movement is increased with the expanding entropy of catalytic action. 2. A "kinesin motor."[40] Some proteins form track-like surfaces called microtubules for other proteins to walk on. The structure forms two "feet." One foot takes energy from an ATP molecule to detach from the track. The other foot uses ATP energy to bend the whole protein forward, which forces the dangling first foot forward, which then attaches to the track releasing degraded ATP as ADP. Then the other foot repeats the cycle. The direction is determined by the combination of the random motion of the water molecules on the "dangling" and the deterministic drift toward lower energy.[41]

The protein pathways form a cognitive system. The protein selects its partner with a choice determined by its interpretation of DNA information. But without having to always directly refer to DNA long term memory, the concentration or accumulation of a certain molecule, reflecting the recent environment, can be compared to the current concentration. Bacteria have been shown in fact, to measure the rate of change in the amounts of a signal, performing a kind of calculus.[42]

The catalytic loop is at the heart of the "cognition cycle," beginning with a computation of the proportion of matching substrate boundaries.[43] Beyond catalytic closure the entire system of molecules is connected. One computation may affect changes in many molecules. **Cellular pathways are logical not spatial**. They go in all directions, even through one another. The first amino acid in the protein sequence serves as the locater, but the address is logical and timed by the cell. Proteins are like microprocessors serving as conditional logic gates in regulatory loops.[44] They might be looking for a match that itself has been created by another match, an either/or or not circuit. The more loops, the more internalized the meaning. The precise meaning of the original information becomes increasingly the precision of the cell. Meaning is amplified in protein groups based on function.[45] Relations with similar molecules with similar histories are formed.

According to Mai-Wan Ho, in her book, "*The Rainbow and the Worm*," protein molecules with similar functionality have matching frequency peaks.[46] In fact, the large size of proteins carry with them a large number of quantized vibrational motions. In quantum entanglement, "random motion contains the seeds of direction," she says, and may be efficiently utilized by proteins. Particle randomness and the thermodynamic gradient that began with the energy of light, combine in the living language of structural function.

In a way, the cell as a whole catalyzes its own pathways. Proteins catalyze new DNA molecules,[47] or control which parts of DNA are read at what time.[48] Repressor and promoter proteins may control an entire set of gene-sentences, called an operon. When a repressor binds to the beginning of the operon, the polymerase protein, which is responsible for DNA transcription, is blocked. For example, a toxic level of lactose may bind to the repressor, change its shape and not allow it to function. DNA transposons insert themselves into other DNA segments. Shufflons are DNA sites that allow for recombinations of codons that diversify the protein expression.[49]

**Genes work in circuits**.[50] At the level of the whole genome inside the nucleus, proteins called histones organize the DNA chromosomes to be read more efficiently. Only 1 percent of DNA has been genetically identified. The other 99 percent may contain controlling factors. The efficiency of modularity is demonstrated by genes working, sometimes differently, in different organisms. Some proteins called inteins splice their own segments.[51] Shapes and patterns persist in proteins independent of changes in the genes.[52] Different patterns may use the same genes. It is circuits of genes not genes alone, that are central to living information.[53]

> **The Point:** In the basic catalytic protein loop, shape changing is causative in the exchange from information to message. Proteins and genes work together in interconnecting circuits, all for the good of the whole cell.

## RNA

The DNA is a template containing all the gene-sentences. The sentences, however, are not necessarily worded correctly for every required meaning. Coding sections are called exons and non-coding parts introns. The function of single stranded RNA (ribonucleic acid) is at the direction of the appropriate protein or protein network, to provide a place for the correctly worded sentence to be translated. This preliminary process is called transcription. Proteins direct the timing of transcription as well as the selection of words. Different combinations of exons can produce different proteins from the same gene. Codon sequences can be read starting at various letters. For example, AAGGCCA can be read as AAG-GCC or AGG-CCA. Proteins called transcription factors can turn genes on or off. Activators and repressor proteins can form networks that mirror certain genetic expressions, with each node correlating with a particular gene.[54] A multiplex is a set of genes made by a particular set of proteins.

A protein called a polymerase moves along the DNA picking off the correct codon-words in the right order, forming an mRNA string. In the meantime, proteins have been catalyzing individual nucleotides with corresponding amino acids both floating in the cell in tRNA (transfer RNA) molecules, a molecule that is half amino acid and half codon.

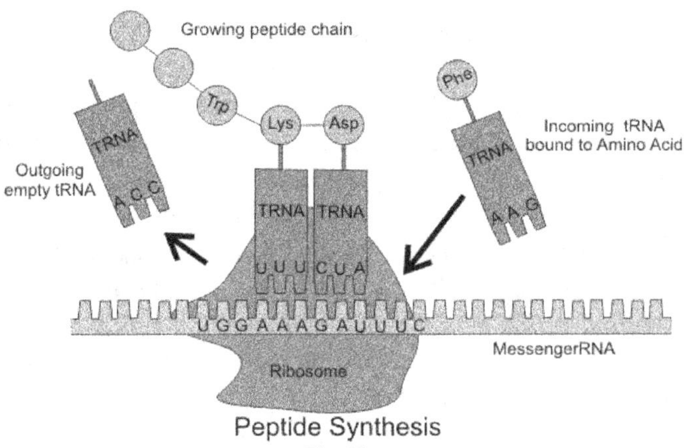


Fig. 13.

The mRNA (messenger RNA) string then runs through a molecule called a ribosome where each codon is matched with the corresponding amino acid via the tRNA molecule. "Peptide" is the term for the growing amino acid chain, that will fold at completion. This process is called translation. The circularity here is obvious. **Proteins are activating genes to express proteins that catalyze the code that will be used to produce the proteins.**

Since the DNA gene-sentences are translated into protein-sentences that in turn instruct the DNA when and how to produce or express the proteins, we have a circular instruction loop, a chicken and egg situation. RNA not only imitates DNA, it can also fold into 3D shapes. In fact, ribosomes are themselves folded RNA molecules. This fact has led some to theorize an "RNA world" precursor to modern life. Adding to the argument is the fact that Thymine (T) in the DNA is replaced with the less stable molecule uracil (U) in transition to RNA. Since RNA exists temporarily, the chances of a mistake are minimal. The requirement for the more stable Thymine in the more permanent DNA template indicates an evolutionary or logical direction from RNA to DNA.

**The Point:** DNA is a database of genetic information that can be read in different ways depending on context provided by protein circuits. RNA serves both as a temporary repository of the contextual meaning and a translating vehicle from genes to proteins. The RNA molecule could have been the progenitor of both DNA and proteins.

## LIPIDS/MEMBRANE/BOUNDARY

The third essential structure of a living being along with genetic memory and protein procedures is the boundary, the cell wall, a permeable membrane. To make the cell an individual, the membrane must be completely made and controlled by the cell, keep the outside separate from the inside, and selectively allow the outside in, and the inside out. The cell surface is a wall of communication, the boundary of "cellular informational selfhood."[55] Significantly, the cell membrane is the only organelle (an intracellular structure) existent in every single living being. It is through the membrane that the self is "downloaded from the environment."[56]

The membrane's most important chemical quality is separation. The origin of the cell wall will be discussed in the chapter on evolution. What is known is that when lipids come together in sufficient numbers they form micelles, which is a bubble like structure that separates oil and water. The energy gradient is stronger than the random molecular motion. The lipid structure self-assembles to the point of lowest free energy as a bi-layered vesicle, formed by two rows of lipids, each with a hydrophilic (water attractive) head pointing out from the wall, and a hydrophobic (water repelling) tail pointed inward towards the middle. In this way they form sealed compartments.[57]

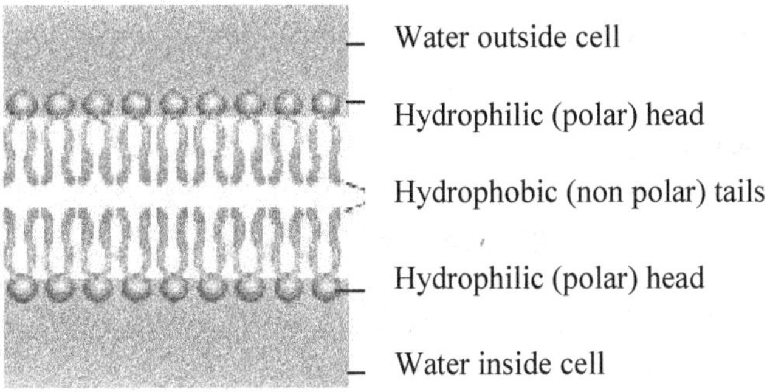


Fig. 14.

Though structured symmetrically as a whole, the membrane is electrochemically asymmetrical. (Electricity is a flow of free electrons. An electrochemical flow is passage of electrons from one atom or molecule to another via ions. An ion is a charged atom or molecule. Hydrogen with an extra electron is a negative hydrogen ion.) The inner monolayer is negatively charged. Asymmetry is the fundamental structural key in converting environmental signals into cellular ones.[58] The charge separation allows for a thermodynamically directional passive diffusion toward equilibrium. This is countered by osmotic pressure which is the entropic force needed to balance the pressure of water against the cell. This expends no energy by the cell.[59]

Countering charge separation is charge transfer. Instead of reaching equilibrium, the gradient direction is maintained by sodium-potassium pumps. To put it extremely simply, the cell normally has a negative charge. Proteins embedded between lipids sense when the charge reaches a threshold, and channels within them will open or shut accordingly. Sodium and potassium ions will be exchanged. Although the pump initially derives its energy from ATP storage, **the fit of the protein shape is so perfect that this exchange will not cost the cell energy.**[60]

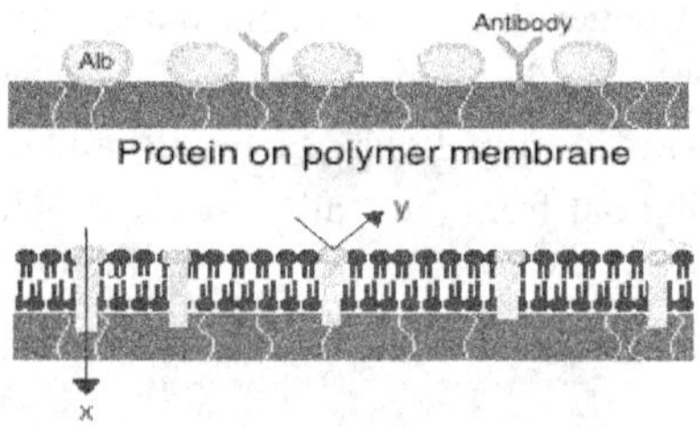


Fig. 15.

Electrochemical gradients across the membrane are environmental signals interpreted by the **selective ability of protein shapes**. The membrane is studded with proteins, up to 75% of the membrane mass, about one protein for every fifty lipids. The shapes and combinations of shapes and their manner of interacting with the lipids vary widely and are peculiar to each kind of cell. Each organism, each kind of cell, senses the world not through every kind of chemical, but through variations in the gradient of a set of chemicals and the relative amounts of each.[61] The cell constantly monitors its own energy state.[62] Some proteins placed in the periplasm, the area between the lipid monolayers, will disable chemicals toxic to the cell.[63] Some cells have regulatory proteins that traverse the membrane and connect directly to the nucleus and gene expression.

Both internal and external stimuli are sensed as bioelectric activity in the membrane. Different kinds of information are integrated by many receptors, resembling the activity of a brain.[64] What stimulates receptors is interpreted by effector proteins, according to the needs of the cell. This is the ground or nexus of awareness. As the manifestation of developing networks, this nexus evolves in some organisms as organelles, which increases the membrane area and total awareness. Since this is the area where initial selection of environmental input takes place, the nervous system is theorized to be derived and developed from the embryonic membrane.[65]

The lipid by-layer is a liquid crystal- half water, half-crystal. Lipids, perhaps spontaneously, exchange places laterally every 10th microsecond.[66] A loop is formed between depolarized open channels and polarized closed channels, creating a wave. Proteins embedded in the cell wall slide with them forming a "moving mosaic." The full lattice surface of all membrane pathways is approximately 50 times greater than the rest of the cell.[67] The cytoskeleton of actin tracks crisscrosses the entire cellular membrane, enabling proteins to carry "cargo."[68] This spatial organization forms **a spatial representation of both the internal and external environment.** The sliding repeating patterns allows the proteins to move on a plane more efficiently than just random movement, with the structure contributing to the overall force.[69] One-dimensional DNA information's translation to 3D molecular meaning, when oriented on a 2D plane, is a fractal[70] connection of the cellular information with the space-time environment.

The flagellar motor on some one-celled animals is activated by a proton gradient interacting with internal molecular calculations informing the cell of appropriate spatial direction.[71] It is also theorized that lipid wave movement is the progenitor of cellular locomotion.[72] For example, when an amoeba extends the cell wall, or pseudopod, by attaching to a surface and contracting to pull the rest of the cell, waves ripple throughout the membrane.

**The Point**: The membrane is the living boundary of the cell. Lipids are structured to separate the outer world from the inner. Environmental information is selected via the shapes of proteins embedded in the membrane.

## METABOLISM

Metabolism is the systematic totality of loops, pathways and networks fundamental to awareness and life. The reduction of the thermodynamic gradient runs through about 10,000,000 chemical reactions per cell per second. In order to stay alive, the cell must replace all its ATP (energy storage molecules) every couple of minutes. Pathways are running in every which direction, timed to receive and produce the correct amounts depending on the ever changing dynamics. The paths even run through one another, sometimes breaking a molecule and immediately repairing or replacing that molecule while adjusting the timing throughout. If there is organized complexity anywhere it is here.

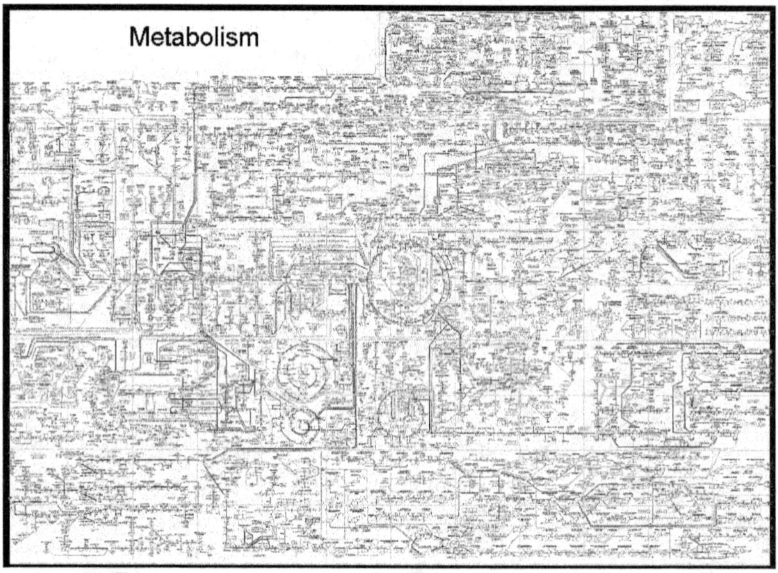


Fig. 16. Biochemical Flow Chart

Material reductionism chases complex details as far as possible. This is not wrong. This is what is revealed when looked at this way. But life is the organized process. It has an in and an out. Free energy comes in and entropy goes out, greater entropy than would otherwise be produced in the universe alone.

The metabolic cycle is divided on the larger inter-life level between plants and animals. Again, stated extremely simply, photosynthesis in plants takes sunlight for energy and combines carbon dioxide and water, creating the sugar glucose and giving off oxygen. Respiration in animals takes carbohydrates (primary source is plants) and oxygen, making energy and giving off carbon dioxide and water. (Prokaryotes and Fungi get energy from inorganic compounds, especially nitrogen, which they may break down into other molecules available for other life forms).

Metabolism in and of the cell, divides into catabolism which is the way in, and anabolism, the way out. Pathways flow in one direction, either anabolic or catabolic. Mirroring environmental information that is first reduced and then interpreted and enhanced with internal meaning, catabolism breaks down food into parts and energy, which anabolism uses to build into products for the cell. All pathways are controlled by enzyme regulators,[73] sensing current needs of the cell. "Cellular sense is a signal spreading and amplified throughout, as numbers of molecules change in various locations."[74]

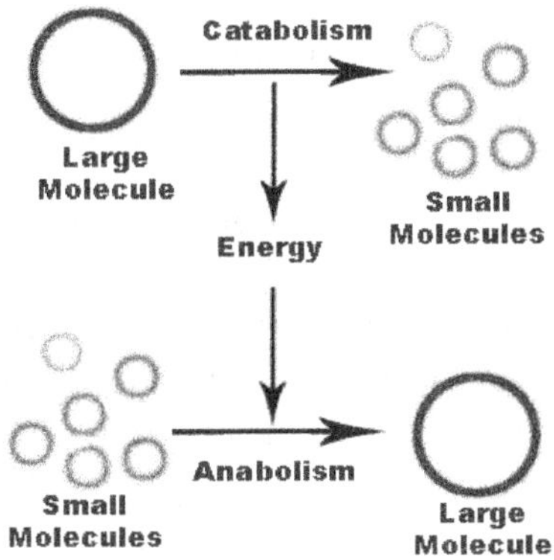


Fig. 17. Metabolism - A mediated energy flow

As life has progressed, sub-loops and patterns have concretized through networks into universal pathways and structures. Eukaryotes, unlike prokaryotes, have a membrane around the DNA. Other specialized areas have formed into membrane covered organelles separated from the watery cytoplasm. There are evolutionary theories on how these organelles first formed, but their functioning determines and is determined by cellular efficiency. The conversion of inorganic carbon and nitrogen sources into biological molecules, as well as other chemical patterns like chemiosmosis, which is the movement of protons across a membrane, and redox reactions for transferring electrons, are also universal to life.[75]

Cell metabolism begins with recognition of food by the protein receptors in the cell wall, the need for which is determined by cell regulation. (Prokaryotes like E. coli bacteria send proteins out of the cell and strip e.g., iron from a molecule and bring the iron back into the cell). The next step is digestion of food into sugar, amino acids and fatty acids and energy. This is accomplished principally in organelles called lyosomes.

Out in the cytosol or cytoplasm, sugar and fatty acids are broken down in a process called glycolysis. Here, like in all catabolic areas, enzymes couple and direct reactions to produce ATP energy storage molecules. Reactions that produce energy are coupled with reactions that require energy.[76] **Because of the enzymes' shape changing, the reactions yield some extra energy rather than lose it all to entropic heat.**

From there, energy carrier molecules enter organelles called mitochondria which contain the central TCA (tricarboxylic acid) or Krebs cycle where catabolism meets anabolism. By reduction from one carrier to another, releasing energy stored in covalent bonds, the Krebs cycle produces high-energy electrons, as well as components for amino acids. Although it is usually diagrammed as a linear circle, it is actually a series of reactions located by enzymes. High-energy electrons are then passed down a gradient in what is called the electron transport chain through the mitochondria membrane in a series of redox reactions, trading protons and electrons, releasing energy to ATP in many small doses and finally reaching the oxygen receptor at the end.

Similar to Mai-Wan Ho's hypothesis, the speed of ATP production is, according to Zeeya Merali in *Discover Magazine*,[77] not accounted for by electrons jumping from molecule to molecule. She concludes instead that the electrons must exist in a quantum superposition across the molecules. Of course, if the living process is explained by electron jumping, we may start to think about a quantum "process." But since living time is outside the wave function collapse, the explanation is molecular directionality- the power of information reception, inherent in the life of the cell.

Anabolism begins with the production of amino acids, monosaccarides and nucleotides. With the energy in ATP manufactured in catabolism, proteins, lipids and nucleic acids are assembled. Anabolism is extremely complex but follows the logical pathways of proteins discussed earlier.

One organelle that is helpful is the Golgi apparatus, which renders random movement more efficient by "sorting" protein addresses. Finally, the uric cycle is a pattern specialized for the final disposal of reduced entropic energy.

**The Point:** Metabolism is the total cellular network, and is mirrored in the general input-output of any multi-cellular organism. The immense numbers of pathways, reactions and molecules and the timing required is almost unimaginatively complex. Life is a network of tasks. Selected energy sources made meaningful by the cell are broken down in the task of catabolism and put back together as the temporary cellular structures for cellular tasks in anabolism.

## STEADY-STATE

Although every level of complexity and individual loop is constantly turning over, and all at different rates, the rate of the living being as a whole maintains a constant state of non-equilibrium above its environment. This steady-state is not instantaneous, or an equivalence of moments, but a running balance in the rate of renewal and decay.[78] Another word for this state is homeostasis. As mentioned in the section on Replication, the organization as such does not turnover while alive.

Although the existence of individual ATP molecules is extremely brief, the total amount per minute is maintained. Regulatory feedback loops at all scales maintain appropriate levels of protein production dependent on fluctuations of need. **This steady-state of many organized processes is measurable as metabolic rate, the energy expended per unit of time.** The space-time structure of the entire organization moves on its own in interaction with its environment. The metabolic rate is not distance per time but energy per time. Moment to moment is matter forming, not lifeless matter in space-time. Steady-state is dynamic non-equilibrium over and against an every-changing environment.

**The Point:** The metabolic rate is a measure of the organic process, which divorced from distance, is a maintenance of form independent of the total chemical reactions. Steady-state is dynamic non-equilibrium over and against an every-changing environment.

## CIRCADIAN RHYTHM

Though there is no one level of causation, the many millions of internal rhythmic loops are synchronized with one another in terms of a centralized rhythm. This rhythm has at bottom the random molecular vibrations in the cytosol, as well as gene expression, and at the top, connection to the periodic variations of the Earth, Sun and moon. **Circadian rhythm operates at every level of every organism yet is individual to the organism.**

The transcription-translation loop is tuned to 24 hour rhythms of gene expression traceable to clock genes.[79] Other clock genes express proteins responsible for feedback loops reflecting circadian oscillations.[80] The TCA cycle, as well as the entire gene-protein-lipid-cell network, are in tune with the overall rhythm.[81] In animals with advanced brains, there is a nuclei located in the hypothalamus that connects through the eye and serves as a master circadian clock.[82] The circadian rhythm serves as a universal feedback component and temporal constraint between the small and the large.

**The Point:** The complex rhythms of metabolism are anchored at one end by the random movement of discrete parts and at the other by large movement in the expanse. Circadian rhythm is the constraint of the parts and whole of time.

## ORGANISM

**An organism is an individual living being**, one-celled or many-celled, but celled. It has a life-span, it begins and ends. It is not itself a species but is a member of a species retaining an individual identity. Atoms of the same kind, called an element, are unique at any given moment in terms of qualities that vary only according to strict physical laws. The vibrational patterns of atoms are interfered with by light waves. Unlike closed material systems which only allow energy to pass through, organisms allow matter to pass through them on the basis of interpretation and evaluation of this pattern of interference.[83] The energy not only passes through, but is stored in organized patterns peculiar to the organism. The energy flow (and the matter converted to energy) are maintained as a macroscopic, dynamic space-time structure above thermodynamic equilibrium. It is the power of the organization itself that enhances the energy and keeps the organized-energy or energized-organization one step ahead of the environment. The stored energy is available for work.[84]

Living complexity, unlike that of a rock, is not arbitrary.[85] It's complexity is specifically organized with changes selected by the organism for its own survival. Though the parts are ephemeral, there are aspects of form universal to all life. The organism is not solid or liquid, but basically a liquid crystal. This allows for "rapid propagation of influences in all directions"[86] blending local and global, part and whole, but moving as one. Universally, crystalline structure is symmetrical. This manifests in an organism either as radial or bi-symmetrical along an anterior-posterior axis. This symmetry acts as an "optical axis"[87] for the self-location of the organism.

Catabolism to anabolism is the principle reversal of direction for thermodynamic energy flow in every cell. These two major pathways, though universally regulated, flow separately. Within each pathway, electric currents flow two ways simultaneously, trading protons for electrons. Reactions are catalyzed, that is, they are directed by enzymes, such that the dissipative energy flow using stored energy is coupled with the non-dissipative processes of recycling information. The rolling balance is symmetrical energy coupling,[88] with energy concentrated at any virtual location. There are no predetermined spatial locations to which the systems have to return.[89] The basic flow scales to multi-celled organisms.

Energy is distributed according to the requirements of the organism. It is not directed from outside or any one internal area. The properties of an organism are functional and structural unities, not directly embodied in the genetic code.[90] The organism's parts are tasks, differentiated by functional integration. Unlike material things, the organism has "counter-rhythmic motion,"[91] that is, complimentary rhythms, on all scales. It is digital- or discrete-like in its genetic code, yet acts in a coordinated continuum. As will be discussed in Part III, it is its own efficient cause by using formal cause to affect its end-directed final cause, a self with a sense of self.

**The Point:** Unlike a single atom of an element or a single molecule of a kind, a single organism, though belonging to a species, is an individual. With its own lifetime, it is its own constraint. It is its own efficient cause by using formal cause, i.e., shapes and patterns, to affect its end-directed final cause, self-survival.

## F. GROUPS

*The manner in which those molecules interact in a holistic ensemble results in something very special*
Addy Pross[1]

### BEHAVIOR

Living beings behave. They interact within their body, with their community and in the historical context of both.[2] Though two organisms, same species or not, may be physically nearly identical, variations occur in the patterns of behavior inside and out. There are organisms with the same proteins- same shape and function, yet share only 20% of their amino acids.[3] They have the same phenotype, but different genotype and/or genetic expression. There is not a direct causal link. Bacteriologists who study bacteria notice personality differences. They can tell them apart by the way they twirl their flagella.[4] **Biological individuality is not the same as genetic identity**. Different proteins are present in different quantities in each cell.

Prototaxis is **the tendency of any cell or organism to react in a particular way to another cell or organism.**[5] Individuals are recognized by other individuals. Some of the receptors embedded in all cell walls are identity receptors.[6] Certain proteins in the wall make others into identity producers. Studies have shown that each bacterium makes a certain amount of molecules that identify themselves or their kind. At a certain threshold what are called "auto-inducers" are detected by the receptors on other cells.[7] At some point the bacteria understand their sufficient group quantity. By joining flagella, millions of bacteria may cling together in a giant mass that crawls across a surface.[8] Others might, at a certain density, all turn on light producing proteins.[9] Enemies are detected in similar fashion. One bacteria might hit and penetrate another with a certain molecule. And close relatives of that bacteria might have a gene that produces a protein that disables the invading molecule. Some bacteria fool others by producing a molecule that appears to be food.[10]

Like patterns of energy flowing in a cell, individual patterns of behavior form group patterns of behavior. At a certain density locusts align in flocks.[11] Starlings fly in synchronous patterns known as murmuration, thousands flying and circling through their own flock. Groups of some species of fireflies flash in tempo, each an oscillator coupling with others. Each follows the other as the other follows it, not in unison but in synchronous rhythm.[12] They are all sending and receiving signals, creating one flowing pattern. From single-cells on up in complexity, organisms are communities that exchange matter and energy as information.[13] They share.[14] Bacteria "communicate and cooperate"[15] in order to survive. As social interaction increases with more complex living beings, self-other awareness increases exponentially.[16]

Interaction becomes infiltration. Species not only live among others, but inside others. There are ten times more bacterial cells (100 trillion) resident in the human body than human cells (10 trillion). There are a thousand distinct bacterial species active in human intestines.[17] Vitamins K and B12, essential to humans, are manufactured in the gut by bacteria.[18] As ameboid protoctists are "integrated colonies" of bacteria, so humans and other complex organisms are colonies of "ameboid beings."[19] Membraned organelles are possibly the remnants of once whole one-celled organisms. Organs and various comparatively stable patterns in multi-celled organisms are evolved organelles.

**The Point:** Individuality is enhanced, indeed achieved, in interaction with other selves of the same and different kinds. The whole-part-whole dynamic circularity in metabolism manifests as population-individual-population.

## MULTI-CELLULAR

**Multi-cellular communities are a collection of individuals that can act as a single entity**. For the good of the community, divisions of labor arise when particular functions require attention.[20] As patterns concretize in a hierarchy of complexity, specialization demands a certain surrender of self. This can even require self-denial. On the other hand, cells not only cooperate, they compete. Nutrients are limited.

With specialization, distinct cell types develop. Cell types differ with similar proteins being expressed in different genetic patterns. Proteins in the cell wall may vary in the extent of their range in cell-to-cell communication.[21] Symmetrical flow between cells of the same type becomes one-way in cells of different types. Relations of part-whole become hierarchically asymmetric.[22] Cell types differ in size.[23] Complexity manifests as an increase in the number of cell types as well as genes, although not in direct correlation.

Whether same type or different, each cell is distinct. In deference to the whole as well as to its own survival, each senses the limit to its own growth.[24] Density in cell numbers is sensed in interaction with density of extracellular molecules. Some are created and some are depleted by individual cells, therefore depriving others.[25] The cell number is in effect coded into the extracellular molecular concentration.[26]

Cell growth occurring in tandem with cell numbers coordinates with cell division. Cells stop dividing at a certain cell number.[27] Cell size must keep pace with cell division. If growth is too slow, the cells get smaller each time they divide.[28] Multi-cellular organisms control the whole organism's growth rate through reproduction in one cell.[29] Eggs may grow to a large size without dividing, but balance it after fertilization by dividing many times without growth.[30]

The physical limit to all cell size is a diameter of about .02cm. Since surface area increases with the square of the diameter and volume by its cube, any larger and the membrane would be unable to pass through enough nutrients for the cell.[31] Other ultimate macro constraints are temperatures of about 150c and water at less than .6 Ph.[32]

The physical limit on the space between cells is about 20 angstroms in order to maintain both separate genomes and intercellular information sharing, or "cellular informational selfhood."[33] Intercellular communication is served by the hormones between cells, cell membranes and generally the protein, cyclic-AMP, that passes the information to and from the inner cell. Cells as wholes, are the senders and the receivers. Timing regulation between c-AMP and the whole cell determines information flow, cell differentiation and "organism level phenotypic complexity."[34] It is not the genes alone. Energy and forms alternate, as electromagnetic energy carries information between molecules (about 3-4 angstroms), and molecules themselves carry information over longer distances. "Chemical and electrical gradients throughout the intercellular channels are balanced by the merging of information."[35]

Cognition begins with the one-celled organism converting high-entropy messages into low-entropy meaning.[36] Cognition expands as the colony achieves a "collective sensing,"[37]with cellular specialization allowing the colony to act as a single, "dynamically stable" entity.[38] With multi-cellular specialization, basic information control, like enzymes within the cell, continues locally at the specialized cellular level.[39] A sponge is multi-celled with millions of cells which are physically connected, but are not specialized and do not share information. A sponge, therefore, has no more intelligence than a one-celled organism.[40] **Specialization requires cooperation**. Cooperative behavior requires communication. As bacteria in a colony communicate, a collective memory develops. The colony learns from experience as genetic differentiation occurs.[41] Efficiency of skillfulness increases with the spread of patterns of loops, helping the colony defend itself and increase access to food.[42]

Each member interprets signals according to its own individual makeup. Meaning is contextual.[43] Each message by itself contains just a small amount of information, but each cell is already an information system which expands each message in its response.[44] Inter-cellular loops include intra-cellular loops which all form in tandem as individuality creates group strategy.[45]

Self-regulating signaling loops, coordinating function, individualize the living being and form an "elementary mind."[46] A "body consciousness" arises in a stable awareness of change. "Sentience, intercommunication and memory," the hallmarks of consciousness, form a bodily nervous system.[47]

As a multi-cellular being becomes larger with more cells, the need arises for speedier transmission of signals. Certain cell membranes develop the ability to send signals as electronic pulses along the wall. These are nerve cells, whose pulses can be transmitted at 100 meters/second.[48] With only one nerve cell, the hydra can coordinate its entire body almost seamlessly. Inter-cellular communication with nerve cells is ten thousand times faster than with chemical signals only.[49] But the bodily system of communication remains, contributing to the intercommunication network. The body has memory.[50]

**The Point:** Other selves become specialized partial selves within a greater individual. Group communication becomes more intelligent with better survival skills, which include but are not limited to the greater communication speed of nerve cells.

## ORGANS

As network complexity increases, specialized patterns become more consistently structured. Specialized organs allow for greater coordination of signals and cooperation of cells, which means the size and detail of the sensed environment, externally and internally, increases. But the network, not the genome, determines organ size. The genome size itself is independent and remains at the cell level. **There is no genome organ**. Cells differ as to which genes are activated and at what time. Organ size, like colonies of cells, must maintain a balance between its cellular replication and its cells' death.[51] The purpose of an organ, like any other living entity, is determined by what it does.

Cells within each organ are synchronized. Groups of genes in different cells within the organ are active or idle at the same time, expressing proteins in sync.[52] Different cells act as oscillating pacemakers. One cell may activate a group of others to open effectors or receptors. A synchronized group may activate a larger group. Like inter- and intra-cellular loops, a hierarchy of activated loops within loops of cells establishes itself.

The function of the heart is to pump blood, while the blood dynamically changes the heart surface.[53] A healthy heart rhythm has a fractal self-similarity not a serial beat.[54] The multi-level loop called the Hodgkin cycle is mutual feedback between the cell-level parameter of membrane ion potential and the opening and closing of the ion channels by proteins. With the heart, all cells are pacemakers. Cardiac rhythm is the pacemaker network itself.[55] The heart is a biological clock. It differs from a mechanical clock in that its shape, its form, is at one with its rhythmic movement.[56]

**Synchrony occurs between organs.** Rhythms in different organs are coupled. One may be dormant yet keep the "beat" while the other is active. Respiratory and heartbeat frequencies are harmonically synchronized.[57] In the intestine, nerve and muscle cells oscillate rhythmically, pushing waves of digestion.[58] With some tissues like cartilage and bone, the material is built by other cells and then self-assembled, similarly to lipids.

Synchrony occurs at the entire body level by the cycles of sun and moon. Blood pressure changes, stroke and heart attack frequency and bodily systems of every sort occur in correlation with the time of day or month.

> **The Point:** With organs, groups of cells within the multi-celled organism are specialized functional structures, with the organs' form remaining through a change in the life and death of constituent cells. Organs are synchronized within themselves and with each other in a organism-wide fractal-like rhythm of complex timing.

## HORMONES

**Signaling proteins in the extra-cellular matrix become hormones at the organ level**. Hormones produced in the cells of one organ or gland or brain nuclei flow to other organs. Bacteria in the intestine also produce messages in hormones. Other hormones are neurotransmitters like serotonin and vitamins.[59] But **the receiving organ's cells interpret the message, not the hormone or molecule itself**. It is read at the membrane gate. A single hormone's message may e.g. spread into a pattern of signals in a liver cell. The same hormone, like cortisone, causes the production of a different set of proteins in a liver cell than in a skin or muscle cell.[60] Entire moods like exuberance and pessimism are affected by testosterone and cortisone respectively. In tune with and part of the whole body rhythm, information about the organism's daily functioning is maintained by intercellular signals. Hormones act dynamically, as first they spread to receiving cells, then genes in the cell produce proteins that enhance the message. Like loops within loops, waves within waves of hormonal messages carry the living body forward.

> **The Point:** The organism's rhythm includes inter-organ messaging by proteins called hormones. The messages are translated at the cell membrane and then the meaning is interpreted and enhanced within the cell. Like genes and proteins, the same hormone can be interpreted differently by and in different organs. As always, the information exists apart from the material structure.

## IMMUNE SYSTEM

Every cell in the body is part of the internal and external information environment.[61] The immune system's function is to distinguish invasive cells and substances called antigens from those that belong. In other words, **its function is to distinguish the whole organismal self from the non-self.** This is, again, accomplished by informational shape. White blood cells called lymphocytes produce millions of different proteins called antibodies, one kind of antibody to uniquely fit each kind of antigen.

The strategy to find the various perfect connections is to randomly resequence its own genes. The cell adds or subtracts nucleotides between genes.[62] Similar to the continuous selection activity that we'll find in the brain,[63] the nonlinear dynamics of selection that is found throughout cellular life, operates on the level of the immune system. The ability to think at all levels is based on the ability to answer the questions already set up,[64] and to recognize change that requires new answers.

Beside antigens, the immune system responds to products of the organism's own metabolism and to bacteria in the intestines that are intrinsic to the organism's self. It recognizes and then ignores them. Bacteria themselves continually read and respond to the environment in a way advantageous to the host body. Both the immune system and the organism's genome is extended by resident bacteria.[65]

> **The Point:** Certain cells are specialized to recognize and destroy material that is non-self. Bacterial cells that belong to the organism are

recognized as self, and they, in turn, recognize their host's cells. The selection process involved is universal to the whole organism.

## G. EVOLUTION

*There is no fundamental difference between man and animals in their ability to feel pleasure and pain, happiness and misery*
Charles Darwin

### THEORY

Is life one thing? Life itself is what all living beings share. Life is what it is to be alive. By living being, we mean an individual living being. The environment, the outside, includes other living beings. So, another way of looking at Life itself is the totality of living beings. But, there is a basic universality in enzymes and general metabolism and cell structure and similarities in body shape in all life. Living beings are all alive but they are not identical. Physical science looks at life in two ways: Biochemistry is a material reductionist approach from the bottom up, that is, from physical parts. Darwinian Evolution is a material deterministic explanation, that is, of cause from before.

In 1859, Charles Darwin (1809-1882) published, *On the Origin of Species by Means of Natural Selection, or the Preservation of Favored Races in the Struggle for Life*. Darwin does not start with life vs. non-life. His theory of evolution is an explanation of living difference. But the individual is a species. Darwin's theory is an explanation of how species differentiate, how they survive as species. The basic idea, is that through competition for resources, by better adaptation to an environment, the fittest species will survive. In other words, survival of the fittest means those that survive will be those that change successfully.

In addition to Darwin's theory about the selection of species, the modern theory of evolution includes the individual unit of life- the cell, first discovered by Jakob Schleiden (1804 -1881) in 1838; the unit for generational passage of information- heredity, discovered by Gregor Mendel (1822-1884) in 1865; and the DNA structure of genetic information discovered by James Watson (1928- ) and Francis Crick (1916-2004) in 1953.[1] There are two basic processes: Variation of random mutations and non-random (deterministic) selection. Evolution arises from the interaction of the two.[2] Between molecular particles and environmental determinism, life organizes in a way not explained by thermodynamics alone.[3]

Let's start with the idea of growth. What gives rise to a change in form? An individual organism grows from an embryo or seed into an adult and eventually dies. A species begins (has its origin) at some point and, very often, becomes extinct at another. Life flows through a change in both species and individuals within a species. But traditional evolutionary theory, as we've said, sees the individual as a species. The word used in evolutionary theory is "populations."[4] In this way individual experience can be ignored in favor of evolving sub-living units of genes which give rise to changes in traits. Structure determined by function in the individual is function determined by structure in evolving populations.

If species are evidence of successful evolution, what then does success mean? What is Life growing towards? Is the manifestation of time in an individual living being applicable to evolution? Is evolution matter through form(s) or form(s) through matter? In this book, matter is traced through a growth in complexity that provides a complimentary well of information available to living awareness. Is life successfully progressing to greater complexity and greater awareness?[5] Is complexity the same thing as variation? Variation alone would lead to an increase in entropy, a dissipation.[6] Sometimes selection by the environment in and of itself can cause the species, the line of descent, to go extinct.[7] **Variation plus selection is the organizing factor that leads away from equilibrium.**

Information derived from selection is stored in genomes whose variation randomly results in traits adapted to overcome the limitations imposed by genomic information.[8] In order to successfully adapt to the current situation, life "adopts and disposes the past."[9] This rolling circularity does not necessarily result in complexity if the complexity does not fit.

Evolution is more the result of the way information is organized. Function is sometimes limited by the integration of form and environment.[10] Evolution (and organization) can lead to simplicity as well as complexity.[11] Complexity can emerge from previously evolved simple components. Simpler organisms can arise from more complex predecessors or a complex accumulation of factors.

Modularity, the retaining of structural form for different functions, is life's method of efficiency. **Evolved network motifs and metabolic structures are the modules, the middlemen between genes and traits.** The same modules appear repeatedly in different evolutionary lines. For example, proteins in animal eyes have different functions in microbes.[12]

Within the self-demarcated cell wall, distinguishing self and other, with personal memory applied to current interaction, the primacy of the individual is established with awareness. Life is looked at, and looks out, through the senses of the individual. Some argue that Darwin did indeed see selection as operating at the individual level.[13] Adaptation is value to the individual. Unlike a physical particle, a physical biological individual is a package of traits.[14] Traits continue and change through individual replication. It is the distinction between a single organism and its environment that is really the origin of competition.

If an aware individual's purpose is to survive, does life in general, that is, the totality of living beings, have the purpose to survive or even a purpose at all? Does success mean progress? Does Life improve?

Life is open to the future. According to Stephen Jay Gould (1941-2002), as environments change, their diversity produces "possibility spaces."[15] Niches in the environment, like tasks in an organism, open up to be filled by the most adapted species to that niche. An unknown future determines creative solutions.[16] But do species, let alone the entirety of living beings, anticipate? Some say that species as a whole have stopped growing. What continues is "the richness of its order."[17] Since the self-ordering of matter culminates in the order of an organism, it seems logical that order itself is evolving.

Network motifs are "information processing functions."[18] In an autocatalytic individual, selection is more controlled from the inside out.[19] Life orders and is the ordering of thermodynamic disorder. But order is not the same as intelligence, if intelligence is defined as the ability to solve problems with some sort of interpretive cognition. Bacteria are the best at what they do with what they have to work with. Is every successful species intelligent in its own niche? Or is it the members of the species that are intelligent in their specie's niche?

The evolution of evolvability is an increase in knowledge. With an increase in knowledge comes an ability to get beyond survival. With time left over from surviving, living beings have time to play, to imagine. Knowledge not only helps anticipate and create solutions, but manipulates the anticipation.

Selection is cognition. Individual living beings selecting what they need to survive, collectively combine their intelligences. Evolution itself is "a form of consciousness,"[20] but is evolution Life's consciousness? Choice is at the heart of life and awareness. And choice is accompanied by feeling. If human bodily structures are traced to the beginnings of life, why is that not true of subjective feelings?[21] The human mind has evolved from elements found in the simplest life forms.[22] Mental processes themselves reflect and are reflected in evolution.[23] Scientific theories from relativity to quantum mechanics are determined by evolution. Prediction is necessary for survival. Individual human awareness, expressed in modern society, understands itself as arising in evolution. Intelligence is a tool.

**The Point:** Biochemistry explains life from the outside. Evolution explains life from before. Standard evolutionary theory individualizes at the level of the species. Random genetic change and deterministic selection combine for species to change. The progressive order of matter implies the organism. Order itself evolves.

## MATERIAL CONDITIONS

Living beings create their own time. The environment, including other living beings from the point of view of an individual living being, is part of the world that the living being is traveling through and that travels through the living being. The time created and the memory set down, implies something that came before. **The material world that preceded life is part of life's understanding of itself.** This is not an anthropomorphic view in which the universe is seen as made perfectly for life to arise, or that the universe we live in had to be this way in order for life to arise. If awareness begins with life then living awareness is aware of what makes sense. What makes sense is a universe of which life is aware.

The universe that life lives in is expanding. Thermodynamic energy encased in a particle/spacetime expanse, chaotically produces structures. Chaos can be seen as material selection. From entropy comes both complexity and variation from which evolution selects order. As more chemical elements run through living beings, the more elements scatter throughout the environment.[25] The second law of thermodynamics and evolution are both parts of the same process providing information for awareness.[24]

Living biochemistry communicates by way of functional structures that fit together. Molecules in living beings can spatially interact because of chirality. Shapes that are mirror images of one another fit together like clasping hands, a left and a right. They are asymmetric. All organisms have left handed mirror image amino acids and mostly right handed mirror imaged DNA and RNA sugars. Why amino acids are left handed rather than right is a question, as chemical experiments have shown the handedness to be random. But for life, one or the other is necessary, and having both conveniently divides information storage from processing. A symmetrical universe at the elemental level, becomes asymmetric in the emergence of living awareness,[26] and symmetrical again as observed bodily traits.

Yet how did chirality arise? One theory, stated very simply and probably insuffiently, is that at the beginning of the universe matter/anti-matter coalesced into matter causing beta particles in the weak force to have a left spin, which when combined with polarized chiral sunlight near the horizons, affected biosynthesized molecules with a handed bias.[27] In non-linear autocatalytic chemical reactions, mutual or circular feedback between thermodynamically entropic and orderly growth processes amplify the asymmetry.[28]

Whatever may be the case, **asymmetry is required to produce information from formless "round balls" like atoms.**[29] Lock and key reactions must be specific because the molecular configuration carries the information. The molecules are spatially three-dimensional and interact only with complementary shapes.[30] Chirality means a molecule can be restructured as a mirror image. A mirror image of a right or left hand is not superimposable in three dimensions, like two right hands are.

The carbon atom is ideal for creating three-dimensional structures. As mentioned earlier, a polysaccharide chain, called a pyran ring, is a hexagon made of five carbon atoms and one oxygen atom. The pyran ring forms a template used for modular construction, as it attracts more carbon and oxygen atoms linking in a series of rings. Because of the chiral carbon center in amino acids, the protein chain is asymmetrical, allowing it to bend three-dimensionally. It is ubiquitous in life, creating a pattern of hexagons with the most efficient use of space.

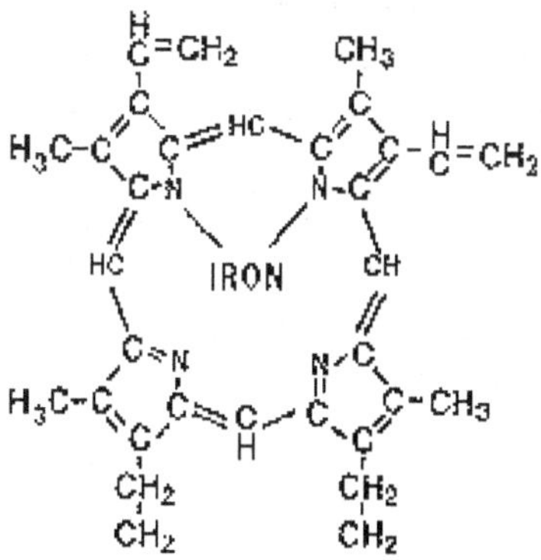


Human Hemoglobin

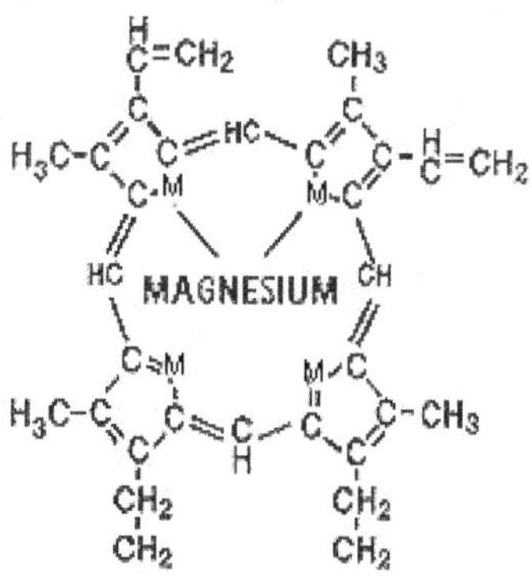


Plant Chlorophyll

Fig. 18. Identical except for iron and magnesium

The Earth's atmosphere, consisting of 78% nitrogen, 21% oxygen, 03% carbon dioxide, and many other naturally occurring elements, plus water vapor, is required, used and defined by life. In fact, the Sun, moon, the Earth's rotation, and the solar system for that matter, are all perfect compliments to life. Nitrogen, an essential component of amino acids, DNA and RNA, is theorized to have been in the atmosphere since Earth's beginning about five billion years ago. Water vapor and carbon dioxide could have come from volcanoes. Methane (consisting of carbon and hydrogen) could have come from inside the Earth. Life itself in the form of photosynthetic cyanobacteria could have produced oxygen as a waste product in quantities over billions of years, enough to produce our current atmosphere. In the stratosphere, molecular oxygen as ozone protects life from ultraviolet light. The carbon dioxide/oxygen, mineral and other cycles that travel through the atmosphere, plants, animals, oceans and soil are the general environmental loops.

One possible scenario is for temperature and sulfur gradients to have been established in volcanic vents beneath the oceans.[31] Comets filled the water with organic molecules that the vents would heat.[32] Reactions of hydrogen sulphide with iron spontaneously yielded energy[33] as iron sulphide bubbles, the progenitors of membranes, emerged. Bubbling alkaline liquid flowed into acidic oceans, producing alkaline/acid proton gradients.[34]

Of course, the principle material condition for life is sunlight. Carbon chains and rings set the stage with electron boundaries able to capture photon energy. Chemistry became biochemistry when cyanobacteria first used this ability. Chlorophylls opened up to quantum energy[35] in photosynthesis and brought in photon information,[36] which life has used as the basis for all subsequent organization. This is what living awareness mirrors.

**The Point:** The material world that preceded life is part of life's understanding of itself. Life compliments the structure of matter. The matter flowing through form is understood as and by shapes. Asymmetry in molecular structures is necessary for parts to fit together and thus pass information as shape qua shape.

## ABIOGENESIS

Abiogenesis, and not Darwinian evolution, is the theoretical description of how life emerged from non-life in the first place. Some say that life cannot come into existence if it had never existed in some way before. It must, therefore, be part of a continuum of increasing complexity[37] uniting chemistry and biology.[38] Chemical selection continues as biological selection.

Metabolism, says Dorian Sagan, is simply a more efficient way to decrease the thermodynamic gradient and increase entropy.[39]

On the other hand, evolution has an organizational history. And the flip side of history is evolutionary potential.[40] Metabolic efficiency increases in evolutionary time. **In matter, molecules may self-assemble but in living beings, molecular assemblies replicate.** In other words, life is the birth of the self. Molecules are not selves. A living being is the self flowing through the molecules. Self-organization arises from externally driven organizations.[41] Matter is real. It is real in its use to survive and in its use for life to explain itself. Living systems not only flow through matter, they turn it into information, flowing in circles, capturing energy and history. In living beings, molecules catalyze their own production[42] but at the direction of the whole living being.

Life began when a thermodynamic loop was closed, leaving autonomy as such to survive.[43] Metabolism likely preceded the development of cells and DNA.[44] The core circle established the constraining template for all life to come.[45] Because RNA can have both DNA and protein elements, it is theorized that an "RNA world" came first. The first metabolic molecules are theorized to be proto-RNA molecules that can spontaneously link together forming ring-like chains called "rosettes."[46] To go beyond self-assembly, a molecule does not simply make copies of itself. Starting with a two-molecule system the molecules catalyze each other. The system as a whole copies itself.[47] **Instead of the loop becoming longer, the loop itself multiplies.**[48]

At first, the pattern would be closed to alternative pathways and have only unchanging persistence.[49] But self-replication requires a context to be a self.[50] Loops competing for the same molecular environment differentiate themselves by replicating at different rates. The fastest is the fittest.[51]

A different rate establishes the other. But the self in the context of that other needs a boundary.[52] Charles Lineweaver gives a definition of the basic first living being: "A molecular autonomous agent is a self-reproducing system that is able to do at least one thermodynamic work cycle...in some boundary, a liposome, and [with a] capacity to make at least one discrimination, food or not food, and to act upon that discrimination."[53]

A self must be thermodynamically isolated. As we stated earlier, lipids dispersed in water form an aggregate called a micelle, with hydrophilic heads on the outside and hydrophobic regions pointing towards the middle of a bi-layer. At some point this amphiphilic bi-layer membrane closes into a vesicle.[54] RNA chains develop into monomers which can diffuse through a fatty acid vesicle. These chains, which themselves close and grow, separate strands of lipids, and attract more monomers. The monomers form polymers, increasing pressure on the membrane to grow. Eventually the vesicles split in two, each inheriting polymer loops. The vesicles with polymer sequences that replicate fastest survive the most.[55]

Originally, these membrane enclosed, looping polymers were attached to the alkaline vents. What sets life free is the ability to establish its own chemi-osmotic gradient, which it does with the DNA genetic code.[56] RNA can contain both DNA and proteins. Chemicals needed for both are plentiful in the vents. As nucleotides and amino acids split from RNA they remained combined by code.[57] The more precise is the code, the more precise are the amino acid combinations, and the more complex genotypes are produced. At first, gene transfer would occur directly between distinct cells. Cell division, or mitosis, made this more difficult.

Gene swapping can cause some genes to be parasitic. Genes that separated from or lost their attachment to replicating ability could use other genomes. These parasites are viruses.[58] Though they are carriers of information, they are not sensitive like a living cell. They are strings of RNA or DNA enclosed in a protein casing. Phages, which are viruses that attack bacteria, have been described as "lying in wait" for bacteria to swim by. This is like saying a mouse trap lies in wait for the mouse. Viruses are not themselves alive. Phages are more like scary robots. They were, however, instrumental in separating the original genomes of archaea and eukaryotes from bacteria.[59]

Vitamins are molecules that helped RNA carry out reactions. They are now made by proteins. Vitamin B12 is made by as many as 22 different proteins. Vitamin B1 helps proteins extract carbon dioxide from other molecules.[60] Speciation occurred in part when some species lost the ability to make proteins and relied on other species. It takes many species to make the vitamins needed by all.

**The Point:** The thermodynamic advance of individualized complexity in matter is logically necessary to establish a continuum to life, but not sufficient in establishing the individual self. Molecular chains self-assemble into loops. A loop captures energy until it divides into more of the same kind. To continue, a closed molecular loop divides into retaining and procuring functions. Nucleotides and amino acids split, yet remain connected by code. Amino acid and nucleotide chains continue to grow enclosed by lipid membranes permeable by proteins. Competition divides kind into kinds, first kinds of loops, then kinds of cells.

But Life begins when molecules become codons inside a membrane, becoming a successful individual cell, looking out and forward at whatever came before, selecting what it needs to continue.

## INHERITANCE

The inheritance of traits from one generation to the next is, following abiogenesis, the primary factor distinguishing living order from nonliving. Thermodynamic free energy that drives chaotic order in matter is further ordered in life by the genetic language communicating information within itself and between beings.[61] History is established. But as we have shown, the genome does not in itself direct the order of change. **For change to occur over evolutionary time, or even between generations, or sometimes within an individual lifetime, the information in the genome must vary.** Over time, genomes mutate. Within the immense cacophony of cellular metabolism, unrepaired chemical bonding errors occur. These errors can be caused by random quantum fluctuations in bonding processes, or possibly damage during or before molecular connections, all resulting in translation errors. Either way it is **informationally a mistake, a miscommunication**.

Sometimes mistakes can be beneficial to the continuance of the species. Proteins expressed by a mutated gene may work better in a new or same environment. The cooperation among proteins and genes may be more, not less, robust.

There are approximately 200 genes common to all current species.[62] It is theorized that many genes have been mutated out of existence since the beginning. The genome, which is the total DNA in any species, is much larger than the gene set. It is theorized that smaller genomes mutate faster. There is "a constant rate of spontaneous mutation in DNA-based microbes."[63] Mutations are perhaps buffered by a larger genome and are collected in the growing "junk-DNA" non-gene area of the genome.

A gene is any DNA string that expresses a protein. A gene can express hundreds of proteins. Humans have about 20,000 distinct genes and up to a million proteins. Complexity of traits does not correlate with the number of genes. Since 4 codons translate into 20 amino acids, the room for error is vast.

**The Point:** There is no one-to-one correlation between individual genes and the structures made by proteins called traits. In the circle of information between genes and proteins, mistakes can be made. The factors are many and random. Random changes in genes are sometimes inherited by offspring.

## SELECTION

For change to occur, some random variations must last longer than other random variations. At the same time, without variations, selection has nothing to work with.[64] Selection is the name evolution gives to the "force" that causes certain variations to endure, that is, to be fit.[65] If individual living beings provide the genetic variations, it is the environment that selects. The environment includes everything in the external world from the available thermodynamic energy to other organisms, to the living being's own changeable internal structures. Though genes themselves are not informationally causative, whatever is produced in the ongoing process of living is available for possible change. Molecules which work better (or worse) are available for selection.[66] Beside its own desire, the constraints provided by the material flowing through and interacting with and being changed by the living being are the limits of what it can do. What is available for selection is dictated by all these constraints.[67] Though ostensibly it is only the genes that are inherited, the entire metabolism is selected.

**Selection is reduction of information**. But if variation is always reduced, then it would seem evolution would be convergent. Unless evolution is the survival of the simplest and the fewest, that is not the case. What is lost or reduced is useless information.[68] Selection filters out mistakes.[69] The fittest variants are the correct information. What is correct is what is meaningful as useful to each member of the species.

If the environment doesn't change, the information in the genome "catches up" to the information in the environment. The genome and the environment are then at equilibrium. Evolutionary choice becomes certain and variation diminishes.[70] It is thought that if the mutation rate increases, the capacity of the genome to retain information reduces accordingly.[71] Therefore, since the nature of information is not limited to a collection of distinct genes, reduction leads to an emergence of "functional generality."[72] Throughout environmental changes, comes a simplicity in maintenance of metabolic networks[73] but with variable functionality. Beaks with different uses are still beaks. The same structure will have multiple functions. And **functional freedom has allowed the same functions to evolve in differ-**

**ent structures**. "New" is relative to the individual species in that particular environment.

> **The Point:** Random variation spreads out from the individual. Functions and their ephemeral structures built from flowing matter are constantly adjusting to a changing environment. Environmental selection reduces. Information becomes more certain with fitness. Functions themselves can then be maintained apart from any one species. Over evolutionary time, lasting functions can have different structures and lasting (that is, recurring) structures can have different functions.

## ENVIRONMENT

The environment, defined as everything outside the boundary of the individual, has a few types of change which are often overlooked in discussing Darwinian selection. The farthest away is the universe, and its thermodynamic push outwards toward entropy, while gathering temporary complexity along the way. The sun, moon, solar system and stars of course, provide an existential boundary but connect to life through circadian rhythms. Next comes astronomical events like meteor crashes and radiation. Then there are geological events like moving plate tectonics, volcanoes and changing climate. Epidemics and diseases are also significant factors. Depending on their structure, species either survived or perished, or given enough time, adjusted or new species evolved. The dynamic competition for existing resources is greatly affected by calamitous unforeseen random change on the outside.

Outside of the immediate environment, there is a material world of change that doesn't directly interact with evolution but only enables "evolutionary opportunities."[74] Non-living change is arbitrary.[75] Prediction is imprecise, if not impossible. Closer to the individual species, the ecosystem (also called the biosphere) not only changes but is itself evolving. The ecosystem includes both the animate (all other living beings) and the inanimate (such as the atmosphere and carbon cycle). Though occasionally affected by cosmic catastrophes, the ecosystem is not simply "willful roots and indifferent soils,"[76] but an evolving interaction between all species and climate as well as among species.

**Niches are to the ecosystem what tasks are to the individual organism**. But instead of being themselves structures, niches are possibilities made available to evolution by environmental change. A niche is actualized by a species. The species in turn is constrained by the niche.[77] Like an enzyme and its substrate, it fits without selection being a causal factor.[78] And out of all possible environmental spaces, only some niches open up.[79] Evolution simply uses those that are available.[80] Niches enable selection to occur.

Traits can provide a niche for other traits. For example, The swim bladder simply provides a place for, but does not cause, bacterial evolution.[81] Which molecular products that are inherited are subsequently selected, depends on which functions become available.[82]

Life as a whole is an interplay of complexity, in both organisms and its surroundings.[83] The reduction of complexity occurs in both the structure of organisms and the organism's knowledge of both itself and the environment. Behavior variability increases with learning about the world and decreases as the organism catches up, all depending on comparative complexity and change.[84] Species are shaped by other species, while at the same time they are part of the environment of others. All information comes in from the environment, but evolutionary change occurs with the intelligence involved in sensing the environment.[85] All species as a whole, in fact life as a whole, evolved to survive.

**The Point:** The ecosystem co-evolves with life. Beyond the ecosystem, the world randomly changes regardless of life. Both due to interaction and without, change makes available niches for species to fill. Behavior variability and learning increase with environmental changes, and decrease as a species fits into a niche.

## TRAITS

Traits are the bodily elements of species distinction. They include everything from metabolic networks to body types. Species have a genotype and a phenotype. The interaction of traits with the environment is both caused by and is a cause of genome structure.[86] **Acquiring a new trait depends on the capacity to acquire it.**[87] The retina, which is traced to primitive light sensitive areas on the skin or cell surface,[88] has appeared separately at different times in evolutionary history. It is not a straight line. At issue is whether or not evolution of traits is change with improvement. Some say life, or at least some lines of species, developed better methods of degrading the thermodynamic gradient. For instance, mammals use more oxygenated air than reptiles.[89]

Some traits which group together, have been traced to locations in the same genome area,[90] but traits do not map one-to-one with genes. Selection operates across the genome as a whole. Genes, like proteins, work in circuits.[91] Genomes are genetic programs with many variables. **Metabolic circularity absorbs, that is, tends to ignore, mutation**. Genome size itself is not evidence of trait complexity. Salamanders, for example, have larger genomes than humans.[92] Genes in only one strain of one species of bacteria (E. coli) have increased, while the genes in common in all strains have decreased.[93] Both general body plans and basic metabolism have remained relatively undiversified.[94] At the smallest scales humans and fungi are virtually indistinguishable.[95] "Generative entrenchment" is the term used for robustness of cyclic processes.[96] More reliable structures mean less need for genetic variation. Groupings of genes have been found to be more similar in genomes of separate species than are found in the same species.[97]

Though there is a concept called "microbial IQ," which rates a genome by the number of sensory systems in the phenotype,[98] a phenotype fits successfully when the genome contains the information about the environment that the species needs to succeed, plus the information needed to adjust to a change.[99] The retina is a development of a visual system that can better distinguish details.[100] But that information is still a reduction of the sensed environment. If the environment fits, the environment sensed is different for different species.

**The Point:** Phenomenal traits, or structures, do not map one-to-one with genes. Traits can develop independently in different lines of species descent. Basic traits like body plans and metabolism can remain entrenched in many lines of development. Genome information survives to match the environment in which it finds itself.

## ADAPTATION and SURVIVAL

Traits vary. Some fit their environmental niche better than others and therefore survive. This is called adaptation. Random inheritance is by chance and occurs first within the individual offspring. Selection by the environment, which includes the totality of living things outside an individual, is deterministic.[101] **Adaptation is the solution that arises out of the interaction between individuals and the environment, between random inheritance and deterministic selection**. The solution is creative in the sense that it is a new variation that fits. Creative solutions are dependent on the availability of a genome that produces sufficient variability for a changing environment and an overcoming of constraints. Behavior variability is itself a trait.[102] Historical stages of evolution show that with time, adaptive mutations increase,[103] as adaptations build on what is already there.[104] The ability to vary behavior requires information about possible, future environmental states.[105] Information remembered in the genome about possible responses, enhances predictive ability.[106] Failed predictions mean failure to survive.[107]

From the point of view of general evolution, the exact usefulness of a trait is highly unpredictable.[108] Out of many traits available to use within a particular niche, it is the individual organism's task that actualizes the trait and the niche. The use is understood after adaptation. "Not only do we not know what will happen, we don't even know what can happen."[109] By surviving, the organism is creating itself. To survive, it adapts.

**The Point:** To adapt, the functionality of behavioral traits accompanying a given genome must be varying enough to cope with environmental changes. What traits are needed and how they will be used is not evident until they are used. They are used by individuals. Behavior variability is itself a trait.

## DIVERGENCE of the SPECIES

As discussed earlier, cell size is limited by the ability of its surface to sustain its volume. To grow and survive, one-celled organisms had to colonize. As colonies grew, cells specialized and multi-cellularity appeared, because coordination rather than strict self-interest usually proved to be more competitive.[110] The first functional separation of cell types were germ-line and somatic cells,[111] that is, reproduction and survival, or gene inheritance and trait selection. Genetic inheritance, though variable, essentially goes on forever, connecting all life, while traits are temporary but interacting. As we've said, cell type differentiation and genomic complexity are not correlated.[112] Sponges, for example, have few cell types, if any, but very large genomes. **Diversity of species does not correlate with genetic diversity.**

Further differentiation of soma cells formed functional sub-units specialized for greater performance. These subcomponents themselves are grouped in larger units, but "interlaced with redundancies," not totally separated.[113] Components starting as proteins, present in the earliest eukaryotes, like SNARE's (sensitive factor attachment protein receptors), important in cell differentiation,[114] homoedomains (areas in a protein affecting development) in body plans, and kinesins in molecular "motors," subsequently diversified within their own types.[115] **Modular structure has enabled life to be both complex and simple.**

There was an explosion in kinds of species during the Cambrian Period, about 530 million years ago. Yet bilateral body types, theorized to be particularly adaptable to change and even necessary for diverse species to build upon, preceded this explosion by 150 million years with comparatively little change. But environmental change during the 20 million years of the Cambrian, provided adaptable life something to adapt to. In one scenario, rising seas caused shallow flooded areas that enhanced contact between rocks and seawater, which in turn increased ocean toxicity. Exoskeletons, developed to protect cellular life from the toxins, allowed for many different and complex body plans to develop. This in turn increased the food chain and the environment of many competing species.[116]

But no matter the period, species come and go. Ever changing environments are accompanied by a constant changing mix of kinds. The Paramos mountain area in South America is known to be the fastest evolving ecosystem on Earth. The combination of the high altitude and the equatorial sunshine, makes for large discrepancies in the surroundings. Many niches, rapidly changing, provide for many changing adaptive solutions.[117]

Thermodynamic entropy increases with the increase in species[118] yet equilibrium is never reached.[119] Convergence by selection and divergence by inheritance is what is actually competing in the dynamic mix of species and the ever-changing surroundings. The information taken in by each species is eventually too complex or too simple for the environment.[120] The information reduced by the species must fit its environmental niche.

Since the future is unknown, the bias is toward divergence. An expanding universe would be consistent with that conclusion. But divergence can be in the direction of greater complexity or greater longevity. Certain species like the lamprey eel have been around for 350 million years, while complex humans evolved about 200,000 years ago. Life has been evolving for 4.6 billion years. Where it is headed, or would be headed without human interference, is unclear. But recombination of information throughout population varieties has yielded successful basic organic methods.[121] **Since the living being is form limited by constraints, and not by any particular matter, modularity has distributed across all living beings.**[122]

When the species can no longer change or reorganize its information to fit, speciation and/or extinction occurs. But all species change within enough period of time.[123] Groups within species can find different ways to reach the same goal.[124] The tools for survival outlive individual species and reappear throughout evolution. The brains of animals might not be an advantage for plants.[125] Progress is more or less simply keeping up.

Modularity manifests in body and limb differentiation within sameness. The forelimbs of small dinosaurs called theropods, which lived about 150 million years ago, grew large enough in relation to their body to allow the limbs to be used by the first birds for flight. Differentiation between limb and body size are noticeable by example, in the comparison of whales and apes.[126]

Like organisms and species, whole kingdoms are as much a part of adaptation and survival as individual competition between types. Fungi are intertwined in the roots of almost all plant species, functioning as providers of nutrients. Fungal DNA was incorporated within plant genomes to produce the first fruits.[127]

Animals evolve more quickly than plants.[128] Because motility brings with it its own changing environment, it is accompanied by the need for faster and diversified sensing. Nervous systems help the organism to adapt more quickly. **Surface cells,** performing the sensing tasks of protein receptors at the multicellular level, gradually became more useful by centralizing in the interior while keeping extensions connected to the surface. These cells **developed into the first neurons.**[129]

Neurons first appeared in invertebrates. Again, these animals are successful within their own niches. The flea, for example, can jump one hundred times its own height. But the protective shell, spine and skull enables more complexity.[130] The first vertebrates were fish that lived 400 million years ago. Cells near the surface connected with neurons to form sense organs.

After touch, smell was probably the main sense. The eye was initially a simple organ used as a source of light energy, but evolved to capture more information. Sense organs moved to the sides to better process coordination in multiple dimensions, and as such were the precursors to cerebral hemispheres. As the eyes became more complex, so did the brain to process more information. Eventually the brain evolved in warm-blooded animals to be the organ that required the most energy.[131]

The earliest brains included the basic functionality of the modern human. Human intelligence might be considered the most complex of all species, but sophisticated mental skills may have been in place 280 million years ago with the common ancestor of reptiles, birds and mammals.[132] Tortoises and lizards have been shown to use memory and learning, with navigational skills and problem solving similar to mammals, which demand more than the sense of smell.[133] Crows, parrots and others have shown remarkable problem solving skills, meaning they require an ability to alter behavior with new circumstances.

The next step in complex intelligence is division of labor within groups of animals. Memory increases to enable prediction of others' behavior, and to imagine alternative, "competing," scenarios. Even more, the group's survival through reproduction is enhanced when the young are reared over many generations of elders.[134] Since dolphins and cows split 70 million years ago, there have been more genetic changes in the cow branch. With dolphins, elephants and apes, child rearing has slowed evolution in correspondence to increased brain power.[135]

**The number of species is to the ecosystem what the number of cell types is to the organism.** Integration is as important as competition. Complexity of color patterns, odors and behaviors in overlapping traits among species, along with eyes, olfactory organs and brains to perceive them, enabled the sophisticated development of reality and imagination in humans.[136]

But the core ecological network has superseded and outlasted the preservation of any one species.[137] Life is something beyond species. Everything attributed to life, like inheritance and selection, necessitates an individual organism.[138] On the one hand, life is generalized simply as a dynamic yet systematic integration born of form flowing through matter. On the other hand, living beings, even within species, are all distinguishable as individuals, with the more complex the more individual. Yet it is the individual that is ultimately ephemeral, with death inevitable.[139]

**The Point:** Genetic information connects all life. But phenomenal meaning of that information is both temporary in detail and lasting in modular generality. Form limited by constraints, not matter parts, leads to distribution of modularity.

Extinction or speciation occurs when information no longer fits. Diversity of species does not correlate with genetic diversity. Complex intelligence expands with the division of labor in groups. But intelligence is always passed though meaningful interpretation by an individual. Life flows through all species but is lived in the individual.

## EPIGENESIS

Natural selection generally serves to preserve the changes in species populations that were gradually generated by previous genetic inheritance. Adaptation is the term for the creative synergy between the two "forces" of inheritance and selection. But what traditional evolution leaves out is the actual point of inheritable transmission. Differentiation in species population begins in the reproduction of the individual. In almost all multicellular beings it is individuals that mate. In order for the environment to select, the thermodynamic energy flow of the universe must first be reversed in the individually aware being.[140] Between environmental selection and genetic inheritance lies the individual cell.

The field of Epigenetics studies characteristics acquired during an individual's lifetime other than inherited gene sequences, such as patterns of gene-proteins loops, DNA that does not directly coincide with genes, other molecules in the cell nucleus and the effects of environmental factors, including the co-evolution of other living beings. Characteristics acquired by the parents can then be inherited by the next generation. Effects can last for generations after that. For a particularly egregious example, exposure by one generation to the chemical, Agent Orange, is now debilitating members of the fourth generation.[141]

The genome is not a static entity. The genome is a "read-write" memory that not only is subject to random mutations but non-random functionality, resulting from interacting with metabolic networks,[142] or not interacting at all.[143] Trait structures and genes do not have a one-to-one relationship. Genes are turned on and off by protein messages from both the organism and the environment. Genes then express proteins, which cause physiological changes, which in turn send signals to the genome. **The regulatory networks of proteins and genes are themselves modified during a lifetime, depending on environmental interaction.**[144]

Epigenetics says it is possible that the genome can learn from experience,[145] that it is actually "sensitive."[146] Protein groups can recombine and grow by copying segments of DNA and RNA. The possible combinations of more than 20,000 genes in the human genome are immense. One genetic network can create more than 2000 protein variations[147] and various genes produce proteins consisting of more than 100 amino acids.[148] Different individuals may have identical proteins expressed by different sets of genes.[149] For example, it has been shown that mice with the same DNA exhibit different behavioral patterns. The genome of two cells in a petri dish may be identical but the physical cells are not.[150]

The genome as such, has a structure beyond just a series of genes.[151] Location in the DNA, along with flexibility in the gene sequence, can affect functionality.[152] The cell nucleus contains molecules beside DNA that interface between the DNA and every other cellular structure.[153]

Chromatin, made up chiefly of histone proteins, wraps the DNA in spools and formats, differentiated by chromosomes. Positions within the histones are modified by RNA and can thereby turn on or off whole regions of the genome corresponding with chromosome placement.[154] Methylation is the name of the process where methyl groups, made up of varying patterns of carbon and hydrogen atoms, switch genes on and off. Ncrna are RNA expressed by supposed non-genetic sections of DNA. All of these methods contribute to the idea that there are sites in the DNA beside coding sequences that regulate cell functionality.

There are direct extra-cellular factors, other than selection, affecting inheritance. All mutations are not random, some may even be "directed."[155] For example, oxygen deficiency[156] and radiation[157] have been shown to affect both particular genes and particular traits. Sugar availability in the E. coli bacteria environment have resulted in genetic mutations expressing digestive processes at speeds not attributable to random change.[158] Developmental stability of a single organism involves the dynamic interaction between the individual and its environment.[159] Outcomes of individual behaviors are intertwined with an organism's awareness of its surroundings.[160]

Outside the nucleus but within the cell, the mitochondria have their own set of genes and proteins, both of which are passed together from one generation to the next. The mitochondrial genes mutate 10,000 times faster than nuclear genes.[161] Prions, which are misfolded proteins,[162] endoplasmic reticulum (membranes used by transporting proteins), microtubules, lipid membranes and chemicals encased in organelles are all inherited without DNA coding. What is finally selected is functionality above and beyond the genome code.[163] **Beside gene sequences, functional shapes survive.**[164] Protein structures do not correlate with body forms.[165] It is evolution of the whole network,[166] not simply gene-trait correlation. Life's functions are both structurally and genomically inherited.[167] **Each species not only differentiates in terms of sexual compatibility, but in how it organizes the world.**

Biological time scales exist not only in evolutionary time of unknowable billions of cell cycles, but at a lesser but still substantial number for individuals, as well as a single cycle for a single cell.[168] Epigenetic inheritance concerns traits that are acquired during a single lifetime and that are sometimes passed on to offspring.[169] Individuals vary within a single species. Selection works on a variety of individuals whose very variety affects each individual.[170] The magnitude of the interactions of any given single generation is itself a possible variable in evolution.[171] Learning that is passed on through generations could possibly arise in some form as inherited behavioral traits. Those that learn, survive.

**The individual is aware, not the species.** The species is part of the environment of the individual. In one organism's lifetime, the interaction with the environment, its awareness, is a more fundamental influence than its genes.[172] Evolution is affected by the choices individuals make.[173] Indeed, it may be awareness of another individual member of the species that is more significant to the individual than the species itself.[174] **Choices made by individuals may favor specializations over sheer genetic reproduction.**[175] Genomic mutations or changes may be selected by the cell. Evolution in E.coli, for instance, has been thought to select for abilities to manipulate its own mutation.[176] Standard evolution theory explains the present by what happened in the past. But learning takes place in an individual lifetime.[177] Learning itself can include creative solutions by various living beings[178] whose solutions become themselves agents of selectivity.[179]

Learning is a cognitive act in which representations of the environment are retained and applied to the predictions required for survival, including the selection of an appropriate mate.[180] Cognitive networks evolve from and by the ability to change.[181] Structures like larger brains for learning or e.g., the social skills of bird calls, are not directly correlated with a genome. Dog genomes are "noisy," as ancient survival skills have degenerated.[182] As was previously mentioned, cognitive capacity increases when the rate of change in the environment exceeds the ability of the genome to cope.[183] At that point the species may go extinct, or at minimum many individuals will die. Some might survive.

Within an individual's life cycle, genomic network expressions also contribute to tissue diversity. Different tissue types with the exact same DNA have different forms and functions. Cell division requires something else to know whether to retain the same or specific difference.[184] Each cell division increases the number of genomic "memory states"[185] so that somatic change in each lifetime outpaces genetic change.

Tagmosis is the term for the functional differentiation of sectional body shapes such as limbs and heads.[186] It is thought that larger skulls enabled the space for cortical development of a greater association area between sensory and motor areas, which in turn allowed neurons to dynamically rewire during a lifetime of learning, leading to greater internal imaginative communication.[187]

Human evolution can be seen to be affected by culture. For example, Scandinavian dairy farming arguably lead to the inheritance of lactose tolerance, despite there being little immigration of other groups. Urban poor in the United States have a lower life expectancy possibly due to higher stress levels. Shortened telomere length, associated with premature cellular aging, is directly linked to stress rather than through genetic inheritance.[188] Richard Dawkins has theorized the concept of memes, ideas as cultural level genes. Behaviors of the surrounding society affect genetic changes.[189] **Traits affect genes**. Interactive thought patterns are the outgrowth of biological networks.

Organisms loose or gain large parts of genomes to and from other organisms. **The interlocking of genomes indicates that an organizational principle exists beyond the sole genome, and certainly beyond random genetic mutation.**[190] Many individuals have multiple genomes. Some have genomes that came from other people outside of inheritance. Some have groups of cells with genomic variations. Twins have acquired the other's blood cell genomes while in the womb. This is a known condition called Chimerism. Fertilized eggs can combine genomes. Fetal cells left in the womb can cause genes to be absorbed by the mother. Neurons with Y chromosomes from the son have been found in the brains of mothers.[191] Microbes in the same environment commonly share DNA.[192] **The genotype alone cannot convert itself into a phenotype**.[193] Random genetic changes feed into an overriding pattern of phenotypes, indicating that evolution occurs within another level of order.

> **The Point:** Species adapt but individuals learn. There are many factors in the cell network beyond the genome that are inherited. Functionality evolves. Functionality, not the genome, is what is internal to the individual. Cognitive ability increases when the genome cannot cope. Greater cognition arises from ability to change.

## DEVELOPMENT

There is a great deal of similarity in the genomes of jellyfish and humans.[194] Trait-wise not so much. Since some beings grow up to be jellyfish and others humans, the process of becoming an adult, or development, bears examination. **Developmental life-cycles may themselves be the central factor in evolution**.[195] Rather than being separate static structures, **traits are processes "integrated in development**."[196]

Traits or structures have a developmental history[197] in a single lifetime. The DNA remains the same but different patterns develop.[198] **Structural modularity constrains the development to certain potentially varying**

**paths.**[199] Developmental time further constrains these paths.[200] The divergence of species in evolutionary time, is countered by convergence of structure in individual development, albeit enormously complex. Throughout all species, there are divergent kinds of limbs and heads. But there are almost universally limbs and heads.

Ontogeny is the process of transforming undifferentiated cells into complex organisms. **The complexity developed from an embryo does not result from competition.**[201] Since the genome normally remains the same, the genome itself has got to be regulated to grow in an organized fashion. Gene regulatory networks are themselves structures. These networks can act to form structures simply by a change in a switch in the network. Thus, **the same basic network can act to form different structures during development.**[202] A protein called the nuclear Fibroblast Growth Factor Receptor 1 has been isolated as necessary for multicellular differentiation.[203] This protein works in a network of other proteins turning genes off and on. If-then logic among genes and proteins can create cascades of proteins in the development of a particular body part. Signaling circuits formed in each cell regulate the specialization of chromatin in the nucleus.[204]

MicroRNAs (miRNAs) help regulate the development of cellular networks that differentiate cellular identity, by switching genes on and off in patterns. The resulting pattern of genes, rather than genes working alone, form networks with miRNAs in establishing the cellular type.[205]

The Hox complex is another genomic (genome as a whole) but not genetic (collection of genes) structure that has been shown to control body plans along the head-tail axis. The Hox gene-protein circuit determines which kind of general structure (legs, ribs) will go where, but does not itself determine details.[206] But **the same circuit produces a different pattern in different animals.**[207]

Cells at the blastula stage begin to move and realign.[208] While still embryos, regions of cells attract others to them.[209] Cells gravitate toward others by sensing pheromones.[210] **The rate of cell differentiation occurs at a slower rate than the increase in cells per type, thus patterning the distribution.**[211] Positive feedback loops combine with negative feedback loops on different scales to produce the patterns that will develop into different body parts.[212]

**The general evolution of a species influences individual development but at the individual's pace.** For example, in the mammalian embryo the brain at first is reptilian, with the olfactory bulb placed in front. Later, the occipital lobe grows as the olfactory bulb shrinks. After that, additional neural layers develop.[213] Development tends to mirror the taxonomical sequence of order, family, genus, and species. The specific distinctions of the individual within the species are last.[214] Significantly, the head comes first. It arises out of the embryonic skin, the cell surface where the interaction and initial interpretation of the outside initially takes place.[215] There are immensely more neuronal connections than genes, in the human, trillions more. Neurons organize themselves as they interact with the developmental environment.[216] **The flow of information in development from the bottom up is also top down.**

If parts of the organism are removed, development in many instances will continue with the missing structure re-developing.[217] There are many examples in fully grown animals. Separated sections of a hydra will regenerate. Lizards can grow a new tail. Separated tails can regrow entire worms.[218] **Cells seem to remember where and what they are supposed to be even when they are moved elsewhere.**[219] Wnt is the name of a network of genes and proteins that act like dominoes in building the pathway to a whole organism. This pathway exists in all animals from worms to humans,[220] and involves detailed timing.[221] Stem cells in simpler animals may differentiate into all cell tissue types belonging to a section such as a limb. In humans they are confined to specialized tissue.[222] New cell growth of any kind is regulated by the cooperation of the rest of the tissue or cell type.[223] The body plan of an organism is not directed from one spot or level. **The body plan and the body that is planned are one and the same. In an organism, "form comes into being."**[224]

**The Point:** The organism develops from the seed with increasing individualization, mirroring the generality of living forms progressing through to the species and at last the individual. Throughout the process, the developing individual directs its own parts. The head comes first, developing from the embryonic skin, the original boundary of sensation and selection. The developmental process itself mirrors and is mirrored by evolution.

## REPLICATION

Evolution travels by necessity through a single very narrow passageway. With few exceptions, unicellular beings replicate by the process of mitosis or cell division, in which the chromosomes and genome are duplicated via a complex choreographed process.[225] Unlike individual cells in a multicellular organism, the resulting cells are both whole organisms, rather than cell types. Each daughter cell is not constructed solely from a genome. Functions and information in cellular networks are also divided and contribute to distinctions.[226]

**With multi-celled organisms the information passed from one generation to the next is not a copy but a reconstruction.**[227] The process is sexual and begins with the development of specialized sex cells, each carrying half of the parent's information. These cells undergo, more or less constantly over a lifetime, the process of meiosis that creates sperm and egg cells. This is a complex process in which genes will first be reshuffled on each chromosome, then new combinations of genes will be sorted on different chromosomes. In a human with 23 independently sorted chromosomes, the result is an individual genome sorted from a possible 8 million combinations. After fertilization, the two chromosomal strands are combined in a double strand, meaning the fertilized egg, or zygote, will have a genome constructed from 8 million x 8 million or 64 trillion possibilities. Genomic mutation is not needed for randomness. **An individual is not copied. A member of the species is.**

**Before fertilization the individual has input.** There is environmental randomness in breeding but also choice. There are non-genetic environmental factors during and after fertilization. The entry point of the sperm into the egg determines the first segmentation of the new zygote. In a frog, this will determine the ventral side.[228] Both sperm and egg cells contain non-DNA material. The womb environment contributes to the development of the pre-birth offspring. But at first, the newborn human has no bacteria in its gut. Gradually after birth, bacteria come to make up ten times more cells than belong directly to the human. Evolution is part of the environment to the individual. Nowhere in the planetary genome is there a blueprint for individuality.

The 2nd Law of Thermodynamics is dissipative. Evolution as sheer divergence is dissipative. Evolution is also more replicative.[229] Life as such finds its stability in replication within dynamic interaction.[230] But making more of itself is exponential and therefore, unsustainable. Living cells and organisms disintegrate in their own time with the increasing dissipation of energy. Sex began with death of individuals. Species evolve. Collections of individuals live and learn from one another.

> **The Point:** The whole of life flows through distinct species and whole species flow through distinct individuals. The whole of an individual develops through its parts in the order of life. The whole of an individual life flows through a distinct cell into another individual whole.

# III. AWARENESS IN FORM

## A. INFORMATION

*Life is awareness*
Lynn Margulis[1]

### AWARENESS

**Awareness is the self processing of information**. Matter flowing through form creates a one-directional past-future, organized by human awareness as a flow from ordered energy to disordered entropy. Form freed to flow through matter flows forward in the direction toward the uncertainty of total information deduced as such by human awareness at the intersection. Living awareness receives and orders the information according to its own flowing structure. Through living beings, information is derived from thermodynamic energy.

All living beings live their own versions of the same directional nexus. Each lives in an environment that makes sense to its living structure. Life is at the point where measures of entropic disorder and measures of informational order cross. Informational order is a reduction of the total information available. The living being decodes selected signals and then increases patterned complexity according to its own meaningful interpretation. Nutrients are nutrients only in relation to an individual's metabolic usage. Communication of information occurs at every level and every point in the living organism. The pure change of the universal given is interpreted as information, producing living activity. **Living awareness does not emerge from matter. It is the informing of matter.**

In living awareness, infinitely directional change becomes living time. Time is change with direction. But human consciousness has generally perceived time as change within time, measured by a clock. Clock time is a measure of change in the world compared to change in a clock. Living time is not quantifiable clock time, it is qualitatively smooth duration. Awareness itself changes compared to everything else in the world.[2] And everything else is felt as resistance to a living being's change. Awareness feels itself resisting change by escaping change with its own sameness.[3] **To maintain sameness, it does not simply react like matter, nor merely act. It overreacts. Awareness adds something**.

An individual organism detects distinct positions that continue the same or change according to the organism's own "expected intervals."[4] Movement and object are the background of one another. The environment is the organism's interaction with the environment. Any regularity in the flow towards entropy forms the general background of all information discovered in change.[5] What is discovered is informed regularity. That regularity is a reduction of environmental information but enhanced throughout metabolism. Information is realized in the movement and changing shapes of proteins.[6] Awareness reflects change and sameness of shape interpreted as moving objects. Objects exist in a field of awareness, which John Searle calls "indivisible [and] multidimensional,"[7] as opposed to divisible four-dimensional matter. They are ever forming objects. **Awareness of change arises in a changing awareness**.

Spaces and things are interchangeable,[8] one forming the background to the other. A thing that moves through space is, in general, a "unity formed by distinctions."[9] And distinctions are nothing in or by themselves.[10] Relative differences between any two positions are distinctions formed by the relation. In a living body, distinction, shape and change are manifested in action. The switch in a protein does not exist qua switch until it switches. Changes in protein shape differ at a point in time defined by the timing of the switch.[11] The switch exists only in time.

Information does not reside in a separate information molecule like a Cartesian conscious substance.[12] **From infinite possible unities, awareness creates its view of reality by limiting and re-ordering the flow of information.**[13] Interaction with the environment is a dynamic relation. Living awareness, looking forward into the past, relates the present-to-the-not-yet in distinction from the environmental past.

Though not separate from the environment, awareness still necessitates boundaries. Boundaries include the organism's own cell wall, the specificity of tasks and multiple constraints or limits against or around which an organism self-organizes. Terrance Deacon describes the foundation of living awareness as "nested layers of constraint."[14] A constraint is a relation to other constraints, e.g., volume and pressure, but is also the images and categories created by the organism. Information probability, thermodynamic energy and evolution of self-maintenance are all relational constraints contributing to the dynamic self-relation of the living being.

**The Point:** Awareness is the self-processing of information. It reduces and enhances information according to its own structure. This living awareness is the informing of matter. It turns the thermodynamic flow into a flow of information. Life continually adds meaning.

Infinitely directional change becomes one-directional two-way flow in living time. Unity and distinction, background to one another, in the flow of awareness are movement and object, background to one another.

Boundaries and constraints are necessary for self-organization. Infinity is limited by individuality.

## FUTURE

Recognition of sameness, including the sameness in the regularity of repeated changes, is based on expectation of what might happen, and is thereby projected into the future.[15] With sameness comes the possibility of surprise, of unexpected change. If there were no retained sense of sameness there would be no awareness of change or sameness. There would be no awareness. The difference between the expected and the unexpected arises in the temporal window of awareness. **The dimension of time manifests in two separate recognitions of material information.**[16] Sameness and difference require memory and anticipation. Time means nothing if it doesn't include both. Direction toward a non-existent future is awareness looking forward at the past.

Information is communicated between the living being and itself at all levels and moments.[17] In the words of Hans Jonas, "The forward trajectory of the organism at the cellular level of metabolism is the vital source of the potential openness of time-consciousness."[18] From the underlying universal distinction and sameness, what is projected forward in anticipation, is determined by the structure of the particular organism. The organism is aware of whatever it is built to be aware of. What it is aware of is not confined to past experience. The organism is built to adjust as it continuously compares internal and external messages. Any adjustments are carried forward to the next moment with new expectations by a changed structure.

**The point:** Beginning with the anticipation of sameness and recognition of difference, the idealization of the past in a non-existent future pulls the living being forward. The present moment, facing forward into nothing, falls into the future. The future is the life force.

## QUALITY

Relation is fundamental and, therefore, change is a blend, not a number of points. Lived difference is not countable. Continuity and discontinuity, things and between things, are relations not quantities.[19] In awareness, these relations develop into shapes and patterns of shapes. **What is selected in and by awareness are qualities.**[20]

The rolling moment or "transient present"[21] is awareness of identity and difference in the world and in itself. But the dualism is not awareness as separate from externality. Matter flowing through form while form flows through matter develops in living awareness as internal and external flowing past one another. Awareness opens and is the opening of the relation of matter and form. Change within sameness is not numerical distinction. Time is awareness of a "progression of qualities."[22]

**Everything is a flow of qualities that can only be counted after awareness selects sameness.** Awareness artificially quantifies what is fundamentally quality. Quantity is a quality. "It is through the quality of quantity that we form the idea of quantity of quality."[23] Number assumes finite edges around that which is counted. Equations, i.e., relations of equality, are superimposed on relations of identity.[24] Positions and motion assume numerical jumps superimposed on a continuous background. If matter in general is the selected quality, there is only infinite quantity. If form is the selected quality, then quantity reveals infinite forms.[25] Living awareness is itself fractal, that is, similar shapes with infinite sides and sizes.

Living interaction with the world is a forming of "maps," says Terrence Deacon. Maps are patterns seen as spatio-temporal relationships. Maps are applied to recognized coherence or regularity. "**Discrete jumps are mapped on a continuous background whose continuum is in turn mapped onto discrete places.**"[26] Parts in continual time are places in space.

Life is matter informing itself in a continuous process that maintains and extends recognized order.[27] **Patterns within external matter become meaningful content within form.**[28] This process is continually pulled forward by an idealization, which is the expectation of possible similar patterns.

> **The Point:** Qualities are selected out of meaningless matter into meaningful form. Quantity is a meaningful form. Sameness and distinction applied to position and motion, is the fractal awareness of shapes repeating within themselves.

## B. MEANING

*Meanings and intentions exist throughout the whole of the living kingdom*
Elliot Jaques[1]

### CHOICE

When microbes are presented with different possible nutrients, they choose one.[2] They choose between nutrients and non-nutrients. Food means food only to the particular organism for which it is food. To a protozoan certain chemicals mean food.

"**Life is matter that chooses**."[3] More precisely, it is living matter that chooses. Choice itself is "built in." Matter has meaning only to and with life. **Survival requires choosing food**. Survival is the fundamental choice.

To survive, the next choice is between two requirements: Safety or hunger. **Choice introduces effort**. It takes effort both to choose and to avoid the wrong choice. To choose not to eat when hungry in order to avoid danger takes extra effort.

**The Point:** An organism chooses. Survival, the choice to live, is the fundamental choice. Choice determines meaning.

### MEANING

**Matter in itself has no meaning.** Living relations, or tasks, point out of themselves to that which is not yet. Meaning is the residue of a dynamic relation. What is recognized is similar, not identical to what is anticipated. A difference in similarity is an exclusion of other possibilities tied to the anticipation. Similarity is incomplete.[4] It is identity provided by a not yet future. It is ever incomplete. Life is always catching up. In biochemical, thermodynamic terms, difference is negentropy, the amount of entropy locally exported from an isolated system to maintain local energy.[5] In pure formal or logical terms, difference is negation. Matter and form, separated and recombined in life, project sameness and difference into a meaningful future.

**Similarity is not in anything in the world, but exists in the awareness of everything.** Meaning is virtual.[6] The taste of cheese is not in the cheese.[7] Something tastes like cheese. The incompleteness of similarity, the absence of what is anticipated, implies selection.[8] The similar that is selected is seen as an equivalency. This is called "isomorphism." "Everything that's true about a designated part of one object's structure is true about the other's."[9] The "truth" in an isomorphism is meaning to the organism. **The meaning is selected similarity, judged as equivalent.** Equivalent parts are assumed for the whole.

Meaning arises in the interaction with patterns that comply with anticipation. The patterns are recognized when the regularity continuously compliments the organism's actions.[10] Differences are what makes a difference for an aware individual.[11] It is what makes sense. Meaning is making meaning. Meaning making is felt in completing the similarity to fit anticipation. The individual feels itself in the absence that it fills. This primary feeling is what Hans Jonas calls "urge."[12] **Meaning is the content that is felt**[13] **as the individual unveils the context**[14] **discovered in likeness.**

Meaning becomes more elaborate as experience elongates or widens.[15] With experience, **regularity becomes habit**. In habit, the living being continuously interprets within the context of the past.[16] The possible provides context for the probable. Skill and learning provide the context of pattern through time and enlarge the range and therefore, the freedom to select. The more complex the patterns become, the more the living being is engaged or "present" with the world.[17] Possibilities become actualized with experience.

Generalization based on specific facts is the beginning of the scientific method. But organizing the given is more than material cause and effect.[18] Information alone does not increase content. **Content increases with meaning.**[19] Experience organizes complexity and meaning expands into recognized categories. The living being modifies the category by recognizing resemblance. Recognition is "filling in." **Modification is formal causation,**[20] in other words, formalizing matter.

Living awareness is awareness throughout the entire organism. Interaction between functional parts at all levels are caused by the not yet of difference. The organism as a whole is "dynamically engaged" with its surroundings.[21] Bacteria are observed to distinguish between two sugars, then actually change their metabolism to use the chosen sugar.[22]

Patterns in the environment are recognized and therefore, meaningful to the degree to which the organism uses them. Categorization matches the species. **Awareness applies categories to structures**. The organism is free to be aware to the extent of its own structure. That extent has leeway, for **meaning ultimately belongs to the individual**. Aside from an amorphous "something else," the individual is not aware of what has no meaning to that individual.

**The Point:** Matter itself has no meaning. Meaning is a relation with that which is not yet. The living being turns information into meaning by selecting similarity, but interpreting it as equivalent. Meaning making is felt in completing the similarity to fit anticipation.

## VALUE

Looking at the world as matter only, science organizes that world according to timeless laws. Life, on the other hand, operates according to norms. **Organisms make sense of the world based on rules, not physical laws.**[23] The future stands before living beings like a perpetually unfinished canvas just beyond reach.

To act according to need requires recognition of what is needed. **What is needed is always needed more or less than something else is needed.** The difference is recognized. What is recognized is the goal to be attained. As such, it has value to the particular living being at a particular time. This means the value is relative. An organism has priorities.

**Selection contains a variation of value determined by the organism's abilities.** "Even a bacterium has a good in this sense, in virtue of its proper functioning, whereas a rock does not."[24] The organism responds from a selection of obtainable possibilities similar to recognized categories. The action carries with it an "ought"[25] to do the right action, the successful action. **The correct choice is the good choice.**

Value is universal in that it governs all actions. As such, it is purpose.[26] **Purpose maintains identity of meaning in the living being.**[27] Purposeful exploration discovers information about the selection. Purpose drives the process of informing with correct generality. This generality causes from ahead. In the words of Kant, "the action of the cause is based on the idea of the effect."[28] Purpose informs every choice.

**The Point:** Value arises out of successful choices. Value prioritizes choice. It is the purpose of a choice. Purpose drives the informing of matter and maintains meaning. Choices are caused by the idea of what is good in the future, its value.

## ABOUTNESS

Values and purposes are universal, but life is expressed through the individual living being. The nexus of form and matter creates past and future, but is lived directly at the nexus. The feeling of want, of urge, does not take place in the past or the future. It is in the moving present. The world is experienced as something to be reached[29] but the reaching is a continual, immediate distinction between the organism and its world.

Each living being exists within a relationship with its own structured environment. Each living being has its own "interests," such that within its context, things are recognized, and structures become categories.[30] Matter is similarly categorized according to the interests of members of the same species. But the direct immediacy of the world's presence to the living being is not available to others. The feeling is one of acting at the center. Inside the moment, individuality is the most universal of all that is. **Awareness is about everything else.**

> **The Point:** In the temporal space between form and matter, awareness is about everything else. The moment is a viewpoint in a space-time whole. Every task is for something else.

## LEARNING

Structure in the environment is categorized by correspondingly changed structure in the organism, which is in turn applied to newly recognized structures, the recognition of which is an expansion of categorization. **"The metabolic causal circle develops into the dynamic progressive interchange of learning"-** Gregory Bateson.[31] Stimulus gets reinforced by response which is then reinforced by previous response.

Evolution is the adaptation of populations of living beings to their environment. Heredity is the adapted result embedded in a species, which then is both the starting point and the constraint that the individual brings to learning. Individuals then form habits which habituate in species. The combination of these evolutionary levels is a recursive circularity. **Contained both within and contributing to evolution is the "sub-routine" of heredity-instinct-learning-habituation**. These sub-levels, though distinct, tend to merge.

The difference then, between instinct and learning is one of scale, not kind. Habituation, which has even been observed in snails, combines learning and memory.[32] They "get used to" previously un-encountered situations. Learning from experience can be observed in the simplest of living beings.[33] Paramecium learn the quickest ways to obtain food.[34] Single cells can generalize from specifics.[35] Information not categorized by the cell is passed over. What is retained is the meaningful reduction.

At the smallest scale of a second or less, cell structure alters. Other bodily processes with larger time scales can make the alteration permanent.[36] At the genetic level, the difference between association and habituation is partly attributable to different genes. But with the combination of these genes comes a repurposing of the genes in learning. Genomes "move beyond themselves."[37] Learning "tunes" the metabolic connections.[38]

**The advantage of habit is efficiency**. Learning once learned tends to be automatic. Repeated experience, like practice, reinforces the internal categories. Habits become skills. But skills must keep pace with the surroundings. If the environmental structures change more rapidly than the organism can categorize them, survival is minimized. Living beings learn what Susan Allport calls "good biological sense."[39]

**Learning is itself a circular process**. Learning transfers an interpreted part of the environment to memory.[40] Association and habituation are expressed in the present moment as recognition and recall. At intervals occurring at a rate set by the complexity of the organism,[41] the current recognitions of sameness and difference are compared to the recalled past, and changes that happened in that past. **Learning begins with the felt surprise of a difference.** The body is aroused by attention[42] when something is unexpected and therefore, new. Categorization, the collection of similarities in a whole, deepens as new content is added to the recognition of what is similar.

Categorization also connects categories in time and place, "the order of information arranged by the living being."[43] Cause-effect relationships between categories are recognized, associated and remembered, and recalled when recognized again. Recognizing cause and effect contributes to survival.[44] Association predicts one event by another.[45] A frog notices a fly at one position and flicks its tongue at the predicted position. The prediction extends the content derived from the initial comparison.[46] To the frog, the fly's motion has a particular speed.

Active learning takes more energy than habit. Openness to the future provides that energy. There is more information in possibilities when they are actively engaged. Actuality is compared to future possibility. **With the future wide open, possibilities are larger than the actual.** When anticipation is surprised by the new, awareness actively adjusts to erroneous expectation. The new is interpreted to be about something more, which is then predicted to be related to what is recognized.[47] Difference is detected within a developing coherent pattern. Memory is "projected to the future."[48] **The new flows in from the past through the future.** The not-yet becomes both learned and recognized to have been learned. The accumulation of certainty pushes and is pushed by the feeling of both attainment of the new and satisfaction of the correct prediction.

**The Point:** The metabolic causal circle develops into the dynamic progression of learning. Categorization deepens as new content is added to the recognition of what is absent. With an open future, possibilities are larger than the actual. Openness to the future provides the necessity to learn.

## C: INTERSUBJECTIVITY

*The sign rather than the molecule is the basic unit for studying life*
Jesper Hoffmeyer[1]

### INTERSUBJECTIVITY

The individual subject is not alone in a world of so-called objective matter. There are other individual subjects.

1. Other.

There is an old adage, "It is better to give than to receive." **A giver is a subject, a self. A receiver is an object, an other.** The matter aspect of life is an object. An object's movement and shape is relative to the point of view of all space-time positions. A living being can observe its own body as well as other bodies from a particular position. That is the body as matter only.

2. Self.

The subject is not an object.[2] What the world is to first-person experience is not experienced by another. There is no way to deny what another claims is their personal experience. The feeling of self is not an objective quantity measurable by others.

**The feeling of self feels like a belonging to the self**. But the feeling of belonging is also a feeling of possible engagement with the world including others. Ours means not theirs.

Subjectivity begins as a self-relation, where the subjective experience within the body is differentiated from that which is outside the body.[3] If we rub our hands together, the feeling of rubbing and the feeling of being rubbed do not coincide. The feelings alternate. Both feelings are subjective, passively, then actively, objectifying. Our description of our own experience feels to us to be objective.[4] We are our own other. To itself, a living being is both observer and observed.

3. Other as Self.

The recognition of certain others as other selves or possible other selves is universal throughout living beings. There are three molecules that have been observed to mean "quorum-sensing" in a species of bacteria. One type of molecule means a close family member, another means the same species, another not the same species. If none are present the bacterium feels itself to be alone. The bacterium counts the respective amounts to know what to do next.[5]

Other as self is called "**theory of mind.**" Beyond the other's body, there is something that indicates to an individual that another individual is present. Another individual means another self whose actions mean something to both selves. The "sense of me" is closely tied to the "sense of others."[6] In Psychiatry, mental illness, when not reduced to brain problems, can sometimes be seen as caused by interactions with others,[7] where the self goes so far as allowing itself to be defined by others.

Alva Noe wonders whether we can trust or love others if we do not believe that they are another self.[8] For some people, maybe all, purpose in their own life is tied closely to what others will do after they die. The meaning in our lives survives with others.[9]

4. Self as Other.

We cannot imagine our own non-existence or death as nothing. We can only imagine our self in other bodies.[10] The view of time as mutually external moments mandates that we imagine a body after body going on forever. We see our self as an other.

Empathy is more than the ability to sense oneself in the other. It is **also to sense oneself as an other for the other.**[11] We cannot see the other as only an other, if we see ourselves as an other that affects the self of the other. Our bodies are acted upon from the other's point of view, like we act upon the body of the other. Sensing oneself as an other for the other is probably a sophistication of human, or near-human, consciousness.

5. Other Selves.

Inter-subjectivity is the awareness of all individuals in a group that the others are also aware individuals, and that the objects of awareness are to some degree common to all. The quorum once analyzed by all neighboring bacteria is an inter-subjective agreement among bacteria. The collective awareness in a close group, or the whole species or other living beings sharing an environment, together shape the possible universe of meanings understood by any member. All behavior has an inter-subjective aspect, a **shared point-of-view**. The world common to all participants is a world with common meaning.[12] "The self and other arise together" as Evan Thompson puts it.[13]

Objective space-time, whose relative positions require non-simultaneity when viewed, is inverted in the inter-subjective world of aware living beings. **In any living moment, our self, other selves and all meaningful**

**things exist simultaneously.** Other selves are not relative to a living individual's viewpoint.

**The golden rule is not only theological but ontological.**

> **The Point:** 1. Other: A receiver is an object, an other.
> 2. Self: Subjectivity begins as a self-relation. We subjectively feel ourselves to be objective.
> 3. Other as Self: The recognition of certain others as other selves or possible other selves is universal throughout living beings.
> 4. Self as Other: Empathy senses oneself as an other for the other.
> 5. Other Selves: Inter-subjectivity is the awareness of all individuals in a group that the others are also aware individuals. Groups shape the possible universe of meanings understood by any member.

## COMMUNICATION

**Communication is living information.** Information is coordinated in mutual behavior. Behavioral communication does not require a nervous system.[14] Coordination does not require language.[15] Before a thought verbalizes, it can be understood.

Awareness of the other's self, in theory of mind, means awareness of the other's intentions.[16] In order to survive, living beings must be aware of why others act, what is their "agenda."[17] In order to understand another, one must to a certain degree know what the other is going to say before they say it.[18] **Communication is the state of being "inter-subjectively open"[19] to the next moment. Communication is pre-conscious awareness.[20]**

Communication is a mutual reaction to a shared experience, the relation with the other already present.[21] Conversation is not simply taking turns talking. Listening involves "reading body, tone and rhythm,"[22] all with anticipation. Pairs of living beings form special bonds, a private "singularly organic"[23] language. Knowing together shares an "interval of awareness, mutually new and therefore, nonlocal."[24] **A shared present has no spatial distance.**

The subjective is first-person, the objective third-person, but communication is primarily second-person. Internal dialogues with an imagined partner are the first inklings of social awareness.[25] Society, then, naturally begins with two people. The theory of other minds is instinctual in babies. "Infant gaze" experiments have shown an ability by infants to anticipate their mother's intentions.[26] Baby and mother's responses to one another are close to being simultaneous when looked at from a third-person perspective, but simultaneous from theirs.[27] **The interaction establishes a greater self.**

Pairs are a source of creativity. Creative dialogue drives the new with merging perspectives, agreeing and disagreeing, yet understanding how and why. Their interaction flows together. "Reading" behavior and "reading" minds are closely related.[28] The ideal for Socrates was to mutually build ideas. An other mind is perhaps the one "thing" we do experience directly.

The addition of a third person establishes rules or structures that reduce flexibility in favor of social order.[29] All living beings communicate in building their intra- and inter-species societies, called by Chesher, "larger units of being."[30] Meaning in dialog becomes group purpose in group decisions.[31] Decisions become collaborative with goals developing within group communication.[32]

Within a group like a school of fish or flock of birds, the meaning is synchronized, where individuals all move according to awareness of another's position at all times. Organic structures, like neuro-chemicals, circuits and hormones have all evolved similarly in many organisms for the purpose of controlled social interaction.[33] **Biological rhythms** from the smallest cellular oscillations to the largest circadian rhythms of the sun and moon **create unified space-times within which communication can be said to be instantaneous.**[34] The vagus nerve, shared by all vertebrates, controls organ rhythm and general homeostasis. Vagal tone has been observed to be tuned to vocal frequencies.[35] We are "hardwired for sociality."[36] Evolution selects for co-operation.

**The Point:** Communication is the state of being "inter-subjectively open" to the next moment. The self in the other is already there.

## SEMIOSIS

Semiosis is the process of **creating meaning through the use of signs** by living organisms. The Kantian Whole, the autocatalytic cycle at the core of life, is an anticipatory process. Protein catalysts represent anticipation. They are the signs that connect the two ends that otherwise might not be related. Since the catalyst exists before the connection, it contains within itself the purpose for parts selected.[37] Meaning arises out of anticipation. The meaning is preserved in the self-renewable structure of the cycle. The

**matter changes but the meaning remains**, allowing for future encounters to be recognized.

The evolution of morphological, i.e., structural, diversity accompanies a deepening of meaningful communication.[38] The capacity to express, understand and replicate messages is universal to life. Parts are tasks. **Tasks are messages**. Within the body, components are restricted to the good of the whole. Whole individuals are open to the world and their own future. Individuals are in charge of themselves. **Evolution is, in fact, the gradual expansion of "semiotic freedom**."[39] Self-determination arises with communication.

> **The Point:** Semiosis is the process of creating meaning through the use of signs by living organisms. Beginning with cellular tasks, the capacity to express, understand and replicate messages is universal to life. Evolution of the use of signs is an expansion of freedom.

## SIGNS

Throughout the organism, molecules are symbols of activity beyond themselves. They mean something else. "Reading and translating processes" make the organism a "self-reading text."[40] Beginning with single celled organisms, chemicals from both outside and inside symbolize conditions that require particular responses. **Meaningless molecules become meaningful symbols.**

Matter and form flowing through one another create living connections between the two. Signs and objects presuppose and mutually imply each other.[41] The immediate representation becomes an object, which then is a sign, a pointer to some other object, including the relations between objects. Categories are processed along with their recognized instances.

Living beings' urge to continue is also a "desire to separate."[42] Internal duration is projected into spatialized clock time through the intermediary of signs. A sign means an object. It is not the same as the object.

> **The Point:** Throughout the organism, molecules are symbols of activity beyond themselves. In separating the sign from the object that it belongs to, the future is anticipated.

## SIGNALS

Signs in the context of communication are termed "signals," or "symbols." Communication theory generally assumes a minimum of sender, receiver and a signal that conveys the transfer of meaningful information about things or events, separated by space or time from one to the other.[43] Signals can be conveyed in all senses, including those foreign to human senses such as electrical currents and other vibrations and wavelengths.[44] Most organisms communicate by distinguishing chemicals.[45] Species in the same environment can locate themselves and others with signals that are unique.[46]

Signals increase the chances of survival, as groups communicate in searches for food, mating and identifying themselves and others. Successful signaling occurs when the mental sign or representation of one organism is replicated, or usefully approximated, in the representation of the other organism. Interpretation of signals, the meaning conveyed, assists successful collaboration.[47]

Though signaling on the one hand is an evolutionary trait, it also has a purpose in individual lives.[48] In any successful communication, what is signified should ideally be unified in meaning. But the signal is external to both sending and receiving organisms. Even a particular behavior by an organism can be distinguished from the organism itself. **Because signals are independent entities, they can be combined in any way.** Signals can mean different things in different contexts. The receivers themselves can change the context.

For improved usage, the symbol must be a discrete unit with a distinct meaning, but able to be subdivided into smaller discrete units. **Meaning deepens with the ability of signals to become more complex**.[49] New symbols are created with increasing precision in the description of the environment. The order of a string of symbols must be repeatable. Eventually, symbols are combined with consistent structures or rules. **As with all living systems, recursion develops when sub-routines or clauses are inserted in the main string**. An example in human language is "Hand me the nails (that Dan bought)." A further recursive call might be "(with the money his father gave him)." Recursion has been observed in chickadees' and starlings' calls and prairie dog chatter.[50]

Symbols along with the empty space or "noise" between symbols, together constitute the pattern which must be recognized for accurate meaning to be conveyed. The noise is the carrier or container, or background of the message. Symbols "shape" the noise.[51]

Communication can in some instances be synchronous, yet actually involve sending and receiving signals. According to Steven Strogatz, this is true of all biological oscillators.[52] In one species of firefly, for instance, the oscillating flash cycle can speed up or slow down depending on when it senses the flash of another, which then influences its flash in turn. The multiple effect of all the fireflies together is a repeating pattern, not simply a synchronous on and off. A sign in isolation is binary or digital. But symbolized patterns are not simply "yes" or "no."[53] **Distinct signs are interpreted, and become meaningful, as smooth, analogue processes.**[54]

**The Point:** Communication requires signals between sender and receiver. Because they are independent entities, signals can be combined in any way. The more complex the signals, the more meaningful the message. Symbols and the meaningless "noise" in between, form patterns to be meaningfully interpreted. Patterns of distinct symbols and continual noise are living communication's coherent reduction of the all pervasive point and continuum.

## LANGUAGE

**Language is signaling looped upon itself**. It is communication taken to the next level of complexity. Signals convey information about specific events or objects or feelings about them. Language is not limited to a particular place and time. It both conveys information about an objects categorization and is itself the way things are categorized.

Syntax is the formal system of symbols and the "rules of manipulation."[55] Manipulation allows for the ability to combine general categories with particular objects, what linguist Noam Chomsky (1928-) calls the "active structure in hyperspace." Semantics is the meaningful content abstracted and conveyed.

Semantic content begins at the cellular level where chemical signals trigger restructuring in the receiver.[56] The autocatalytic cycle is open. Content is what diverges. The general applicability of syntax gets applied by semantics. One object is valued instead of another as an instance of a category. In language, valuing creates deeper meaning in the assigning of names, as names are open to changing application. Language is potentially open to infinite generalities about any object it can name. **Syntax is the limit. It forms the basis of how the world is structured.**

Language builds upon itself. The categorical abstractions, and especially the recursive loops, add to the independence of users. The largest recursive loop articulates recursion. The organism itself and even "self" can become a linguistically distinct object. "I" becomes the ultimate subject as object.

Since the self is an object formed in language, it is socially located. It is a self among other selves. Language and/or symbolic thought do not develop in social isolation.[57] The philosopher, Ludwig Wittgenstein (1889-1951), famously concluded that language is necessarily public.[58] Linguistic socialization is embedded in human thinking. For instance, Russians identify distinct shades of blue faster than English speakers. The French language identifies objects as feminine or masculine. French people will imagine feminine objects to have a higher voice.[59]

**The future directedness of living awareness is reflected in language.**[60] Language is "inherently teleological." For example, the prepositions "to" and "for" imply both future and purpose.[61] Sexual selection is symbolized.[62] Utterances imply responses.

Language is an integral part of physical adaptation. The moving point of interaction with the world expresses itself in verbs. Emotions in language connect with the body.[63] The FoxP2 gene has been linked to linguistic abilities in all vertebrates.[64] Specific languages have evolved along with other traits to fit the organism's environment.[65]

The language of Prairie Dogs has been studied extensively. Prairie Dogs do not simply convey how they feel, but what they know. They have a sound for e.g., the category of predator. Like grammatical syntax, the components of the sound are arranged in different ways for different types of predator. Adjectives of color, size and shape are symbolized by different frequencies of sound. Like a sub-routine, color can vary, while size and shape remain the same. An adverb for variations of speed is indicated by the rapidity of calls.

Body language, of course, is more universal than spoken language. Though it tends in many species to be more instinctual (genetic) than learned. The honeybee's ability to signal with complex dances is most likely instinctual. Yet the messages are sent and received with the same circle of awareness found in all languages. Language may not be necessary or sufficient for conscious awareness, but a state of learning is. When awareness is instinctual, even in humans, it is unconscious. That there is a tiny point of learning or surprise in a bee's message reception, seems reasonable. According to Terrance Deacon,[66] there are no language structures unique to human brains. Language is a living process, not a thing.[67] But written language, unique to humans, lends itself perfectly to objectivity. It maintains meaning beyond living communication.

Beyond sounds and signs that simply communicate, humans make art for its own sake. Music, for example, has syntax starting with scales and measured rhythms. There are many musical languages. But the semantics are emotional. And the emotional satisfaction largely rests in the fulfillment of expectation. In playing and listening to music, the present, future and past are linked. Complex temporal patterns can be heard with three-dimensional depth. The underlying pattern recognition of language ranges from emotion to the level of strategy in areas outside of music, like math and investing.[68]

**The Point:** Language is signaling looped upon itself. Syntax is the formal system of symbols and the "rules of manipulation." Semantics is the meaningful content abstracted and conveyed. Language is potentially open to infinite generalities about any object it can name. Syntax is the limit.

## D. THINKING

*Knowing is the action of the knower; it is rooted in the very manner of his living being, in his organization*
Humberto Maturana[1]

### THOUGHTS and IDEAS

"Stored memories and desires, like hunger or thirst, are added to information about the world, and actions are the result. This is the case for all sorts of animals, not just humans. It is thinking, at the most basic level," J. Gorman.[2] It begins with metabolism.

A thought is not matter, yet it is an object. **Thinking is the act of separating form from matter, established by form flowing through matter.**[3] A physical object when given away is no longer possessed. A thought cannot be taken. It is shared by being understood, and in the process increased.

Rosenblum and Kuttner see the uncertainty principle inherent in thought.[4] Like position and motion with the observation of a quantum particle, thinking about a thought changes "where it is going" and thinking about "where it is going" blurs the precision of the thought. **Thoughts are simultaneously retained and expanded**. Thinking combines the general thought with the specific thought, making another thought. Thinking, like all living tasks, involves a degree of randomness,[5] which allows an open layer of possibility beyond what is current. The randomness means ideas can generate ideas, some of them new.[6]

Ideas are complex objects in thought that may or may not refer to the world or to other modes of thought. They might not refer to matter, yet still be objective, e.g., a circle, or they can be possible yet not actual or completely impossible. Ideas as such, are timeless. To paraphrase Hans Jonas, the blur of the given world is incorporated in the time of thinking and externalized from the moment as timeless objects.[7] Thoughts themselves can become timeless objects for the abstract thinker. Ideas can be thought without referencing a specific example.[8]

**The Point:** Thinking manipulates every sign. Ideas are signs freed from any particular object. Thinking freely manipulates ideas.

## DESCRIPTION and EXPLANATION

Descartes' "I Think Therefore I Am" is generally considered to have removed all distance between the thought and the thinker. As such, the thinking is always immediate. Consciousness is not extended. But even Descartes recognized the difference between duration of thinking and measurable motion. Motion, for Descartes, is dependent on duration, but not vice versa.[9]

**Thinking in duration, unlike the motion of objects in clock-time, is relational not relative**. Thinking is foremost description, that is, it describes relationships between parts and wholes.[10] Description is at each step an increasing elaboration of the comparatively syntactical or rule based smaller category.[11] Descriptions are not extensive material parts. They are intentional, which means descriptions can separate from any particular part-whole system.[12] Descriptions are meaningful because they are isomorphic, "a plural thing made singular and re-pluralized" differently,[13] a similarity reduced to a same with a new similarity. Descriptions are the stories that all organisms share through language. "A logical argument and a story are two ways of putting fragments in proper relationship and guessing where the whole sequence leads and how it gets there."- David Gelernter[14]

Explanation adds assumptions. Descriptions as explanation are tautological. A tree is a tree because it looks like a tree. As Aristotle might have asked, "Why does a tree look like a tree?" Explanation adds its own facts to the described facts.[15]

Imagination separates the remembered descriptions and explanations from the original encounter and uses them to anticipate the future. We don't anticipate infinity, but it exists implicitly inside the manipulation of ideas.[16] **There is an inherent incompleteness in thinking.**

**The Point:** Descriptions are meaning that living beings share through stories. Explanation adds its own facts to the described facts.

## REASON and KNOWING

What is present in the intersecting flow of matter and form is neither the ultimate particular nor the most general category. What is known is what has been successfully categorized. Living knowledge is embedded within its own body and environment. Knowing is recognizing change within itself and corresponding change in the environment. "To live is to know" - Humberto Maturana.[17]

**Reasoning is a relation that connects what is known so far, with what will or could be known.**[18] What is known as useful extends to what could be useful. Deliberating is a self-conversation comparing the relative validity of alternatives. "Grasping a concept" has the feeling of coming to know the solution.

Intelligence is the scope of ability to apply, and if need be extend, knowledge successfully. It is the scope of expectation and correspondingly correct anticipation that exceeds what the organism's physical structure would normally allow.

> **The Point:** What is present in the intersecting flow of matter and form is neither the ultimate particular nor the most general category. What is known is what has been successfully categorized. Reasoning is a relation that connects what is known so far and what will or could be known.

## BELIEF

The first form of **belief is mistaken reason**. What reason assumes will happen may turn out not to happen, although it was possible that it could have happened. The "subjective theory of probability" is the degree of certainty in the original belief in the reasonableness of the projection.[19] A living being does not always know its limitations.

The second form of belief is magical thinking. This is belief based on reasoning with no or little probability of outcome. The Cartesian split gave birth to **the belief by some that form exists on its own without matter**. Intentionality not only can be ascribed to material things, but believed to exist on its own, invisibly. Purpose is imagined to exist inside external events. Intuition is a belief that events can be known, or felt as present, before they are encountered. The process of knowing is replaced by the known.

If knowing takes place only within thinking, then thinking about thoughts is a tautology signifying nothing. On the other hand, the idea that everything is matter denies that thinking itself has an effect on matter. The thinker's feeling of effort, says the materialist, causes his belief in his effort to be causal, to believe in free will. That thoughts can be causative is a magical belief for the materialist.

The third form of belief is faith. This is the belief in belief. Belief is, in fact, the superior or correct way to know. Belief can never be mistaken. And if it is, it is the fault of the believer. **Belief in the next moment expands to the eternal.** The living urge to continue into the unknown becomes a "desire for transcendence."[20]

| **The Point:** Belief is reasoning that form is without matter. |
|---|

## TRUTH

**Truth is the object of knowledge**. We know the truth. The Cartesian split makes it a problem for subjective thinking to be a knowing of objective matter. Where does a thought leave off and an object begin? Inanimate systems are explained by laws. But the idea of truth is an attribute of life, not matter. **Truth is not the object of faith in the unknown, but faith in the knowable**. The truth is that which can be explained but has not yet been explained. The truth is the truth so far.

The belief that the future is knowable contains the idea that it is completely knowable. Truth is a standard by which knowledge competes. When knowledge temporarily catches up, when the parts and whole match, the feeling is that of directly grasping the truth.[21]

The experience of truth is also the experience of untruth, that which is other than truth.[22] The content is remembered differently than the original encounter. Categories of content are malleable. The truth itself remains independent. There is always something else.

Of course, we also believe we know the truth. Truth is ultimately its own ground. There is nothing else that it stands on. In that way truth is like faith.

| **The Point:** Truth is not the object of faith in the unknown, but faith in what can be known. Truth is always a relation between form and matter, that is, an attribute of life. |
|---|

## E. OBJECTIVITY

*There just does not seem to be anything like "function" in physics and chemistry*
Alex Rosenberg[1]

### LOGIC

**Categorization has not only structure, it has order**. Things are not only parts of other things, but they become parts of things in certain sequences. Matter alone has no direction. Form without matter is timeless. Living awareness applies forward direction to bi-directional time. Form flowing through matter categorizes matter with timeless logic. There is order and an ordering.

Logical relations in material things arise in awareness as cause and effect. But physical causation differs from logical implication. In logic, A implies B, B implies C, A implies C. The sequence is both necessary and sufficient. In nature, A causes B but C might also cause B or C might cause B instead of A. The cause is not contained in the causal relation but in the general laws of nature. But nature is both random and endlessly complex. The only absolute certainty is that one thing comes before another. Sufficiency is open. Necessity is discovered. Contradiction in logic is absolute, in cause non-existent.

Form alone and matter alone are devoid of choice. The intersection of awareness contains within itself neither logical implication nor analyzable cause and effect, but instead, the "intentional logic of reasons and purposes."[2] The intersection introduces immediacy. Awareness formalizes matter, but immediate experiences can never logically contradict one another.[3]

Living beings insert valued choices and selections within "a logical frame over time."[4] **The application of logic to cause is reason**. Inductive or bottom-up logic generalizes from specific observations. Objects are categorized. Deductive or top-down logic begins with a generality assumed to be true and members of its "class" necessarily true if the general is true. If a category is assumed to act in a certain way, then it is reasonable to assume new objects similarly categorized will act in a certain way.

A tautology is a statement that says the same thing twice, e.g., "I am here while I am here," or something that can't be proven or disproven, e.g., "I will be there or not be there." This is similar to circular reasoning where the premise is also the conclusion, e.g., "The reason he is not here is because he hasn't arrived yet." Neither case contains any meaning. They are locked inside themselves. There is no time.[5]

Teleology is cause, reason or explanation for something by its own purpose. The function or purpose of a seed is a tree. The logic of self-causation is opened up to what it will be. Teleology is tautology in time. Something is caused by its own future. The tree causes the seed. Self-reference is necessary but not sufficient. Purpose requires implementation.

**Efficient cause and effect is in time, imposed on matter by timeless formal logic. The imposing, itself, is a procedure**. A procedure has input and output, similar to a living being. In a computer, the instructions are carried out in time. The output is caused by the input. But the instructions themselves are timeless. The output is logically implied by the input. However, somebody (or something) from outside the procedure writes the instructions, inserts the variables and starts the program. Time is from the outside.

INPUT -> PROGRAM -> OUTPUT

To meet a user's condition, the program might form a repetitive loop, circling until the condition is met. The loop and therefore, the program, is completed to the satisfaction of the users of the program, who have set up the condition. The condition in the loop would be expressed by formal logic, e.g. if x = y then proceed.

INPUT -> PROGRAM -> LOOP -> PROGRAM -> OUPUT

More complex conditions can be imposed in the form of loops or subroutines. For example, before continuing, check to see if x > than some number. A call to a subroutine loops in a circle adding 1 to x until something else, then the procedure returns to the original loop. Each call and the called loop is a function. It has a purpose.

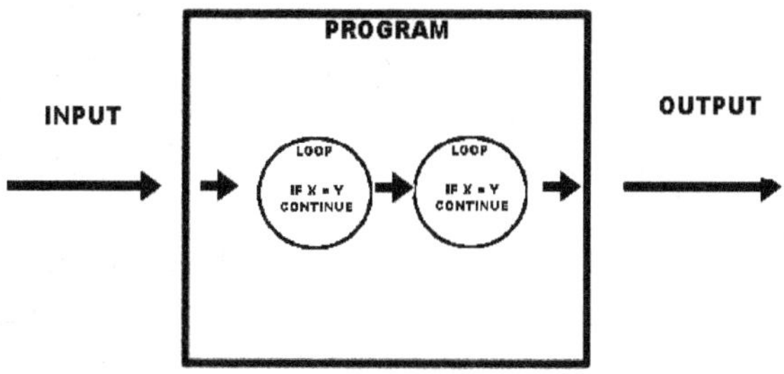


Fig. 19.

Recursions are loops or functions that call themselves or are in a cycle that self-references.

Fig. 20.

Three loops, a, b & c in a cycle of self-reference     Loop c calls itself

A recursive loop must contain within itself a "base" condition which throws the procedure back to the major loop or to the end of the program. Otherwise the program will continue, never reaching the final output, in other words, an infinite regress.

**Recursion**

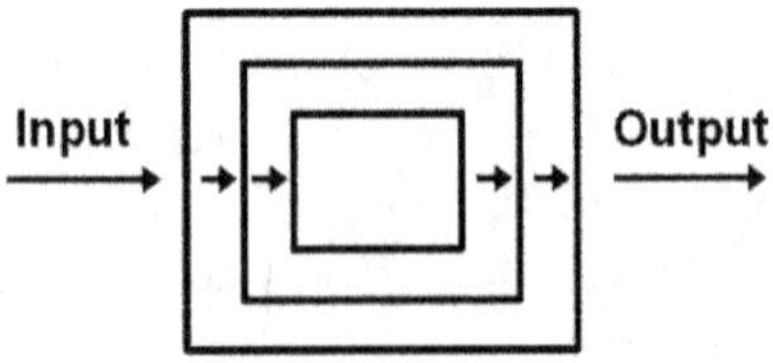


Fig. 21.

**The logic or reasoning of awareness itself is recursive**. Material causation occurs in a time imposed on it by timeless formal causation. The intersection of form and matter is the present time. The input is before. The output after. Living awareness is a self-referential loop at the intersection.

But with life, the starting point is never the same. It never quite catches up to itself. It is also ahead of itself. It is its own future. The whole is a function calling itself. The output is the next input. But that circle, as logic, is nothing but sheer timelessness at the core. Tautology is iteration of the same starting point. It is true. If it is true then it is true. But in time, If it will be true, it is true.

Recall and recognition are generally considered to be divisions of memory. But recognition and recall in the present are an active categorization. An object is recognized as belonging to a recalled category. In living flow, objects are recognized and categorized indefinitely. **Object and category are the boundaries of recognition.**

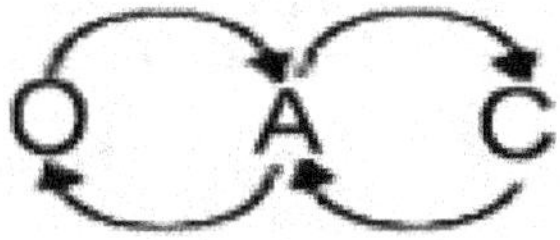


Fig. 22.

Here the flow is a self-referential closed set of loops representing matter object O, formal category C, and the recognizing/categorizing act A in the middle that recognizes O and categorizes C.

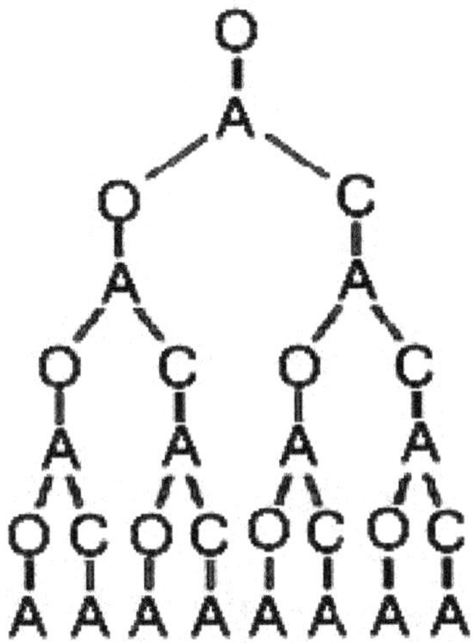


Fig. 23.

This chart follows the logical path. The object O is categorized by A as category C. A recognizes new object O as category C. A recognizes C as including new object O. The categorizing of the object and the recognizing of the object as a member of the category simply accumulates. Awareness A is represented by a static row occurring separately. **The flow is described by an infinite regress.** This is as looked at from the outside of the living procedure.[6]

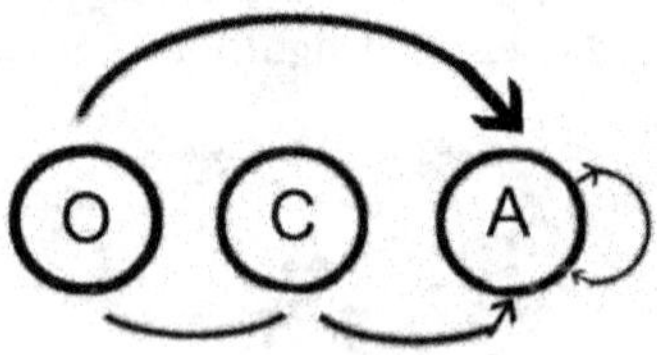


Fig. 24. In this diagram, A is recursive. It directly refers to itself. A is "looking" from the inside.

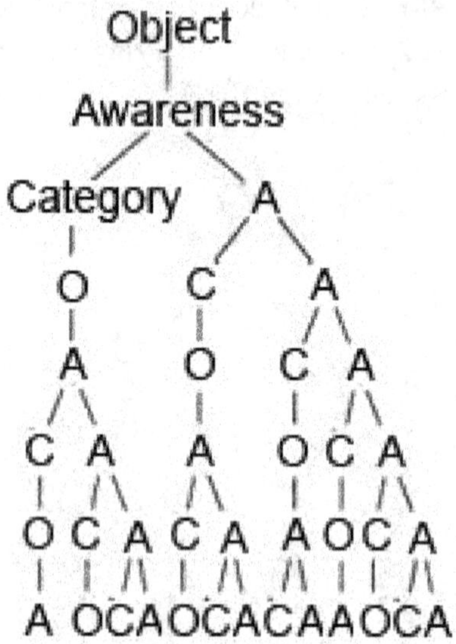


Fig. 25.

**This chart follows the recursive flow**. Awareness is represented by A. Object O is recognized by A, which then categorizes it as C and categorizes C with a new object O. The recognizing/categorizing process A is always partly closed off as excess from object O and category C. The recognizing by a living being is the meaning left over, represented by the residue of A towards the right. The residue also produces an asymmetry. **Meaning tilts living awareness in one direction**. Living awareness is a circular flow directed forward and outward.

The aspects of awareness are not separate, but are organically related with no clear individual moments. Moments blend. Both charts form a "tree" diagram, mirroring the certainty in the present moment. The certainty in the second chart is never absolute as time is ever fluidly moving. The leftover meaning lies in the next moment.

A recursive procedure programmed on a computer, however, simply repeats the same object and category. The remainder is caused by an imposed condition. A living procedure expands the input. It is its own categorization.

But to recognize the recognizing, the entire process has to be outside of itself. Each moment A is the same and not the same. The call by A to itself finally "pops" at the contradiction. Kurt Godel's (1906-1978) incompleteness theorem showed that in a mathematical system there are always axiomatic statements that cannot be proved or disproved. Applied to quantitative objects and qualitative categories, it shows that it is impossible to be completely accurate and completely universal. **Sheer recognition and sheer categorization are unobservable limits**. The object and the category are always observed in merged moments. Roger Penrose says that because we can see the algorithmic contradiction revealed by Godel, our awareness is something real but outside of logic.[7] Logic and reasoning can never completely explain themselves.[8] From a next moment vantage point, both contradiction and necessity of logical entailment are "seen" or intuited.

In communication, "Information economizing" is an evolutionary advantage.[9] The less time the receiver needs to understand the message the better. **"Logical depth" is a measure of the effort needed to logically convey the meaning.**[10] The depth of the message must be unraveled by a shared context which need not be logically explicit. The intuitive leap is felt as an agreement. But the agreement is always open to another logical step carrying potentially more meaning in the receiver.

**The Point:** In time, parts and wholes are ordered as categories and objects. The logic is procedural. Seen from the outside, the procedure is a regress. From the inside, it is recursive. There is a residue of meaning that drives or pulls the logic forward and outward. In the living procedure, the procedure is input.

## MATHEMATICS

There are a few different ways in which mathematics seems to exist.

1. Mathematics is a separate realm. Mathematical objects such as a circle, a straight line or a numerical system do not exist in the observable world.[11] They have no direct instantiation in matter. Because they are timeless like logic, they are not causal.[12] Mathematicians themselves admit that mathe-

matics exists in a separate Platonic realm.[13] Kurt Godel said that mathematical concepts "form an objectivity of their own, which we cannot create or change, but only perceive and describe."[14] In some sense they are both objective and necessary. Mathematical truths are agreed upon by all.

The symbols in mathematical systems are not related to phenomenal experience.[15] They are not governed by meaningful shapes. "2" does not look like two. These systems can loop within themselves, cutoff from external meaning.[16]

So, in what way do they exist? Where does mathematics come from?[17] How is it knowable? Do these immaterial objects exist in their own right or are they creations of awareness? They are in some manner **the necessary application of pure form to itself while interacting with concomitant quantitatively understood matter**.

2. Mathematics is the only realm. A further ideal is that the universe itself is one mathematical object. Everything is not merely described by math but it is math. Everything in the universe is made of elementary particles whose properties are purely mathematical. Electric charge and spin, for example, are just words for what are in actuality numbers and equations.[18]

What is sensed is just an illusion in a mathematical universe. Reality is beyond the senses. Language is only metaphor. Physicists describe the world outside the brain mathematically while neuroscientists describe the brain as a computing machine.[19] The mathematical universe contains immense amounts of information which ordinary language cannot begin to cope with. Logically coded mathematical language is the only one capable of describing the inherent "reciprocal relationships between things."[20] **Existence is nothing but equations.**

3. Mathematical truth is applied to a true mathematical universe. In this view, mathematics is both transcendental and transcendent, that is, it's realm of existence comes both before ordinary experience and beyond it in pure matter. The "things" whose relationships mathematics describes, are numbers and points all defined by their relationships to other numbers and points.[21]

However, **Changeless, timeless mathematics is applied to motion, that is, things in time.**

Between the integers 1 and 2, is an infinity of decimals or fractions, that is, parts. Science applies math to matter in causal time. Scientific theory both informs and is informed by observation.[22] The language of purely theoretical mathematics becomes applied mathematics in science.[23] Computation by brain or machine takes place in time.[24]

Matter is quantified. Quantified matter is measured. But measurement is approximate. Precision reaches a limit. What is predicted is a point just beyond observation.[25] Any system's whole is beyond the sum of its parts. Whole and parts are approximate in their interconnections. Their relationships are merely "parameter fitting."[26]

Experiments, though, occur outside mathematics, which means that the language of science is not mathematics. It is language.[27] Sequences of numbers are not motion. Their meaning is in the approximation of measurement. Matter is not an exact mathematical double.

**Theoretical mathematics is pure form, waiting to be applied to matter as physics in the future, when life catches up.**

4. Mathematics is an explanation. "What turns a state vector of integers into an electron?"[28] What is real and what is math? That determination is not found in mathematics itself. Truths found only in the mind are superimposed on experiential time. The relativity of simultaneity shows that three-dimensional geometry is only a mental construct.[29] We live in a now that is not the time of measurement. We can only imagine three-dimensional images superimposed on a four-dimensional space-time. "Mathematics only expands the world, but does not envelop it completely."[30]

Godel's incompleteness theorems demonstrate that **the limits of knowledge are in the interaction with mathematics or maybe within mathematics itself.**[31] In geometry, a theorem is a logical consequence of the axioms and rules. The rules must be listed in a particular order which is not itself part of the rules.[32] This order is selected in an outside decision process.[33] Axiomatic truths cannot be proved or disproved within mathematics.[34] They will always contain un-decidable propositions.[35]

The famous "liar sentence," "This sentence is false" is false if it is true, and true if it is false. A Godel sentence adds mathematical provability to truth: "The proof G says, 'The proof G is not provable in theory T.'" If theory T proves G, then the proof is false. If theory T cannot prove G, then the proof is true.

The problem is the paradox of self-reference. Sets that contain themselves must be able to change their own rules, or better, the rules change the rules. This is sometimes called a "strange loop."[36] The loop self-references itself by being different. That which is looked at closely appears like discrete parts and jumps, but smooth from above.[37] **Measurement is about precision that is not experienced.**

5. Mathematics pervades life and awareness. Is there a necessary connection between mathematics, life and awareness? Pure mathematics does have both mysterious physical and mental manifestations. A prime example is the Golden Ratio. Certain numbers divided by certain other numbers will equal or approximate 1.618. For instance, a sequence of numbers called the Fibonacci sequence is created by progressively adding the two previous numbers. Dividing each number by the next, progressively yields the Golden Ratio. For example - 0,1,1,2,3,5,8,13,21,34,....13/8 = 1.63 , 21/13 = 1.615. The higher the sequence the closer the ratio comes to 1.618.

Or, 1 + 1/1.618 = 1.61805 which can go on forever adding by 1 and dividing into 1 and always coming out the same. Self-similar mathematical relations disappear into themselves...

$$\varphi = 1 + \cfrac{1}{1 + \cfrac{1}{1 + \cfrac{1}{1 + \cdots}}}$$

Fig. 26.

The Golden Ratio approaches but never is a whole fraction. **It is the very essence of limit.**

But equations acquire clearer meaning with an image. In geometry, divide a line into two parts, with the longer part divided by the smaller part equal to the whole line divided by the longer part. For example- a line 10 inches long divided into 6.18 and 3.82 is 6.18/3.82 = 1.618, 10/6.18 = 1.618, that is, a+b/a = a/b = 1.618.

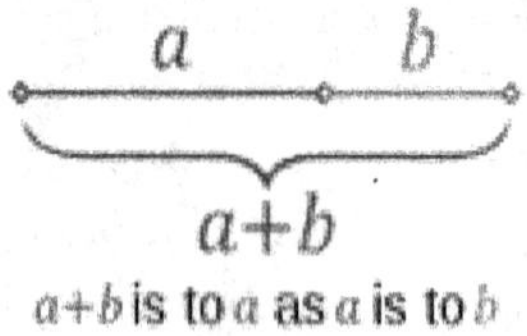


Fig. 27.

From there can be derived the Golden Rectangle....

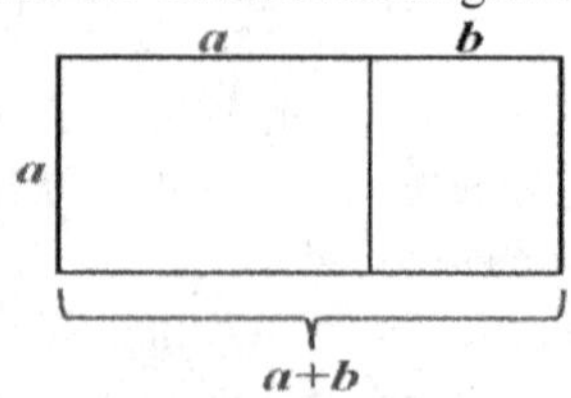


Fig. 28.

And from there the Golden Triangle.....

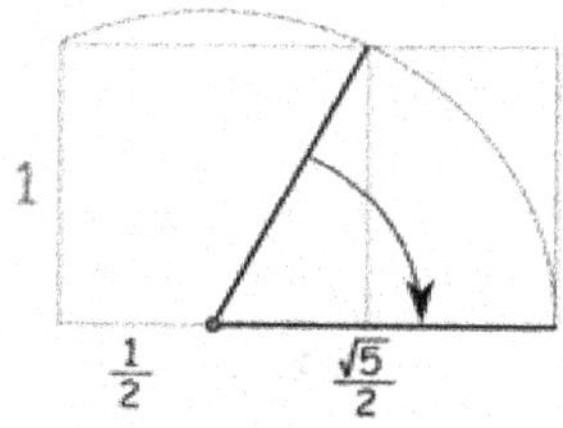


Fig. 29.

the Golden Ratio....

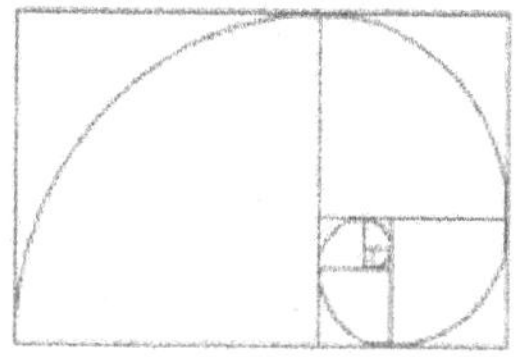

Fig. 30.

and then to life....

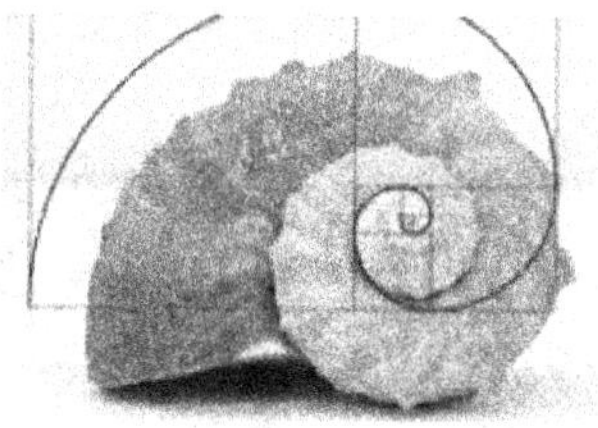

Fig. 31.

A plant with spiral angling with rotation (each new cell a turn) of a fraction made with two successive Fibonacci numbers. The result is something round with no gaps....

Fig. 32.

With .15 for example, it doesn't work very well. Eventually the pattern stacks up with gaps....

Fig. 33.

Many people believe the Golden Ratio is the most pleasing of all shapes. And it is found in architecture throughout the world in buildings such as the Parthenon.

Fig. 34.

Aristotle thought the Golden Mean is the desirable middle between "excess and deficiency." Keith Devlin says in "The Math Gene," that we are "wired for mathematics." If you take any polygon and keep increasing the number of sides, eventually a circle appears while remaining multi-sided. The circle is a limit of a polygon. The realization of this, says Devlin, is "exhilarating." The brain's pleasure centers are affected.[38] **Somehow infinity is grasped in felt application to the finite.** The mathematical limit of reason is felt at the absolute present in living awareness.

**The Point:** Quantitative matter is pure quantity in form. Pure quantity in form, applied to quantitative matter is approximate. Quantity seen in matter is felt as truth.

## SCIENTIFIC METHOD

Before the 17th century, information about the world was largely confined to the senses and the ability to imagine and reason with what the senses revealed. Objective truths, on the other hand, are ostensibly those that exist beyond the senses and are therefore available to all, independent of subjective opinion.

The scientific method discovers objective truths. Understanding of reality is gleaned from making consistent predictions about the outcomes of events. Physicist Neil Degrasse Tyson (1958-) summarizes it this way- "Do whatever it takes to avoid fooling yourself into thinking something is true that is not, or that something is not true that is."[39] Results of experiments that prove the prediction must be agreed to by an increasing number of scientists until a majority of scientists claim an all encompassing truth called a theory. But the process is eternally open-ended because every conclusion must be falsifiable, that is, in some way new evidence could always be found that would disprove the theory. That Thor causes forest fires cannot be proven or disproven.

**The basic scientific assumption is that the world has an "intelligible order"[40] underlying the chaos of the senses** and that science can reveal its correspondence to a "logically uniform system of thought."[41] What exactly constitutes that "correspondence" is the question. Does science merely describe the world or understand it? Is the order itself an explanation?[42]

One answer is that a scientific law is an explainable description. Newton's Law of Gravity describes motion and gravity. Einstein's Theory of Relativity explains them.

The scientific method can be laid out in six steps:

1. Asking a question about a natural phenomenon.
2. Observing the phenomenon.
3. Hypothesizing an explanation.
4. Predicting a consequence of the phenomenon.
5. Testing the prediction with an experiment.
6. Coming to a conclusion that either a new hypothesis is needed (if it is falsified) or that it must be confirmed by others in further testing.

**Inductive reasoning** goes from the specific to the general, or in other words, bottom up. After observing numerous instances, the general event that would include those events is hypothesized. The necessity of their inclusion in the general is assumed by an intuitive leap. If each specific statement, or observation, is true, the conclusion or hypothesis might not be. The general form is: x, y and z all have quality w. Therefore, all X, Y, Z (all members of the general pattern) also have w.

Charles Darwin observed variation in Finches that corresponded to the geographical isolation from each other on different islands. He hypothesized a common ancestor.

**Deductive reasoning** starts with the general hypothesis and examines logical, possible, specific, observed outcomes. It is top down. Predictions are made about what the outcomes should be if the hypothesis is correct. If there is a common ancestor, then if every red bird has a long beak, this is a red bird, therefore this bird must also have a long beak.

**The Point:** Objective truths exist beyond the senses and are therefore available to all independent of subjective opinion. The scientific method discovers objective truths. Originating in successful survival, understanding of reality is gleaned from making consistent predictions about the outcomes of events.

Inductive reasoning goes from the specific to the general, in other words, bottom up. Deductive reasoning starts with the general hypothesis and examines observed outcomes that are consistent with the general. It is top down.

OBSERVATION

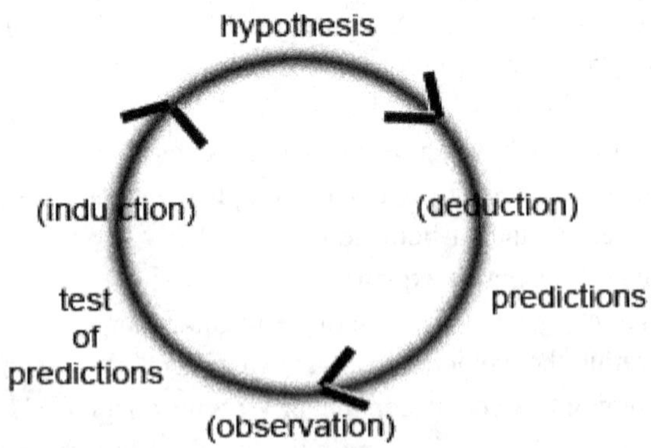


Fig. 35.

The first thing we notice in the above graphic is that **the scientific method is circular, albeit expanding. "Observation" occurs twice: Observation**

**of the phenomena and observation of the experimental results.** As results are more widely confirmed, the general tendency is to make further hypotheses about the phenomenon or hypothesis about generalizations of the phenomenon, leading to an encompassing theory. Experimental results become more and more part of, or even the whole of, the next observations for which to make an hypothesis. As scientific tools become increasingly sophisticated and accurate, the phenomena studied become progressively quantitative and statistical. In quantum physics, point of measurement itself rather than the scientist, is argued to be the "observer." Observations by the senses, which are both subjective and qualitative, are not as objectively comparable as are quantitative measurements.

An extremely primitive hypothesis uses just the senses, e.g., "If the plant gets sunlight, then the plant will grow." Compare that to, "If the wave length of light affects photosynthesis, then different wave lengths will affect the rate of oxygen production."

Modern physics has gone so deep into the material that even space and time are something other than that which is sensed to be real. We cannot directly experience four-dimensional space-time.[43] Science has revealed that a particle's mass is inversely proportional to its radius (the average variation of the probability cloud), because the component probabilities cannot travel faster than the light speed of the particle. As the mass and energy become larger, the volume becomes smaller, that is, closer to the point of light. In other words, the world of quantum mechanics and space-time are so far beyond what we experience that real space and time may actually be what we sense.[44]

**The Point:** The scientific method is circular, albeit expanding. It starts and ends with observation. Observation is increasingly becoming measurement and statistics.

## CAUSE

Inductive logic may lead to an hypothesis and deductive logic may determine an hypothesis's prediction, but the world that the scientific method reveals is not logical, it is causative. The if...then of logic, including the leap of induction, is timeless. Logical conclusions are necessary, they are not predicted. If... the hypothesis is the cause...then... the predicted is the effect. The very fact that the hypothesis must be falsifiable means that it is not necessary.

**Cause and effect imply time.**[45] The effects observed in the experiment must happen after the cause. Objective time is thought of as spatial or occurring in space, yet **causation is not seen. It is quantitatively connected to effect.**[46] Relations between measured phenomena are somehow entailed by causality.[47] The before "shapes" the after.[48] In other words, cause is the meaning of different effects.[49]

Causes contain within themselves the potential, but only the potential, for effects to occur.[50] The reasoning is inherently uncertain. It is always possible that the future will reveal a different cause for the same effects. Each experiment is also performed under a minimum of some new initial conditions. Sameness of cause is assumed in particular experiments and then surpassed in conclusion. The circle expands to include wider and wider generalizations.

> **The Point:** The world that the scientific method reveals is not logical, it is causative. Cause and effect imply time in matter. Relations between measured phenomena are assumed to be entailed by causality.

## LAWS

As the scientific method expands scientific knowledge, causes eventually become laws. Like causes, laws of nature are not logically necessary. Laws may be materially necessary, but are logically deniable.[51] They are timeless, yet part of the world, but they are **sustained neither by logic nor observation.**[52]

The First Law of Thermodynamics is the law of conservation of energy. The total energy of an isolated system can change form but cannot be created or destroyed. The Second Law implies that the energy of the entire universe irreversibly dissipates in entropy. This is the asymmetric arrow of time. If the First law, and other laws, are fundamental, then the universe, space and time, are symmetrical. If the Second Law, which is the only asymmetrical law, is fundamental, then they are not. The predicted relation of symmetric, timeless laws to events operating in asymmetric time are approximate. **If laws are part of the world they must be otherwise explainable.**[53]

**The Point:** Laws themselves are not explained and therefore, not part of the world. They are purely formal. The First and Second Laws imply contradictory concepts of time.

## TIME

Time is more fundamental than laws.[54] What laws describe is motion-in-time, which is spatialized time, not time itself.[55] In physics, time is measured by the timeless function f(t).[56] "The most elemental observation we make is that nature is organized by time."[57] But observation itself takes place in time. What is observed is public. The observing is private.[58] Yet the experience of time passing is assumed to be universal, which of course, conflicts with the relativity of simultaneity. **If the present is real** and the future is not yet real, and the past no longer real, **then a universal simultaneous now transcends the past and future measurable time.**[59] It is in the present that the timeless law and the space-time event are unified. Encountering the quantum realm, the present is where "the superposition ends and the wave function collapses."[60]

**The Point:** It is in the now that the timeless law and the space-time event are unified.

## MEASUREMENT

In physics, observation is a measure, that is, **an association of a quantity with an event**.[61] The observing is not measured and is not explained by either relativity or quantum mechanics.[62] The physicist might say that the act of observation is whatever is most convenient to describe under the circumstances. The particle collision itself, the effects on the cloud chamber, the visible light from the cloud chamber, the retinal image or the electrochemical activity in the scientist's neurons are all events that could be said to constitute observation. If the observer is the measuring, these are all candidates, and the choice is arbitrary. The observing act is not an object. Objectively, the point at which the observation takes place is the measuring instrument.

But the experience of observing is not a material event. **The observing observer is the point through which events flow.** From the observer's perspective an event in time and the observing of it, is a single occurrence.

Simultaneity is relative objectively, but simultaneity is the core of subjectivity. Qualitative observation applies a quantitative law.

At the quantum level of objective matter, properties of matter are potentialities whose likelihood is determined only after many experiments.[63] This means that the outcome of any one experiment has no rational reason.[64] The cause is pure potential. The experimenter's selection of the quality measured and the time of measurement are needed for the event to take place. The specific choices of the scientist and how they affect the results are not explained.[65] The interaction of the observer and the wave-function collapse is not explained.[66]

> **The Point:** In physics, observation is a measure. Objectively, the point of observation is the measuring instrument. The act of observation cannot be found in matter. The specific choices of the observing scientist are not measured.

## THEORY

We have noted that laws are descriptions and theories are explanations of why the phenomenon exists and what causes it. Laws describe measured phenomena. But if theories are explanations of the laws, then they are explanations only of the descriptions but "not what the descriptions are descriptions of."[67] Explanation should be the explanation of why the cause produces the effect.[68]

**Theories are timeless abstractions that continually become more encompassing and abstracted in time.**[69] The two widest theories are Relativity and Quantum Mechanics. Events in relative space-time are measured from the perspective of each other. Quantum mechanics takes perspective directly to the edge of measurement.[70]

It is explanation, not theory, that connects the law to the events. The meaning of objectivity to the observer lies in the explanation. But to many scientists, language is not applicable, or falls short of grasping the mathematical truth.[71] The Higgs field, a candidate for a unifying entity that marries gravity and quantum particles is explained thusly - The Higgs field "absorbs the lines of force, much as water vapor will absorb light from a lantern on a foggy night."[72] It is the isomorphism of metaphor that provides meaning. But what is the meaning of "absorbs?" I asked a nuclear physicist what it means and got the answer that the Higgs particle is like the most important person in the room who attracts all the others. The Higgs particle (or field) is a "fundamental ingredient of how nature works."[73] But the "how" when resolved to the most elemental objects doesn't go beyond energy differences.[74]

William James (1842-1910) held that a true theory must have an "informational structure" that can causally link mathematically described objects and experienced actions of the observer.[75] Perhaps this structure is the structure of "absorption." In any case, "neither mathematical beauty nor agreement with experiment," nor language can prove that theory reveals any truth beyond theory.[76] **Science agrees on objective, measurable results, but not on their meaning.**[77] Useful predictions are those that work but do not explain. All theories are approximate, albeit effective.[78]

An explanation would have to go beyond comparing the timeless with the cause-effect events and explain what is it about the cause that contains the effect.[79] Separate events connected by timeless theory leads only to further material reduction,[80] where the qualitative universal arises out of mathematically equating quantifiable properties.[81] Two qualities or objects can be meaningful in terms of one another, but even the smallest object cannot be contained by one description.[82]

**The Point:** Separate events connected by timeless theory leads only to further material reduction. The meaning of objectivity to the observing human lies in the explanation, that is, what the theoretical description is a description of. Qualitative universals only hint at quantitative relationships.

## MODEL

Experiments and observations are necessarily subsets or simple models of the universe. A complete model of the universe would contain the entire universe. Each subset is one step in an infinitely incomplete set, eventually covering ideally all that is.

**The model is an isolated system within which successful predictions are made.** In some cases models can distort dynamic reality by establishing static reference points to measure results. For example, the "phase space" invented by Henri Poincare (1854-1912) visualizes position and velocity simultaneously as a single point moving in time on top of a timeless space. Models not only reduce but simplify. Simplifications apply to all possible configurations.[83] The most general theories derive from the "simplest of phenomena."[84] **It is assumed that the most thorough and**

**accurate knowledge of the universe is derived from the "simplest and most unified physical laws."**[85]

> **The Point:** Models are an isolated subset of the universe within which it is assumed the most thorough and simple laws discovered are universal.

## EVOLUTION

Many, perhaps most, scientists categorize living beings the same as matter, that is, a "consequence of the equations of particle physics."[86] For them, neither matter nor man's understanding about matter, are based on any evolutionary origin.[87] Yet, **standard evolutionary theory remains deterministically causative, with the present explained by the past.**

Though prediction in science generally simplifies patterns quantitatively, **Darwin's theory of natural selection and common origin is largely qualitative.**[88] It is difficult to isolate the model environment for any particular organism. And the causative entities, in addition to multiple environments, can be a gene or groups of genes or groups of organisms. Statistical studies are done on rates of genetic variation, but the gene-trait correlation is far from exact.

**Evolution is a theory that predicts the future of its own theory**. Predictions in evolution are validations that pieces are consistent with the puzzle. For example, genes thought to have caused or been correlated with certain extinct traits are predicted to be found in current species and are then indeed found. It is hypothesized that proteins found in a bacterium's flagellum would be qualitatively similar to others within the bacterium, and indeed, this turns out to be the case. This similarity shows the path of the evolution of the flagellum. Similar but different fossils are found to be in two eras and environments. A prediction is made that a transitional fossil will be found in an environment and era known to be in the middle, and sure enough, the transitional fossil is found.

Compare this to hypothesizing that dirty hands cause infections and predicting that if surgeons wash their hands, the number of infections in patients will be less. Variables can be isolated and the cause directly locatable. In physics, chemistry and molecular biology exact chemicals and conditions can be replicated in isolated systems.

For evolution, it is not that a certain bacteria cause disease, but that certain bacteria evolved to cause disease. The unquestioned assumption of evolution, the settled fact, is that different species have a common ancestor, not that they share common molecules. The actual theory is that the fittest traits, derived from random genetic changes, survive the longest in changing environments and in competition with others. **The causal entity is not a thing but a dynamic interaction called selection.**

Evolution travels through reproduction. Reproduction goes through the "bottleneck" of single cells reproducing single organisms. All of evolution passes through the reproductive fitness of individual organisms. The totality of evolution is the totality of individual actions. The theory is circular-the fit survive because they are fit. But the tautology is in time, fitness adapts to change. **A selecting subjectivity is both cause and effect of objective selection.**

The ability to generalize and make predictions based on those generalizations has evolved or changed with the corresponding organisms. Learning is a form of adaptation.[89] Adaptation itself adapts. In some cases, survival depends on having an awareness of another's changing knowledge and intentions. The individual's "theory of other minds" is an advantage.

Individual organisms have eventually evolved into humans who look at their own evolution. The different ways that different species interact with and categorize the world are themselves categorized. Subjective awareness, the "inner point of view," has evolved with and within life.[90]

Evolutionary epistemology is based on the idea that human knowledge is itself the result of adaptive selection.[91] The capacity to formulate a theory of evolution, indeed objectivity itself, is therefore, an adaptive advantage.[76] **If reason has evolved, then the subjectivity that reasons has evolved with it. Selection is what life does.**

> **The Point:** Evolutionary theory explains life as being caused by its past. The causal entity is not a thing but a dynamic interaction called selection. All of evolution passes through the reproductive fitness of individual organisms. Subjective awareness, the "inner point of view," has evolved with and within life. The theory of evolution is a product of evolution.

## PSYCHOLOGY

Material reductionism, or objectivity, applied to subjectivity reduces subjectivity to matter[93] and therefore, assumes the quantification of thought.[94] From the outside there is only the material body and the manner in which that body behaves. Mind does not exist.[95]

Psychiatry has largely abandoned treating the mind as a subjective realm of causative thoughts, and replaced it with the brain as an object to be treated with chemicals. Mental illness is a brain disease. If that is true, then all mental occurrences, both psychological and social, are brain events.[96]

Neuroscience, the discipline that studies the brain directly, often looks at the brain as part of biology and therefore, connected to other parts of the body.[97] Mental phenomena are looked at not only as a brain mechanism but as a biochemical mechanism. Of course, either way the inescapable conclusion is that the body/brain causes behavior and the person is not responsible. This conundrum has led some scientists to the idea of unifying the brain and the mind as one inseparable thing. "Our mind is a set of operations carried out by our brain."[98] The brain/mind is inseparable like a computer, with hardware and software. Our thoughts are programmed and run on the brain. The programmer is our genetic predisposition along with our ongoing evolutionary causation.

**Material reductionism attempts to understand conscious awareness by correlating conscious states with physical structures**. The problem is that mental phenomena are not structured in a physical way.[99] Locating mental causation like billiard balls hitting one another is futile. Subjectivity is in the present. Subjectivity is a biological process. Individual points of view are acts of subjective viewing. Instead of combining brain and mind, the necessary combination is subjective first person and objective third person.[100] The brain (and body) is the mind looked at from the outside.

Ross Douthat lays out the following argument: Morality depends on the existence of multiple subjects. A subjective analogy to objective bodily health would be the feeling of happiness. Morality would be the idea that "well-being" should be achieved by the most subjects. If there is no way of knowing this, then it could also be said that there is no way of knowing that health should be achieved by most subjects. The desire to understand what disease is, and how to alleviate it is a value of objectivity, no less than the desire to alleviate unhappiness. An objective science of subjectivity should be possible.[101]

How do we explain explaining? If the universe is mathematics, are the instructions of the mental program equations packaged in quanta?[102] Distinctions and correlations may be the transcendental input. But the subjective loop is open to the future. The present subject is the center of perennial motion. Like a Godel sentence, logic and matter have to be looked at from outside. Timelessness is the object of experience in time. **Between the first and third persons, there is a second person, anticipating itself.**

**The Point:** Psychiatry now treats the brain as an object. But mental phenomena are not structured in a physical way. Subjectivity is a biological process, not an isolated object.

## F. BIOCENTRISM

*The relative positions of physics and biology become interchanged*
Robert Rosen[1]

### INTRODUCTION

Biocentrism is the position that life and consciousness are not material things described through physics and chemistry, but that **the material world is truly understood through life and consciousness.** This understanding must first be reached by life and consciousness and understanding itself. The biocentric explanation in this book is largely based on the arguments laid out by Robert Rosen (1934-1998) in his book, "*Life Itself.*" Rosen believed, as do I, that life has to be explained at the level of the present individual organism, not in terms of evolutionary past. In later sections, I'll try to show graphically how his theory fits with both Hans Jonas' form-through-matter and Aristotle's four causes.

| **The Point:** Biology is not a subset of physics. Physics is a subset of biology. |
|---|

### SCIENCE AND EXPERIENCE

The methodology of science is measurement. By quantifying nature and measuring it, physical science has attempted to bypass what is assumed to be imprecise perceptions in order to get a direct explanation of reality.[2] Observable events are mapped to numbers. The logic goes both ways: Induction generalizes quantified propositions from instances and deduction finds instances based on quantified propositions.[3] To make sense of this, reductionist science understands the world in terms of quantifiable parts. And with a direct correlation of parts and ideas, there is no loss of information. However, an infinite universe can only be modeled by finite subsystems. And since measurement depends on precision that is precise only so far, both model and measured event are abstractions from reality.[4] Scientists constantly correct measurements and guess what is missing.[5]

What is abstracted at the limit of precision, and beyond the scope of experiment, is a categorized recognition. Physical events are made meaningful in experience. Meaning is what is abstracted in living awareness.

Meaning, intentionality, thinking, and other subjective entities are qualitative with no direct correlation with material entities.[6]

Biological explanation has to be more than cataloging. If experience is part of nature, then how nature experiences itself has to be explained. How can we understand not just the models but the subjective modeling of objective nature? The part-whole relations in space-time matter can be visualized geometrically. The part-whole relations in living experience are different.[7] The structure of experience is a dynamic biological process.[8] This process applies timeless categories on measurable objects in time. But subjective time, the rolling present, is not itself measurable. **Observation is not subject to equations.**[9]

**In addition to the limit, the act of limiting has to be taken into account.** Infinite sums have to be limited in order to be meaningful. In doing so, the living being encounters the ever-converging discreteness and the ever-expanding continuum.[10]

> **The Point:** The limits of objective measure are made meaningful by the limiting process of subjectivity.

## MIND AND BODY

Mental illness has been reduced by psychiatry to brain dysfunction, and by extension, mentality has been reduced to the brain. But **the brain is alive**, which means that abstract thoughts, attitudes and values are grounded in the body and brain, not reduced to them. What is important and felt in any living being is tied to the structure of its particular body, which includes its brain.[11]

The problem of course, is that importance involves purpose. Reductive materiality applied to the body looks for objective causation.[12] But **mental life and bodily life are both biologically interacting life.** Biological explanation must include both "immanent purposiveness" and the "continuity of life and mind."[13] Individual biological systems can be observed but cannot be predicted like material objects.[14] **We can know everything about a living being but not what it will do next.**

Alternatives to biological reductionism include holism, which looks at the total integration of mind and body as a single entity. Downward causation is acknowledged in holism, as larger phenomena can affect smaller phenomena, all within a self-contained purposiveness.[15] Dynamic interaction, however, "reveals holism to be a larger level of reduction,"[16] as individual holistic entities are still external to one another.

Another way of combining the mind and body is monism, which assumes that physical states of the brain are also mental states. Since these physical states are matter, all matter, alive or not, is both physical and mental. In other words material elements are both passive and active in and by themselves.[17] There is no essential difference between life and non-life. Rocks are, therefore, in some way aware.

> **The Point:** The brain is alive.

## TELEOLOGY

In studying life we are studying ourselves. And in studying ourselves we are studying our studying. We have noted by now, how both life and awareness are circular. With self-explanation the circle widens. **Nature explaining how nature explains, means objectively understanding subjectivity. This means looking at the subject/object from the inside out.**[18]

The fractal self-similarity that connects quality with quantity, exhibited in the idea of a finite area with an infinite circumference, is reflected in a universe aware of itself through a virtual infinity of living beings.[19] Each living being is imbued with a goal to survive. Striving toward that goal implies action. Awareness is not static. "Knowing is the action of the knower."[20]

To be objective, teleology cannot be confined to separate individuals. **Universal awareness is present in a universal now, through which individuals move.** A theory of everything must include the awareness of that theory. The self-organization of matter into awareness requires its own kind of explanation. Objective reality and objective purpose must be transcendently understood in a functional, not merely a structural, model.[21] Epistemology and ontology, awareness and the beings that are aware, meet in a bio-centric biology.

> **The Point:** To know life is to know knowing. A theory of everything must include the awareness of that theory.

## SYNTAX vs. SEMANTICS

There is no physics of meaning.[22] **Quantitative science and qualitative narrative are two distinct ways of knowing.**[23] Strictly logical or mathematical knowledge is syntactical. Quantitative particles acting according to external causes are modeled logically by meaningless symbols whose "cause" are the rules of axioms.[24] But the rules, as pointed out by Rosen, are implemented by the human selection process that must list the rules in a particular sequence. In other words, the input values and their order are chosen. The selection process is outside the formal structure. Godel's Incompleteness Theorem shows that semantical knowledge about numbers is outside the syntactical numerical proposition. There is always a "semantic residue" of meaning.[25] The general linearity in syntactical modeling ignores the semantic context.[26]

**The point:** The standard modeling process is syntactical. Modeling life must include semantic meaning. Qualitative narrative imposes causality on randomness.

## ORGANISM VS. MECHANISM

The subsets of the universe modeled by standard science are mechanisms.[27] Systems constructed of mutually external parts are described by symbols and a "a set of rules governing change of state."[28]

The bio-centric view holds that **the relations between parts are more fundamental than the parts.** These relations are themselves processes, that is, relations in time.[29] Their modeling requires an additional limiting process. An organism, unlike a mechanism, is a whole greater than the sum of its parts. It is not a mere summation,[30] which can be modeled without semantics. **An organism is informational, not mechanical.**[31]

**The point:** The physical models of the universe are mechanisms. Living beings are organisms. Because an organism is informational and not mechanical, the relations between parts are more fundamental than the parts.

## COMPONENTS

**The parts of an organism are functional tasks, which Rosen calls "components."**[32] As such they are not distinct parts totaled in a summation, but are qualitative, that is, "non-fractional."[3] They blend. The component is not in a one-to-one correspondence with its material. The structure is continuously replaced by different material and appears only on demand of the context. The function itself is both defined by the context and affects the context. The function is defined by and defines many other components. The component's properties can be defined by their use, top-down by a larger whole.

Because it is form moving through matter (and matter moving through form) an organism is essentially process. Because the flow at the intersection creates direction, the components are inherently asymmetric, as in protein structure. The three levels, genes-to-proteins-to-three-dimensional structure, do not arise out of separate parts but through the direction of form. Emergence of functional structure arises through "morphogenesis."[34] **Asymmetric components both communicate ahead and are "seen" from ahead within the greater context.**[35]

> **The Point:** Components are functional parts, i.e., tasks. Like all life, they communicate forward into and with the greater context.

## TIME and RECURSION

Matter systems consist of separate parts in time. Formal or logical systems consist of symbols in sequence. Causality in matter is generally considered to be asymmetric, that is, one-direction, past-to-future, or at least past to present. Differential equations applied to matter first assume the flow of time and then relate functions of continuously varying quantities (e.g., velocities) to their derivative (e.g. rates of change in those velocities).[36] The future is entailed, that is, necessarily connected, by the present. But this can only be expressed by an endless recursion- measuring finer and finer by looping back onto itself.[37] Though human choice determines where to stop measuring, i.e., where or when the limit to precision is, the relational connection of the things that are measured is assumed to be causal. **The flow of time is represented by the recursion in matter.**[38] In other words, catching up to the flow of time approaches but never reaches the ultimate measurable material point.

**With formal logic, time is mirrored by an irreversible sequence of steps.**[39] Recursion involves selecting a value to start and then applying the same process repeatedly to each new resulting value. As stated earlier, a function is a relation between a set of inputs (the domain) and the set of outputs (the range). A recursive function calls itself. The output is entailed, i.e., necessarily connected, by the input. But the function can end only by an independently selected stop value.

Something as basic as addition can be represented by a recursive equation, e.g., starting value $a_1$, $a_n = a_{n+1}$, f(1) = 2. Addition as a function is an infinite summation. The total sequence of steps starting at 1 approaches but never reaches 2.

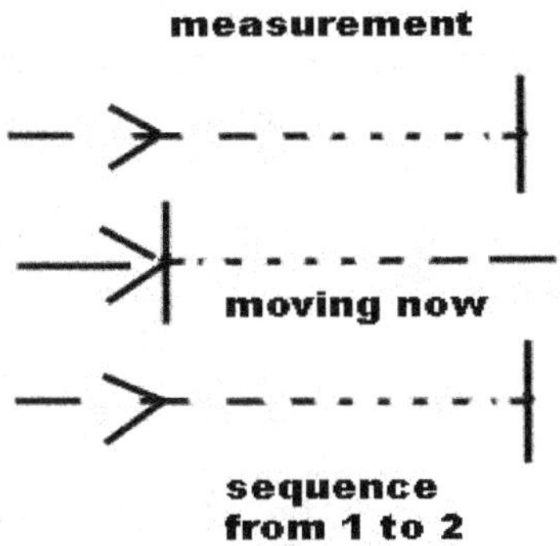


Fig. 36.

**Top: Measuring space-time matter never reaches limit.**
**Bottom: Formal sequence never reaches e.g., 2.**
**Middle: Moving now is the limit of awareness.**

**The Point:** In time, both measurement of matter and formal logic are recursive. Awareness is both the start point and selection of the end point of each measurement and logical progression.

## MODELS

So what we have are two different kinds of entailment, one- formal, syntactical and inferential and the other- material, mathematical and causal. With the standard modeling relation, the natural system of causality is encoded into the formal system of logical implication and the formal system is decoded into the material system.

Scientific understanding consists of circular mappings of one system to the other. For humans to explain their own biological selves, a model of the modeling process is required. The material and formal systems are both subsumed within an organic system. This organic system retains semantic aspects which can neither be a direct model of the universe with its timeless causal laws nor can it be constructed of the pure syntax of logic. Comparing a causal with an inferential structure is something outside the formality of logic. **The modeling relation of the natural and the formal is itself not formal but creative.**[40]

The organic model does not consist of mutually external parts or symbols, and cannot therefore, be replicated in any one-to-one manner. The parts are relational, which means the causation is relational. **The organism is an organization which models itself.** The relational organization has no one-to-one relationship with its material structures.[41] The organization itself is the reality. As such, it is neither infinitely syntactical, nor an infinite summation, nor an infinite material beyond measure, but is hierarchically structured without limit.[42] It is fractal.

> **The Point:** The modeling process is an organic relation of the formal and material mechanistic models. A living being models a living being that is modeling.

## RELATIONAL EXPLANATION

**A relational explanation not only matches relations between material events with relations between formal symbols, but it explains the action of explanation.** This action is the categorizing done by a living organization. It explains itself. Within its own time it is circular without being regressive. The living system is an integration of the infinite summation of matter objects on one side, syntactical objects on the other with "direct product" of group theory.[43] In group theory direct product takes two groups and makes a third. (Direct sum creates a new object of the same group). Instead of external imposition of limit, relational explanation creates a semantic "residue" of meaning openly limitless and inclusive. Living beings take information and add something.

**The component is not an object but a relation between formal inference and material cause.** The component is a functional relation between the action of formalizing and material events. In other words, **a component is the formalizing of matter in every task**. Relational explanation is a dynamic interaction of the organism with nature, which includes its biological self.

As form flowing through matter, life is not based on necessary formal laws or universe sized material models, but on its own possibility. Reduction explains matter without the organization. Relational explanation describes, as it builds, organizational causation. The organization is not pushing itself from behind or before. Rosen describes this as "immanent causation."[44] The living organization pulls the future into itself and thereby pulls itself into the future. To avoid the infinite recursion in material or formal cause, the causation's necessity is its possibility. Understanding (or knowing) is based on relating the present and past to an expected future. Life, that is, both the whole organism and all its components, points toward the future.

> **The Point:** A relational explanation explains the action of explanation while building organizational causation.

## TOPOLOGY

As we saw with chaos theory, **qualitative "measurement" is topology**. As form separates from matter, the quantification is graphically represented. Responses to the large quantitative data input captured by computers vary qualitatively as a graphic.[45]

**The image of a functional component can be represented by a mapping**, f:A>B.[46] A is the domain, or the inputs to the function. The inputs include the "influence" of the larger system.[47] B is the range, that is, the outputs of the function into the larger system. The particular type of change from A to B is represented by f, the type of functional task. In Aristotelian terms, A is the material cause of B. "f" is the efficient cause of B. The entire mapping is the formal cause. But f is also final cause in that its purpose is to make B out of A.

The mappings themselves can be mapped. **Self-reference without infinite regress, and without formal or matter recursion, can be seen to invoke a purpose not in space-time but in subjective time**. The future exists only as pure anticipation of possibility. The future in itself is ever and always nothing.

Three general functions in a living system are metabolism, repair and replication. These three functions are each entailed by the other two.[48] This system can be represented by a diagram.[49]

1. Metabolism is represented as f:A>B. Metabolic products, B, are used to maintain structures with new material A.

2. Repair is represented as g:B>f. The result of metabolism, B, is used by repair to maintain metabolism, f. Repair components are not externally caused. They are self-repaired within the system.

3. Self-replication of repair components can be represented by h:f>g. Replication is the function in which metabolism, f, provides the components for repair, g.

Starting with form remaining through a change in matter, we can then visualize matter flowing through form while form flows through matter. Change is then represented as time. Matter time flows into the subjective present. If the perspective of Aristotle's four causes is added, and Rosen's mappings are then superimposed, we can create the following chart:

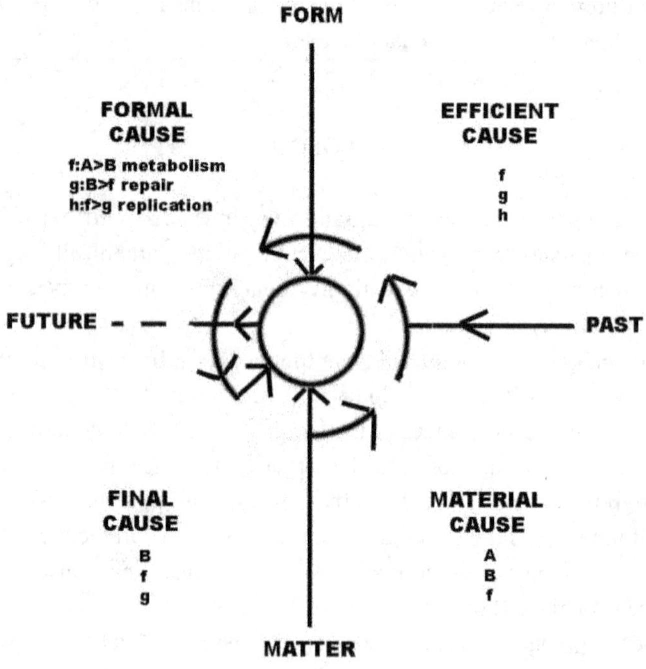


Fig. 37.

Imagine the circle to be the awareness/organism circling around itself at any particular level or component. Matter comes in, or is brought in, from the past. It is recognized, categorized and used for the organism's own future possibility. But awareness is not an infinitely regressive circle in universal time. Purposeful meaning leaps out into another level, which then is the input to another component or functional process which again consists of multiple components. These mappings map all components, not simply the three most general. A hierarchy means components can be the object of components, but recursive in the sense of loops inside loops, each with a semantic residue that "pops" into some other component. The awareness/organism circles around itself by circling through itself.

In a living being none of the four causes exist separately. Material cause by itself would reduce to mutually external parts. In relational causation, material cause is the entailed input. Efficient cause by itself would be what is normally thought of as external cause. Here it is the particular kind of task required by the larger system. Formal cause is not merely the categorization but the formal mapping, a categorizing. Final cause by itself is the output. In relational causation, final cause is transposed to efficient cause, the purpose of the component itself. **The output of one component is the internally entailed input of another**. The final cause of one component is the material cause of another. Internally means that external or clock time is eliminated. The living being is self-explanation, a rolling now.

**The Point:** At the intersection of form and matter in the living moment, components run through material, efficient, formal and final cause in a forward direction bringing each other through the past. Subjective time pops out of its own circularity into another level or functional process or task. Living awareness circles through itself.

## EVOLUTION

Evolution is the historical development of species, not life itself. Life itself is a process of "the momentary context."[50] **Evolution explains where life came from, not what it is.** Life is a functional process, not a collection of antecedent parts. Life does not exist at the molecular level. Indeed, it is the constant rebuilding and self-replacement that allows evolutionary change to take place. The population of living beings as a whole evolves through the circle of awareness to be observed in spacial time.

**The Point:** Life is a functional process, not a collection of past parts.

## G. CONSCIOUSNESS

*Life is the lower boundary of consciousness*
Alva Noe[1]

### DEFINITIONS

Awareness, unconsciousness awareness, conscious awareness, consciousness, attention, experience and feeling are all terms used interchangeably throughout scientific and philosophical literature. There are entire books written on the subject, whole schools of thought, in fact. This entire book is a defining of awareness. Here, I will try to recap and clarify my particular use of the terms.

Conscious awareness and unconscious awareness are the principle demarcations. Awareness, of course, is common to both.

Beginning with the absolute, invisible edge of the future, in the present moment, assuming we are awake, there is attention or not. Attention can remain present in humans as much as 400 ms. or longer, in the sense of "now." This sense of now can contain memory of un-attention or inattention ("half" attention). In the present, i.e., "now," we can attend to remembered past attentions or un-attentions in addition to the remembered present. This could also be called conscious attention.

Conscious awareness contains our organization of what we remember and what we currently sense, or attend to. This also contains our memories of past conscious awareness and conscious attentions. Consciousness is generally used as a synonym for conscious awareness.

Unconscious awareness is everything else. It includes non-attention in the present and the past. It includes everything that the body, that is, the entire organism, senses. It includes any unconscious decision processes.

Experience is generally used to mean everything from conscious awareness to sensation. Experience is the most general feeling. It is what life feels like.

Awareness is the organism itself, which includes consciousness. The body parts are tasks timed to and for the whole. Each task has a purpose ahead of itself. Consciousness is a task.

Feeling is the effort of maintaining the organism ahead of its past.

**The Point:** Conscious awareness and unconscious awareness are the widest tasks or functions of living awareness, each calling the other.

The difference between these tasks and all others is that these tasks are OF the object. Other tasks are FOR other tasks. Being for themselves opens out to being of any object. Being for themselves constitutes the one self which is of the other, in other words, matter.

## BODY

Is consciousness something separate from the material body? Is it even real? As Descartes' "I think, therefore I am" attempted to show, the only absolute certainty that anyone has is his/her own conscious experience.[3] What is certain is that consciousness is involved in what appears to be real. But the question then becomes, "does anything exist outside of our conscious experience?"[4] Is the "outside," generally referred to as matter, organized in a certain way that produces consciousness?[5]

The difference between matter and consciousness, is that consciousness is directed at matter. Matter is the object of consciousness. At the very least, consciousness is conscious of the outside, or something other than consciousness. There are three fundamental entities: 1. That which has consciousness. 2. That which is the object of consciousness. 3. The consciousness in between. Consciousness is primarily a relation between the inside and outside.[6] The "of" of consciousness is open to a relation with anything.

What is the nature of this relationship? There is a sense of boundary between me and not-me.[7] This boundary would seem somehow to involve the body. But the involvement of the body is not passive. The relation is dynamic, involving the body, the outside and the inside. Consciousness is tied to the life of the body, not to the body's inanimate, past matter. Since the living body is one whole relation of relations, consciousness is a relation with and of the body. **Consciousness is alive.**

Life is aware at all levels and in all species, aware but not necessarily conscious in the way humans are. Whether other animals like apes are conscious or bees or even plants are conscious like humans, will be explored in later chapters. As was discussed in the first two parts of this book, **from the end-directed fractal chaos of matter, life has arisen (or arrived) with purpose in individual survival.** Purpose driven meaning connects every component in the body. The difference between consciousness and the body is that consciousness can be directed back at the body as an object.

Body components recognize and choose according to the good of the whole. The organism has sense of itself at the whole body level. Consciousness categorizes that sense. It recognizes its choosing. The value embedded in the living body becomes explicit in consciousness. Meaning is applied freely. Information is integrated into "layers of permeable meanings"[8] which are recognized and categorized again. But with meaning, the meaningless also comes into view. With consciousness comes awareness of a general boundary, that which lies beyond.[9]

Many neuroscientists divide consciousness into two levels. There is the autobiographical self, aware of objects and ideas, consciousness at the high end. At the low end, there is the more body-based emotional consciousness. This consciousness is what Nicholas Humphrey calls 'feeling.' It is the awareness at the very base of all living beings- "the emotional undercurrent of awareness," [10] the urge to survive.

Consciousness is consciousness "of" things. It is a direct relation. Feelings, on the other hand, are "about." The relation is indirect. I am conscious of spiders. I am afraid of spiders. The fear is about something the spider might do. I am conscious of a food. I am hungry for a food. The object of the hungry feeling is not yet, it is a lack. The object of consciousness is present. Urge is not "of" survival, but "to" survive.

Without the presence of the object, physically or imagined, the consciousness of the object is not there. Feeling is its own presence. Consciousness is relational like a Rosen component, but freely floats in its relation to different objects. Feeling is a sheer pointing at the future. Feeling as indirect relation feels the necessity of freedom. "You can't help the way you feel." Basic feelings like happiness or satisfaction need not have an object at all. I feel happy. I am happy. Period. The reasons can be so indirect as to be non-existent. Pain, on the other hand, is completely tied to body feeling. Pain is about pain. To feel pain, like feeling happy, is redundant. But pain is externalized by its own location. I feel pain in my leg. But I don't feel pain "of" my leg. **Pain is spatialized feeling**.

We speak as though we can be conscious of our feelings. I am conscious of my hunger. But is that consciousness something other than the feeling? Or is it the consciousness of the word or abstracted idea of hunger? I am conscious of a hunger that another might feel. I feel my own hunger. I do not feel another's. We can reverse the question and ask "what does it feel like to be conscious," or "feels like to be alive." Awareness at the level of feeling is always present, and like every mode of awareness, points forward, free from the past.

**The Point:** Since the living body is one whole relation of relations, consciousness is the whole relation with and of the body. Feeling is the undercurrent of awareness, the urge to survive, not conscious of survive. The value tied to survival embedded in the living, aware body becomes explicit in consciousness. Feeling is its own object. It is for and of itself.

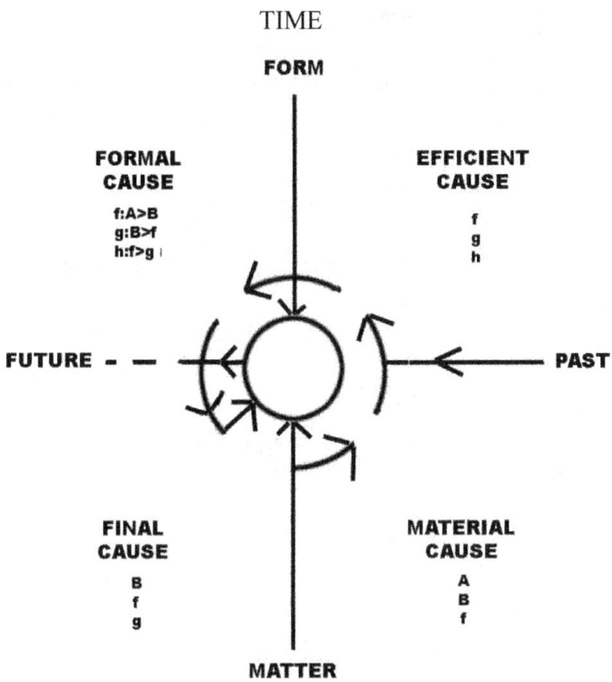


Fig. 38.

The above chart is just like the one before except this time the components are not labeled, because they refer to any components at any level, body or consciousness. It is **a topological bio-centric explanation of the organism explaining itself.** Each component's purpose is linked to other components both wider or smaller whose purpose, in turn, feeds into other components' selections. Instead of copying itself in a linear row of charts that would be tied to spatial time, each component "falls" through its final cause as purposeful meaning, the "semantic residue," which in turn is "material" input into other layers. Each component is a layer in the whole at any moment. Within every moment, the circle of awareness is instantaneous. All four causes are present. Purpose is timeless at the point of now, yet open to, and both falling and being pulled into, the next moment. The recursive flow is the reduction of infinite directions of change to a forward direction. In the forward direction of time, **a living being is formalizing matter while simultaneously, at the nexus, matterizing form.**

Living time occurs when formalizing-matter/matterizing-form constructs three basic relational (and related) components: The functions, Metabolism, Repair and Replication, mirror respectively, the three necessary structures of life, Procedure or Sustenance, Memory and Boundary. **Internal time is relative timing among components**.

Complexity ensues as components are both added and increased in reach. Components are maintained across repeated ephemeral material structures. They are real in their functional relation with other tasks, resolving in the next moment.

**Meaning is purely temporal**. Temporal interaction requires that each task be timed relative to all other tasks and to the whole organism. Each component, from the smallest to widest encompassing loop, brings its future back into its present, while the present, which includes it material input, is pulled into its future. The components all point forward. The whole is pulling itself forward. Living awareness is pointed in one direction- forward.

The whole is the largest function encompassing all the unseen levels and/or components in the chart. But final cause for the whole is a residue of meaning pulling the whole organism forward. The residue of meaning at the level of the whole organism, the purpose of self-survival, is anchored in feeling. Primal awareness feels time. The whole is open to the future. But feels itself in the present. The relational component might look like this, feeling:body->now. Feeling brings the body into the present.

Consciousness is OF something. That something is anything it categorizes. Consciousness is neither matter nor form, it is living. Consciousness is the widest component in the organism. **For consciousness, the future is input.** If feeling brings the body into now, consciousness brings the future into feeling. Since the future itself is open, this means that consciousness brings categories into feeling. It is a relation that is aware of relations. Relations mean something to consciousness.

In dynamic interaction with the environment, varying degrees of novel uncertainty to be categorized, along with the varying flux of metabolic relations, generate the feeling of time.[11] Consciousness extends internal time. The now of feeling, the feeling of presence, is extended to the degree that the complexity of meaning interacts with the corresponding object. The felt now, the sense of now, is a "space of temporal internality."[12] But the absolute timeless now is the limit. **Unlike the limit where measurement ends, the now is the limit where awareness begins.**

The chart at the widest level is the conscious living self. Since the future is sheer openness, consciousness is aware of the future only by means of the past. At the level of the whole, the component is Life: past matter -> future forms. Life is open to itself as Awareness: future forms -> past matter. By looking at the past as if it were the future, consciousness is of other, and of itself. It is "the propagation of thought in time."[13]

**Consciousness becomes aware of an external time when it isolates the past by spatializing the past as pure matter, that is, by separating the past into space and time within which it can measure parts and wholes.** Spatialized clock time comes from only applying form to matter. Internal time is formalizing matter/matterizing form. Consciousness is aware of the experiential process itself. In other words, consciousness sees itself as something to categorize. The act of categorizing itself becomes explicit. Instead of being in the moment and being "of" categories, the moment of categorizing is in and of itself. Consciousness: Awareness -> Life.

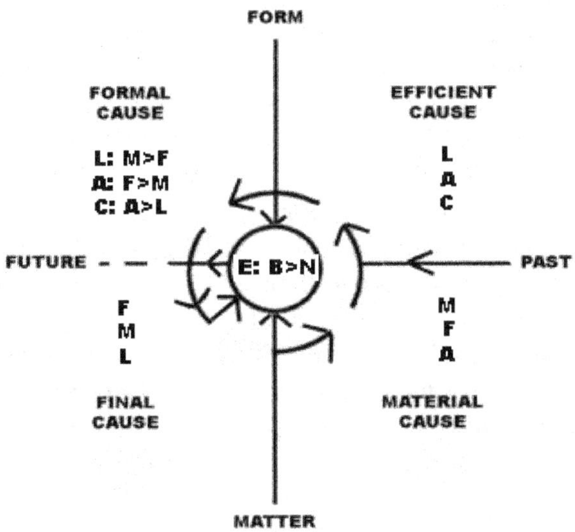


**The Circle of Living Awareness**

Fig. 39.

Life: Matter -> Form

Awareness: Form -> Matter

Consciousness: Awareness -> Life

Feeling/Effort: Body -> Now

Consciousness is a relation with the living body. Studies by Benjamin Libet (1916-2007) showed that the brain reacted to stimuli or a decision about stimuli seconds before the person was aware of his/her making the decision. This has often been interpreted as meaning the body or brain makes the decisions and consciousness is merely a bystander.

Some consciousness studies similarly claim to resolve the continuum - duration vs. discrete-moments question of information processing, by proposing a two-stage model. During a 400ms. period, first the brain unconsciously analyzes objects, while consciousness is not aware of time. The brain presents the results simultaneously to consciousness which resumes the sense of time.[14] In a second study, consciousness of time is measured by momentary object recognition. Consciousness is seen as a passive observer of the body's perceptions.

But if time passes without objects, this does not mean time is not felt. If consciousness does not completely span an objective clock-time, the body's awareness does. Consciousness is a component of the body. **Consciousness anticipates what the brain presents.**

If we look at it from the outside, consciousness is the last step in a material causal chain. If we look at it from the point of consciousness, the decision is waiting in the future. The decision is an anticipation grounded in the next moment. Alwyn Scott calls it "backward referral in time.[15] The future pulls the now. Conscious awareness has a feeling residue that stretches over the extended now. Feeling: Life -> Consciousness.

Matter appears to the organism through the Boundary. The here-and-now splits as the now covers multiple heres. The now is free of the heres. Since the ongoing present has no spatial dimension, yet it points forward toward a non-existing future, there is only an in-out through which anticipated meaning categorizes the past. **The continuity of duration is an ever-moving point at the present edge, directed forward through an in-out into a small-large space-time.** The small, discrete part and the large, continual whole are the boundaries of awareness.

> **The Point:** Living awareness through internal time consists of the relative timing of forward directional relational components in the moment, causing themselves at the durational nexus of form and matter.

## RECURSION AND REGRESSION

A loop is how components act. Loops begin in the cell with negative feedback that rolls into the component's purpose.[16] Information expands with meaning. Living recursion has a meaningful residue understood by a larger loop. For instance, the message "fast" is looked at by a correction loop, "too fast," which is looked at by a correct-the-correction loop, "not fast enough."[17] Of course, loops are layered only indirectly in the hierarchy. Loops move upwards, downwards and sideways throughout the aware organism.

At a whole body level, practice is a repetitive loop aiming for, but never reaching, infinite perfection.[18] **In conscious imagination, the recursive looping is semantic infinity**. In other words, finite qualities can be infinitely recombined,[19] though only one or a few at a time. Loops are not infinitely recursive, they only imply infinity.

A materialist would say that awareness of awareness is meaningless because it is nothing other than an infinite regress. Regress, as opposed to the self-call of recursion, has only the regress as object. Information is beyond the regress. The infinite regress that is applied to the infinite summation of quantitative matter when applied to awareness[20] is halted by the limit at the timeless now. Recursion is an excess of meaning. The fallacy of consciousness being explained by another consciousness placed inside, that then requires another consciousness inside that one, is corrected by the concept of living awareness. The "ghost in the machine" is the next moment. Now is the limit of the infinite regress. To repeat a previous image, this time showing that **regress in the limit of recursion**.

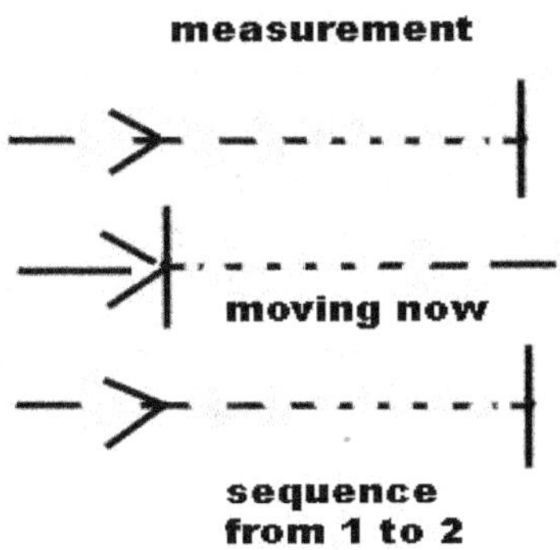


Fig. 40.

**Top: Measuring space-time matter never reaches limit.**
**Bottom: Formal sequence never reaches e.g., 2.**
**Middle: Regress is the limit of recursive awareness.**

At the level of thinking, there can be the endless loop of e.g., "I know that I know that I know...,"[21] but each loop is always another moment. At the limit, the future is absorbed by the now. Supplying its own meaning, the primordial feeling stops the regress.[22] But "the more self-referentially rich"[23] the more complex the system of loops is, the wider the sense of now and the more complex the consciousness.

**The Point:** Regress, as opposed to the self-call of recursion, is an empty, repeated present. The infinite regress is halted by the limit at the timeless now. Recursive looping in conscious awareness is semantic infinity, that is, finite qualities can be infinitely combined, always and continually in the future.

## QUALIA

Robert Lanza lists the "easy problems" of consciousness:

"1. The ability to discriminate, categorize, and react to environmental stimuli.

2. The integration of information by a cognitive system.
3. The reportability of mental states.
4. The ability of a system to access its own internal states.
5. The focus of attention.
6. The deliberate control of behavior.
7. The difference between wakefulness and sleep."[24]

These aspects can all be observed from the outside, or the explanation or description can be recognized by others. The "hard" problem is the subjective- what is directly presented to the organism's consciousness.

When the living felt moment is wide or complex enough, it might be generally acceptable to call it conscious. Categorization is experienced as images to be manipulated.[25] Those images are off limits to objective observation by another living being. The subjectivity of experience, the feeling of experience, is a phenomenon not present in the material world. Neither is it a formal category. David Chalmers (1966-), claims that this phenomenon called "qualia" is a separate phenomenon of the universe on par with mass and charge.[26] Mass and charge are ways in which humans (human scientists) organize the universe. **Qualia are the immediate presentations of the universe to consciousness.**

**The Point:** All objective things are the ways in which life organizes the universe. The "hard" problem is the purely subjective- what is directly present to the organism. "Qualia" is the term for that which is immediately presented to consciousness.

## UNITY

In organizing the universe, physics has run into the concept of quantum coherence. Wave functions are "successions of overlapping simultaneities."[27] This mirrors and/or is mirrored by the unity of awareness in living beings. Internal time is not chunks of space-time. It is more like successions of overlapping nows. It is duration.

Material or spatialized time is mapped from the coordinates of current awareness.[28] Calculus approaches from a past never reaching a more recent moment. Awareness's relation approaches from the anticipated future without ever reaching the limit of immediate identity with the present moment.[29]

Forming-matterizing-awareness generates a "window of presence,"[30] which internally models universal information in a way that is coherent to the organism. Different senses are sensed as simultaneously unified. And there is no reason to believe that different types of organisms don't also unify senses.[31] They may do so in windows of felt unified time that are closer to now's limit.

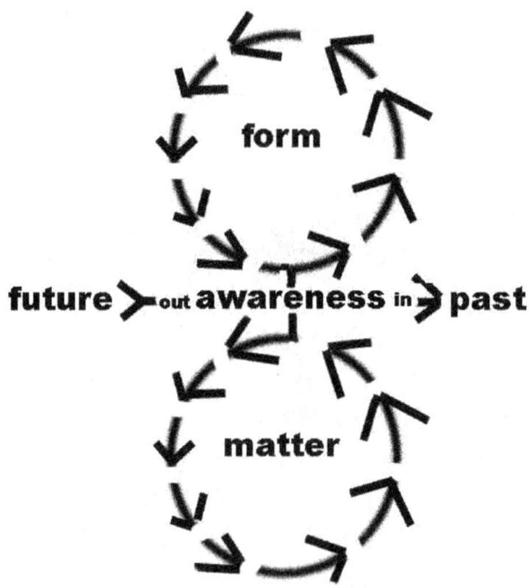


Fig. 41.

Living awareness exists in the point of duration, the point of view at which form and matter pass one another. This point is without spatial dimension, and thus has an in-out that looks with anticipation toward a future which always does not yet exist. Life experiences its own forming of passing matter, and feels as if it is the one that is moving.

The meaning that it anticipates pulls form one step ahead of, or free from, matter. This is what keeps it alive.

Beginning with the blending of qualities, consciousness feels united with memories both recent and past, which themselves seem unified. But consciousness also tends to separate the current moment from past states both internally and externally. Though language, with subject and predicates, divides like an equation,[32] consciousness sometimes remains in a state of pure duration,[33] where the absolute unity of consciousness can rest by riding its own now.

**It is consciousness's sense of its own unity that seeks and finds unity in the universe.**[34] It does this by striving to unify the discrete and continuum that is both reflected in, and discovered in, its own conscious duration that alternates with pre-conscious awareness.

> **The Point:** The unity of consciousness seeks and finds unity in the universe. Consciousness unifies conscious awareness with unconscious awareness. It unifies awareness against that which is not aware.

## EXPERIENCE

There is a knower and a known (of objects both in the world and in consciousness), but also a knowing.[35] **Experience is qualia plus temporal unity. It is continuous presence.** Now slips into the next now. Experience is ontological not epistemological.[36] In other words, ongoing experience is what it "feels like to be alive," not what we know about the world. If consciousness is controversial when applied to animals, the word "experience" should not be, unless we are to deny that animals experience their world.[37] Experience is both the point through which events flow and the point at which the flow of events are felt.

Experience is the presence of what is there that perception or thought or language cannot grasp.[38] In fact it could be said that most experience is in some way unconscious.[39] Experience is lived body action. **Experience becomes fully present to consciousness when "potentiality becomes satisfaction,"[40] when something searched for is found**. The object of conscious attention is explicit, as is also true when a mathematical or logical truth becomes clear.[41] Potential objects of consciousness, both mental and physical, can be unconsciously experienced when the mind "wanders," so to speak, in free association. Recall can bring these objects to "full" consciousness. Modes of consciousness like memory or thinking, or modes of feeling like fear and hope can also come into consciousness as objects. Unlike consciousness, **feeling is its own qualia**.

Structural coupling with the world is manifested in experience as an "ongoing dynamic relation to objects."[42] Experience is not just action, but interaction.[43] Experience is affected by what it experiences, and what it experiences affects experience. Maurice Merleau-Ponty (1908-1961), describes experience as "skillful coping," where the felt intention is already "at" the object.[44] Yet the immediacy of felt experience indicates no causal (past to future) delay between the object as input and effort as output.[45] **Within the experienced span of now, there is a continual alternation between interaction and immediacy.** The temporal extent of the relation varies, but it is ahead of and at its object,[46]

**The Point:** Experience is qualia plus temporal unity. It is continuous presence, alternating between interaction with the whole and immediacy at the point. Experience is alternating conscious awareness and unconscious awareness. Experience is the most general feeling. It is what life feels like.

## SELF

The word "self" is unavoidable. It is used as a substitute to describe an individual anything, including inanimate objects. "That little car has enough power itself to...." Chaos theory talks about matter e.g., a hurricane, as self-organizing. But **self-organizing in life means something beyond alone-organizing.** More than separate and distinct, self means autonomous, that is, in some way it affects its own motion. In a world of objects self is the subject. It is the center of the world. The model of the world that the living being creates has itself at the center.[47] All modes of awareness are relational tasks within the organism, but **the self is the feeling of the whole organism as one entity. The self is that which "has" consciousness.** To be a self is to be self-aware.

From the body to consciousness there are four relational components of self:

Awareness is self-awareness.

At bottom the self is self-survival. Self-survival is common to all living beings. The self is not an ethereal essence that exists apart from the living body, or that directs the body from a special place inside. The organism is an object that "selfs."[48] It acts. It is aware of those objects in the world that have an effect on it.[49] As a subject in a world of objects, the whole organism is an inside. In order to survive it knows how to survive. It self-organizes in dynamic interaction with its environment. **Self-awareness, the feeling of being alive, begins with the urge to survive and the whole body knowledge of how**. When the environment changes more rapidly than can be normally absorbed, resistance is reverberated throughout the organism.[50] Component loops all have a semantic meaning culminating in the feeling of the individual.

**The self is not alone. It is an individual.** A significant difference between the animate and the inanimate is the individual behavioral variations in a group of the same species. Material elements are completely identified by their reactions. Living beings, not only humans, are persons. Individuals announce themselves, in some cases with a particular sound[51] or for example, with bacteria, emit a certain chemical. Total behavior in a living being adds up to personality. Personalities exhibit unique actions. They exhibit awareness of themselves, if not necessarily consciousness of that awareness.

Self-awareness is future-awareness.

Selfing is the process of self-propelling beyond itself. The urge is to stay ahead of environmental events.[52] A self is beyond itself. Conscious and unconscious awareness are aware of something other than itself. Awareness of the environment means an environment that contains a feeling self. The aware self, even without being fully conscious, organizes the world around itself.[53]

Acting for the future, or planning, is the next stage. For example, many animals hide food for later consumption. Behaving in accordance with the seasons is a general precursor to more particular behaviors. **Survival of the self demands prediction of its environment.**

Future-awareness is self-awareness

At some point the self becomes its own object. The self is seen as one more object in the world. Recognizing oneself in the mirror, including realizing a correspondence between e.g., touching an area of oneself and the mirror image of that touch, is certainly indicative of awareness of one's own body as belonging to the self.

Self-awareness in its most abstract phase sees the future in light of the past. As the past seen as future becomes more explicit, the past as past does so also. Consciousness sees the self in its historic and anticipated future perspective with regard to the world. Looking beyond oneself is also seeing oneself as the one who looks.[54] Consciousness not only sees the self as an object but as a subject. The conscious self sees itself as an "I."[55] The widening of now within the conscious moment allows the subject to linger as an object.[56] What is sometimes called the ego, is the awareness of something that had all along been the subject of objects.[57] Self-knowing is the autobiographical self.

**With conscious awareness the idea of self is included within consciousness.**[59] The interacting relational components affecting one another at all different levels in the organism manifests in a conscious self arising along with the consciousness of the self.[60] Consciousness of objects and consciousness of itself as being the subject of those objects form a changing gestalt of background-foreground.[61]

Self-awareness is awareness.

This is not only seeing oneself as the one who looks, but is **aware in the present of the present looking**. This is the self of reflective or contemplative consciousness. This is not abstraction, because awareness of awareness is not a consciousness of the self as an object, but an identity. The identity is felt. The now is narrowed to the limit.

> **The Point:** The self is the feeling of the whole organism as one entity. From the body to consciousness there are four relational components of self. The self is not separate, it is a unique individual. The individual selfs outward to the predicted future of others. From ahead it looks at itself as a past object called a self. The self identifies with itself in the present.

## BEHAVIOR

Just as awareness fills the entire body, the body is present in all awareness. Most of an organism's actions occur below the level of subjective whole body awareness.[62] Even higher brain functions are not always and, in many cases never, conscious.[63] Behavior represents awareness as an observable object.[64] Behavior defines the external space-time location of an awareness within. Space-time location implies motion. Motion does not indicate the feeling of awareness, but the action of awareness.[65] **In behavior, subjectivity can be seen as having a causal effect on matter.**[66]

In an organism, efficient cause is final cause, but final cause is also efficient cause. The future comes into the present as qualia, and as meaning pulls the present into the future with excess bodily energy. The being and doing of awareness are each commonly conceived respectively as the phenomenal and psychological ways of describing awareness.[67] Cognitive science, which more or less looks at the brain as a computer, explains behavior as objective, without subjectivity.[68] Material cause as sheer input is what is seen of the body from outside, the observable doing.

An individual organism exhibits changes in routine behavior upon recognizing new circumstances. Living beings have the capacity to modify their behavior, not simply react in accordance with a genetic disposition.[69] Learning is behavioral adaptation.[70] It is with change that the feeling of what is felt arises. Forward leaning anticipation that accompanies the rolling feeling of living, is surprised by a difference. The feeling of the unity with itself is passive but the feeling of the difference from itself is active. When something comes to attention, unconscious awareness jumps out to a conscious awareness of the difference. **Being and doing form the ground of one another.**

Behaving is a feeling of spontaneously generated action that can only be indirectly observed from outside.[71] Because the behaviors of single social insects are less that of individuals than parts of the individual colony, studies of ant behavior provide a glimpse of internal life. A colony will travel along the fastest route rather than the shortest distance.[72] Each ant moves randomly, yet like the chaotic move of matter toward self-organization, the random converges on an optimum point where form flows through matter. The colony is the self. The living conscious moment is felt as spontaneous action because, as we will discuss in Part V, internal time is velocity without distance.

**The Point:** Behavior represents awareness as an observable object. In behavior, subjectivity can be seen as having a causal effect on matter.

## INTENTIONALITY

Though the feeling of difference can be active, behavioral doing and the awareness of the doing are mostly passive.[73] This can be the case even when something comes to attention and awareness is raised. Action has been previously determined, not by outside forces, but by the organism in all levels of awareness. But occasionally, when a living being faces alternatives that are particularly important, the transition from unconscious awareness to conscious awareness is itself felt.[74] The primal urge, the basis of intention, is brought to the surface. The full causation of the component model is felt in its widest relation open to purposeful meaning. The body self feels itself as its own cause, as internally directional. Making sense is intending.[75] **Intentionality is how causing thoughts and action feels.** Intention is the outward directing of consciousness. Direction is felt as directed.

Concentration is an ongoing decision, an elongated intention, interspersed with relaxed drift in unconscious awareness where detail transiently appears without felt cause. Intentional felt selection links with intentional felt action. But action, like concentration, drifts into a more automatic mode. The detail of the action, for example, the feeling of arms swinging at a given moment while ignoring the feet, changes with focus. The experience of arms swinging can itself with effort be intended. The effort to feel the experience can be felt. Each intention is built upon a ground of "prereflective" felt experience.[76] We can move and be aware that we moved without needing to actively choose.

Feelings also have a range of intensity whether they are felt as intentional or not.[77] A surprise can be accompanied by degrees of fear, for example. A choice may be accompanied or proceeded by a whole range of emotions. The intensity of effort varies with the discrepancy between recognition of novelty and categorization. We might notice at first that something is different and immediately realize that the room has been painted or we concentrate before realizing e.g. a picture frame is askew.

| **The Point:** Intention is the self-causation of directed objectification. |
|---|

## COMMUNICATION

Having a theory of mind means realizing that your knowledge is your own and therefore others have their own. You have your reasons for acting and so do others. Observed behavior is the clue to another's motivations. Enhanced by language, knowledge of one's experience and motivations can be conveyed as information. For complete certainty, not only knowledge but the knowing, that is, the experienced meaning and what that meaning feels like needs to be conveyed. The other's motivations need not only be predicted correctly in terms of observed results, but actually known in the way the other knows them. How would it be possible to communicate not just information but consciousness?

A phenomenological explication of consciousness is beyond the scope of this book, but three general levels can be mentioned: 1. At bottom is the lived moment, the urge to keep going. The selfness of the self should be impossible to communicate. 2. Qualia is the immediate sensation in any moment, for example the sense of the color red. It is the qualitative basis of meaning. The color can be talked about and agreement assumed.
3. Experience, either unconscious or consciously intended is the ongoing organized meaning of objects- both things in the world and objects of thought, including the reflective consciousness of thinking itself.

Unconsciously aware signals, even long sublimated by the organism, may be correctly interpreted by the other. But with conscious signals and language, there is a felt effort of intention to convey a specific message.[78] Although specific, the meaning may involve much of the sender's total experience to be accurately understood.

**The living relation with its passing of meaning is the very essence of communication**. The relational complexity that is the living being is a "communication web."[79] Each loop is timed in the context of all others. The context interprets the information. At the apex in consciousness, timing manifests in an experienced extended now. Meaning is open at the widest loop. Consciousness feels its object.

The immediacy of consciousness and object is experienced vividly in sound. In speaking, we can hear our own voice as an other. If we can hear ourselves as an other, can we hear another as ourselves? The relational act of awareness rolls in and through the immediate grasp.[80] Meaning flows through a door of anticipation. The next moment is open.

Musicians anticipate the next note. People playing improvisational music together report that the music is immediately felt and mutually anticipated. Two musicians know what they are playing although neither have heard it before. Each self is a point of view in relative space-time, but communication between selves can be simultaneous in duration. The same note comes toward both musicians. Both musicians move toward the same note.

Can experienced intentionality be passed between subjective consciousnesses? In the depths of conversation, meaning can be felt as mutual. Speaker and listener merge as they finish each other sentences. The listener anticipates the ending of a story being told.[81] Efforts ride on one another. The listener hears and understands the words that the speaker says as rapidly as it is said.

But do the words mean to the listener the same as they do to the speaker? Meaning is not agreement. It doesn't have to be clear in order to be meaningful. Meaning connects beyond and before. From those particular words both subjects are catapulted to what perhaps might be separate individual choices. Understanding is open to interpretation, but the words while being spoken are not arbitrary gibberish. Sound waves travel in space-time, but the meaning during speaking occurs in living duration. Both subjects hear more than sounds. We may know what words mean separately but not understand them strung together.

An ongoing exchange is the shared object of two awareness's. The meaning is mutually constructed. Anticipation, facing a future not yet filled in, is open intention. **In the shared space of anticipation, meaning is a shared next moment.** In duration, two become one self in the immediate future coming toward both.

Consciousness in the original Greek meant "to know together." For Socrates, true communication is immediate, not mediated by writing. Communication is foremost not third-person objective or inter-subjective. It is second-person intra-subjective.

Information, though, is extended within a group. Simone de Beauvoir (1908-1986) taught that it is inter-subjectivity that allows for freedom. We choose as a group.[82] **Free choice arises not in agreement, but in difference.** Choices are not automatic. Subjective interpretation of shared meaning is demonstrated in further conversation, where freely chosen differences progressively make clearer the ideal, what is "truly" meant.

**The Point:** The relational or functional organization of tasks always resolving ahead of themselves, culminate at the level of the whole organism in communication with another, where meaning is shared in the next moment, and progressively clarified as shared truth.

## H. FREE WILL

*Willpower is itself a biological phenomenon*
Sam Harris[1]

### FREEDOM

Much of neuroscience concludes that both free will and consciousness are illusions.[2] Although some believe that consciousness may be aware of the feeling of free will, they claim will itself is not free. If a living being is explained as merely a rock that moves when pushed, then free will and consciousness are scientifically un-provable.[3] In fact, with that way of thinking, life itself is an illusion because it consists entirely of non-living material.[4] But if life is explained as the freedom of form through matter, then consciousness and free will can be seen as integral to life. Free will is not a quantity that can be measured. On the other hand, either is it a quality like the color red.

**Life is the choice to live.** Choice implies options from which to choose. Life's original free choice to live manifests in the individual, when seemingly limitless, possible categories of form pass through matter's infinite parts. What is given to life is the ability to organize the small and large according to its own selected options for survival. "Behavioral variability" present in all organisms belies the denial of freedom. Those options may not be effective beyond its given environment, but all living beings perform the best that they can under any circumstance.[5] Life is not wholly determined by its environment. Life chooses until death. In fact, **death is the absence of choice**.

**The Point:** By choosing to live, life is freed from its past. We are in between the small and the large as far as we can manipulate the part-whole contradiction.

### FORCE

Newton's laws of motion teach that forces occur in pairs.[6] Like consciousness OF something, a material force is always on something else. It does not exist alone. Stimulus and response are similar to action and reaction, but neither organic causality nor the organism's response is locatable at a point in space-time. Stimulus-to-response is not a mathematical equation.[7]

Bodily causality does not come before awareness, it is integral to awareness. Intention brings forth energy in effort.[8] Life is a continual effort to maintain itself in the world.[9] Far from being a cause arising out of nothing, **the living being causes its world by making sense of it**.[10] The self exists in its own moving, categorizing present, bordered by anticipation and memory. An ant does not crawl like a rock rolls down hill.[11] **It falls into a future which, like gravity, pulls it forward** toward a far, ever unknown, yet certain, point.[12]

| **The Point:** The life force is its own future. |
|---|

## CAUSE

**Free will is subjective causation.** As we've seen, the material world consists of lawfully determined causes on one hand, and random chance on the other.[13] Quantum uncertainty and macro-determinism, the small and the large, point and continuum, meet at the wave collapse. If all is matter, then consciousness is either necessarily determined by prior events or arises out of pure chance. In that case, subjective causation does not exist and humans, and certainly other living beings, are not responsible for individual actions.[14] In fact, we can't even say "their" individual actions. There are no individual actions.

If "free will" is equated with totally random behavior, then choices are arbitrary.[15] A free awareness unencumbered by choice, would have no choice. Pure freedom is meaningless.[16] If the mind or consciousness is a substance totally separate from the material body, and cause is determinate, then all thoughts, all qualia, would be caused by previous thoughts, which themselves have no ability to cause, as they are determined by other previous thoughts occurring further back.

A material machine cannot have will, that is, its own cause, because it cannot supply its own energy. But in the chaos of matter itself, the self-organization of a dynamic nonlinearity arises in a mixture of chance and necessity. The wave collapse is also the point at which objective measurement and consciousness arguably meet. Choice is neither chance nor necessity. Before either, it is free. Un-free choice is meaningless.

**Subjective causation only makes sense if it is aware**. The consciousness that is thought by some to cause wave collapse is literally defined by free choice.[17] The collapse occurs on the behest of a scientist's decision. The conscious cause-effect relation is a relation of conscious free choice.

The random mutation/determinate selection of evolution manifests in the individual's consciousness. Random possibilities limited by the organism are narrowed down with selection by trial and error.[18] Recognition of appropriate responses according to learned criteria is made in the flow of internal time.[19] Freedom is in the present, not in the past. It is a relation of anticipation to immediate occurrence. And since the free act is momentary it can never be repeated. It is, therefore, free of physical laws.[20]

Organisms predict other's behavior and also track their own in order to successfully adapt.[21] This means that, unlike a machine, living action is necessarily accompanied by awareness.[22] Experienced anticipation, the ground of prediction, necessitates action.

Levels of awareness throughout the organism are degrees of freedom. The changing of protein shape is both cause and effect of its own signals, but the meaning is conveyed in a greater context.[23] The most free are the moments when we reflect on ourselves as we act, when we are conscious of the living moment,[24] when we are aware that we are choosing. But unconscious intention and purpose still occur when not made explicit.[25] We usually act without paying attention.[26] Spontaneous movement occurs within and because of bodily restrictions. Controlled timing is relative within a whole body/mind duration.

> **The Point:** The living being causes itself to act by being aware of its choices, and choosing.

## WILL

**Free will combines free choice with action**. Action can affect the body and surroundings, which is felt as physical causation, or it can affect itself in experience.[27] Transitions between levels of consciousness as well as change of focus from one object to another, involve effort. Free will is the feeling of identification of consciousness with thoughts[28] as well as attachment to acts.[29] Free will is what it feels like to be alive.

All modes of experience are action within the ongoing present. Looked at by outside measurement, there are approximately 400 ms. in a human sense of now. The temporal distance between consciousness and its object vary within that span. **The contradiction of the discrete and continuum manifest in the living moment as will and consciousness**. The satisfaction of the binding and the never-ending effort to reach it are entwined. **The effort of will pulls the future so close to the absolute limit of the**

**present moment that the feelings of consciousness and object, thought and act, tend to merge.**

We normally feel "at one" with our willed effort, but sometimes consciousness can be a passive watcher. Behavior and thinking can both be observed.[30] From there a decision can be made which is felt as immediate. Choices based on previous learning are selected and willed, which then leads to more learning. The processes are looped down through the body and back. Not confined to consciousness, freedom is dispersed throughout the organism.[31] Self-monitoring feedback loops culminate in the feeling of freedom.[32]

"I am, therefore, I think" reverses Descartes' cogito.[33] Awareness is more fundamental than consciousness.[34] The unconscious is often a term used for brain processes or pre-conscious bodily processes that provide the choice to act. For example, we urge a particular verbal direction and the words flow out of us. Each word is usually not consciously decided before it is uttered. And those individual words that are consciously decided upon arise from a choice already given to us by bodily awareness. Willful urge is not only felt in overt acts. Every decision to move our eyes, to focus, to think any thought, is not effortless.

To paraphrase a quote attributed to Einstein, "consciousness can choose to act, but not choose to choose." Choosing to choose leads to a regress, but that regress is stopped at the point where choice is presented by the body. Choices were laid down previously in various conscious and unconscious aware acts. Free will is free in the conscious ability to reject offerings by the body. With the unconscious awareness that we live in most of the time, pre-decided goals direct our behavior.[35] Unconscious choices arise from our own history. We can be conscious of those choices well after they've been made. We can choose to be conscious of them.

Benjamin Libet's experiments have shown that activity in the brain occurs about 300ms before a human is conscious of making a choice.[36] But whether that observed brain activity is the actual choosing is open to question. From the point of view of consciousness, the brain activity has yet to appear. **The choice is in the future**. Brain and consciousness, both living processes, roll forward together. We feel the weight of choice and we act. The process is not objectively linear before and after, but nonlinear to, from and around with an ever moving start point.[37]

**The Point:** Will is the effort of choosing to do or think anything.

## SELF

**Essential to free will is the idea of self**. An organism is an individual. It is a self. Chemical reactions occur spontaneously under certain physical conditions that necessarily lead to thermodynamic equilibrium. But the spontaneous action of life is a continual maintaining of itself above thermodynamic equilibrium. Striving to survive creates a constant flow of energy.[38] This creation is self-creation. The self is the agent and agency of its own behavior.

The self is established not as an independent receiver of information, but in dynamic interaction with the world.[39] At the center of this interaction, the self senses the difference between the body and outside. The outside is normally sensed as moving in comparison with the direction of the body, that is, flowing in the opposite direction. If something in this opposite motion changes, it is recognized as other.[40] The self is what feels ownership of the forward direction. And when a living being communicates, it feels ownership of what it is saying.

Experiments on the brain have shown that a feeling of will can be externally imposed unaccompanied by a feeling of ownership.[41] In these studies, a self looks at the act that it is doing as separate from itself. The will apparently becomes an object that is present to consciousness as separate. This is, in a way, an externally imposed state of meditation, where awareness becomes the "watcher."

But consciousness can objectify anything, including itself. We normally can separate our choice to act from the feeling of acting. Free choice as an object can be separated from the effort of will.[42] Freedom, in fact, is found in the ability to distinguish between self and action.[43] Unless we focus on the difference, we don't experience our thoughts or the world as first deciding, then experiencing. Effort is applied by willing to select a choice and by willing to act accordingly.[44] Free will is freedom to choose along with the willful effort in choosing.[45] Separation between decision and experienced action, are united by the act of deciding to act in the ongoing conscious present. Will is present to consciousness as identical to consciousness. Consciousness at the very least, takes a miniscule effort.

There is nothing already there in either the organism or the surroundings that can determine the next moment.[46] Objectively from the outside, immediate future actions, whether human or nematode, are unpredictable.[47] From the point of view of consciousness, future consciousness states are predicted and many are predictable. Intention is prediction. Anticipation is accompanied by expectation. The possible and the actual are each other's ground in duration. **Awareness and choice come before each other**. Theirs is a "tangled hierarchy."[48] The past may push but the future pulls. Life falls into its own future.

Edmund Husserl's (1859-1938) twist on Descartes' "I think, therefore, I am" states "I can, therefore, I am."[49] In active engagement with the world, objects appear as defined by the ability to use them.[50] The qualitative differentiation open to future possibilities transcends quantitative matter.[51] Matter is formed according to the structure of the organism. Purpose drives action by categorizing matter as useful.[52] The self's ability supports and defines purpose.

**The Point:** The self is a choosing organism that belongs to its choosing.

## MEANING

"A living action is an event that can cause other physical events by moving toward a goal,"- Thomas Metzinger.[53] Conscious actions are determined by selections based on future possibilities in accordance with desire. Consciousness feels its own sense of physical causation.[54]

Qualia as they appear are not always consciously chosen, but an action, whether to ignore, explore more closely or move on is based on what the qualia means.[55] Qualia are the qualitative basis of meaning. Meaning drives the categorizing act. The judgment of what serves ones purpose is felt as meaningful. With all relational components, meanings are both result and pointer to something else. Something means something else, within or beyond the recognized category. "Conscious meaning is personal definition."[56] **We are what we mean**. A successful act that goes beyond expectation becomes new meaning. The act causes meaning and meaning causes the act. "The relational excess"[57] open at the width of a living now, causes its own future.

Meaning is not matter. Conscious objects are not matter. Meaning is the world as presented to and by awareness. The living now is pure time. In pure time, awareness looks forward into nothing and out from the inside. Matter in space-time does not have an inside-outside. It is quantified mutually external parts. Meaning is not part of a time-neutral equation. It is a qualitative relation, itself open to be related.

**The Point:** What is meant is the reason to continue.

## INTELLIGENCE

All species are alive. All species are aware. Are all species intelligent? Each species is aware according to survival within its niche. Is human awareness more intelligent than that of other species? One human is more intelligent that another human. Why isn't this true for individual members of any species? Unlike living awareness, intelligence is a variable.

Human intelligence is often defined in terms of malleable language and abstract thought. Does this make humans more intelligent than other species? Not if intelligence is measured by the ability to respond to new circumstances and solve problems.[58] Many non-human animals have been observed to be as capable as human infants.[59]

Increase in choice beyond survival eventually becomes more a matter of preference.[60] Eventually, choices are not between what is simply good or bad for the organism, but are between what is equally good, or equally inconsequential. The greater scope of human consciousness opens to a wider imagined future with more powerful freedom to choose among a larger set of possibilities. Human consciousness can cause events with wider scope than the consciousness of any other living being.[61] But dolphins are better at living in the ocean than humans. Evolution is defined by some as the "evolution of agency."[62] But **if the scope of awareness is relative to an environmental niche,**[63] **then intelligence is more applicable to a range within a single species.**

Species intelligence can be defined as the ability to adapt to a changing environment, and therefore, not go extinct. But intelligence, as well as survival, begins with the individual. Some dolphins are better than others at being a dolphin. If enough individuals are good at it, the species survives.

Consciousness and intelligence are the two sides of the contradiction of the present. Consciousness is at base a binding. Intelligence is at base the willful effort to reach the correct binding. If awareness in any single species is the relation between that which is sensed and the corresponding action, then **intelligence is the capacity for variation** in that relation.[64]

The ability to manipulate beyond the innate capacity to successfully categorize is the core of creativity.[65] Relational tasks are nonlocal rather than linear because they timelessly "pop" out with an overflow of meaning into a wider component.[66] Intelligence means seeing more than what is normal and applying it beyond expectation. Intelligence is creation.

**The Point:** Intelligence is the creative ability of living awareness to survive. Intelligence begins with will to reach a correct conscious binding. Beyond survival, it is the ability to playfully create.

# IV. AWARENESS IN MATTER

## A. NERVOUS SYSTEM

*Brains don't think: they don't have minds; animals do*
Alva Noe[1]

### BODY

Awareness, limited only by scope, is present at all levels of an organism. The widest level of awareness is the organism itself. The brain is an organ within the body. Indeed, the brain, extended by the nervous system, is an organ that reaches throughout the body. The brain is alive. Consciousness is alive. From the point of view of matter the brain explains consciousness. From the point of view of form, consciousness explains the brain. From the point of view of life, **the brain is the way consciousness explains itself**.

The brain is not separate from the body. It is a unifying relational component, a brain-body. An organism is a living totality of functional relations, and therefore, "collapses the two notions of consciousness and awareness into one."[2] Consciousness and awareness are "total organism functions and not purely a brain function."[3] Neither living awareness nor conscious experience can be explained by a quantity of matter parts. The living body and awareness as a whole, and in their components, are each ephemeral structures composed of dynamic functions. Awareness with a brain includes human-like consciousness.

**The nervous system is an extension of the brain**. The brain coordinates bodily actions.[4] **All bodily systems are extensions of the nervous system**. Organs throughout the body communicate through multiple systems with the brain and consciousness.[5] The immune and lymph systems are intimately tied to both brain and bodily neurons, or nerve cells.[6] Respiration depends on constant feedback from the brain down to and back from bodily cells and genes.[7] Our conscious moods and overall health can be affected by the status of bacteria in the gut. The heart even has its own set of neurons, a "heart brain."[8]

The sensory and motor sub-systems work to distinguish self and other, by comparing motions respectively. The vestibular system from the ear to the brain, is the unifying framework through which all sensations connect.[9] Balance and spatial orientation are the general bodily sense. The structure of the environment is reflected in the structure of the body. Different areas of the body are mirrored schematically in the brain.[10] Interactions of and within the organism are mapped in neural patterns.[11]

**The Point:** The brain is a labyrinth of communication throughout the body. Living awareness operates throughout, culminating in full body individual awareness.

## SENSATION

Sensation is to the whole organism what the cell wall is to the cell. It is the boundary function, extending from the surface to the brain. The eternal present of pure duration is bounded by the past and anticipated future. By imposing space over time, it forms the in-out pole of internal identity and external other.[12] Matter parts in spatialized time have no in-out, they are mutually external. **The in-out pole divides quantity from quality, matter causation from living action**.[13]

As science now comprehends it, photons from the Sun hit the Earth at roughly $10^{17}$ photons per second per square centimeter. Neuronal receptors in the retina capture about 10% of the photons reaching the eye. Individual photons chemically interact with individual molecules (actually the opsin protein in the receptor cell), which then spread electron energy by interacting randomly with the time-asymmetrical, uni-directional shape recognition of many molecules, before a patterned digital signal shoots up the nerve. Conscious sensation requires a minimum of five to nine photons hitting the retina within 100 ms. The cell size containing the receptors needed to sample a photon distribution, is two to three times the amplitude of the wavelength, about 1000-1600 mm.[14]

As we have seen, it is debated whether the collapsing wave function is entangled with consciousness. If anything, it is entangled with sensation.[15] Color, for example, is a pattern, a distribution of photons at each wavelength in a given area per unit of time.[16] The light wave pattern becomes shaped molecular information. Other senses detect vibrational patterns of molecules in the air, or against the skin one molecule at a time, and like vision, pass on patterns interpreted digitally in the nerve.

**Distinction among multiple kinds of senses is the fundamental qualitative basis of sensation**. What would be the quantitative difference between seeing and hearing? In the sense of sight, there is no color quality in matter. Photons hitting the eye differ by angle orientation reflected off objects.[17] Vibrations of air have no sound.[18] Pain and pleasure do not differ quantitatively. They are, instead, the fundamental qualitatively distinct feelings of the body.[19] The environment becomes not simply matter parts, but a totality of sensations of "something beyond."[20]

Sensation is an internal information pattern. This pattern is a reflection of differences in change,[21] patterns gleaned against an undifferentiated background.[22] The many sensations throughout the body provide the sense of bodily location. Bodily movement felt as one, against a counter moving or un-moving environment, leads to a sense of ownership, with an in sensing an out. Sense of direction is not a sixth sense. It is the first sense. Living time is change with direction.

> **The Point:** Qualitative sensation is the in-out boundary of momentary awareness. It is the patterning of changing parts and wholes.

## PERCEPTION

Perception refers to the first level of tasks interpreting the patterns of sensation. Various sensory loops[23] feed back to and from perceptual and sensory sub-components. The perceptual feeling of sensation operates within the temporal width of "anticipatory sentience"[24] and immediate retention. Single molecules, each of which can have multiple meanings[25] are translated into neural impulses, each also a symbol depending on the contextual pattern. What is perceived is a single image born of many parts. Sensation detects photons or molecules, but the resolution is perceived as a changing and changeable whole. For example, a color is perceived as the same despite changes in lighting.[26]

**Within the temporality of perception the sense becomes a sensing**. Seeing red is not simply the red seen. It is an action.[27] "Seeing is doing."[28] Resolving the image involves a series of feedback loops continuing until a satisfactory perception is reached. Perception selects what seems appropriate or useful for the sensation.[29] Learning. or past experience affects what is appropriate.[30] But what is perceived is identity abstracted from distinct positions, not yet an identifiable object. Space is never seen. It is the undifferentiated background.[31] This forms the world that we seek to understand. We live within our perceived senses.[32]

> **The Point:** Perception is the ongoing process of turning the patterns of sensation into meaningful identities.

## BRAIN

Over evolutionary time some organisms expanded beyond what the "sensory loop," as Nicholas Humphrey calls it, can capably support.[33] Responses to increasing changes in the environment and within the organism required help.

The human brain contains about 100 billion neurons. Each neuron has 1 to 10 thousand connections, or synapses, to other neurons. There are up to 100 trillion synapses. For perspective, there are an estimated 100 billion stars in the Milky Way galaxy and 10 trillion galaxies in the known universe. It takes about 50 ms. for signals at the nerve ends in the skin to reach the brain. It takes 200-300 ms. for a human to be conscious of a minimum sensation.[34] Sheer distinction has been tracked at 20 ms.

Though the brain is the unifying area of the body, sentience permeates every level. The brain is an organ of the body and communicates with other organs. Its cells, both neurons and other types, communicate among themselves and with others in the body. Proteins and genomes within brain cells send and receive messages in the same way as they do in all the body's cells.[35]

The outer world is reduced to serial photon and molecular encounters at the organism's surface. The quantitative serial pattern is converted to asymmetric protein shapes by sense organs, which sense differences in the serial pattern. Differences are converted into digital neuronal signals sent to the brain by nerve cells for further processing. Multiple signals are reduced by the brain as patterns perceived as images. These images are multiplied and organized by the brain into objects with parts and objects as parts of greater wholes. General sensory information becomes more specific with organization.

The brain constantly makes sub- or pre-conscious decisions.[36] Sensations of various aspects of light are met with over thirty visual areas in the cortex, looping among themselves and with other parts of the brain.[37] Identification of objects arises out of the visual system's recognition of relations between percepts,[38] including shape, color and movement.[39] Components in the brain send and receive signals in circuits of loops with every part of the brain and body, both internally and in interaction with the external environment. Sentient and body states (a state is the functioning of a particular set of related tasks at any given moment) at every stage are not separate.[40] The body state and the perceptual brain mapping of the body state are dynamically united as functional components.[41] In this ongoing looped circuit of signals, the perceptual state predicts what it will do and therefore, "knows" the body state it will produce.[42] Retained perceptions are compared with change or repetition in sensual messages to measure any required correction.[43]

The default brain state, similar to that of brainless organisms, is the expectation of continuity.[44] **Distinct change triggers action**. With a brain, the loop of awareness is elongated and demarcated between the brain-body and the "reverberating dynamism" of sensations.[45] Consciousness requires a time lag to be conscious of sensations. Distinctions are separated and identifiably connected into events in pure time.[46] Retained experience enables categorical recognition of things that can then be counted as separate events in spatialized time.[47]

Sensations and bodily awareness occur at much faster speeds than do brain states. Feeling, therefore, occurs faster than awareness of feeling.[48] Bodily actions are in relations with both sensation and with thought, as bodily feelings reach up to the brain state and emotions reach down to the body state. For example, a fast heart rate and sweating are in a relation with brain states. The limbic system is a system of functional relations between emotions and memory.[49] All brain states, including consciousness, are necessarily accompanied with feeling,[50] as the residue of meaning manifests at all levels.[51]

Sensations, similar to cell membranes, establish the inside/outside of self. Components of the brain identify the series of inside/outsides in time. Body mapping then establishes the self/external world as a temporal self process.[52] The brain monitors its monitoring, and makes adjustments.[53] Causality is upward with multiple sensations and downward with categorical actions based on changes in the sensations.[54]

**The Point:** The brain makes further meaning by combining and recombining percepts into objects and events in spatialized time. The greater temporal distance between sensation and perception allows for more complex awareness of the changing environment.

## B. BRAIN

*There is only energy changing in space and time*
Peter Ulric Tse[1]

### NEURONS

The brain consists mainly of two kinds of cells: Neurons and glia. The glia cells provide nourishment specifically for the neurons. Neurons are to the nervous system what proteins are to all cells,[2] that is, logical processors. They are the cells that as a group unify the body and constitute the basis for consciousness, both human and non-human (those that have neurons).[3] Neurons are the cells that compose the mapping of the body. Most information processed is unconscious

Neurons are different than other cells in three ways fundamental to the understanding of conscious awareness. First, **a neuron is an individual**.[4] The relation of structure to function in a neuron is the inverse of all other cell types in multi-cellular organisms. In other cells, the function remains through repeated instances of the cell. The function is directly in service to the organ and only indirectly to the organism.

The structure of a neuron, on the other hand, remains intact for the lifetime of the organism. (The molecular structures inside the neuron are, like in any cell, constantly replaced by identical instances, which of course, consist in turn of repeated replacements of identical atoms). This is similar to an individual whole organism, both unicellular and multi-cellular. Though neurons generally last as long as the organism, partial structural changes do occur when function demands it, again similar to any cell. Though there are approximately 100 billion neurons in the human brain, each neuron is different. Some scientists had once thought that this means each neuron can be assigned to a different unique physical entity in the world.[5] What was not understood is that it is the neuron's function that makes it unique.

Second, **a neuron, like consciousness, is of something** and not just for. It is of and about the rest of the body, including other neurons. Other kinds of cells exist for the benefit of the organism in changing structure with constancy of function. Neurons maintain structure, whose function is to maintain the function of everything else in the organism. All cells use differences in electric charge to pass molecules through the membrane, but neurons are also able to send an electric charge to other cells.[6]

Third, their relation to energy is different. Cells in the body, including neurons, bring in mass/energy (matter) to store and power functionality one step ahead of entropic dissipation passed back out. But neurons also **interpret and embed information within the flow of energy itself.**

The information is passed **in one direction**. Neurons have an input end and an output end. They consist of three sections: The input end, called dendrites, the output end called the axon and the main body in the middle called the soma. The number of dendrites and shape and length of all three sections vary considerably depending on the neuron's function.

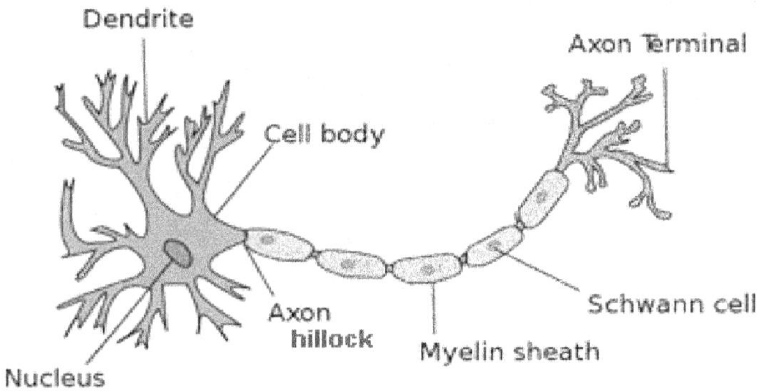


Fig. 42.

In general, multiple signals come in to the neuron for every signal that goes out.[7] Neurons, both individually and as a whole system of neurons, take otherwise unrelated sensory signals and turn them into progressively singular streams of information.[8] Neuronal connections alternate between electric and chemical passageways. Electrical pulses pass through the axon, are interpreted chemically at the synapse between the axon and a dendrite of the next neuron, and are translated back to electrical pulses along that next neuron's axon. The conversion of electric signals to protein chemicals slows the neuronal pulse by about one-half.

The signals propagated along, not through, the axon membrane, are electrical impulses defined not by strength of amplitude or average velocity, but by the variation in frequency; in other words a temporal pattern or shape. This temporal pattern is built upon a digital frame. The channel proteins along the membrane open all or nothing, one after another forming a wave traveling in one direction.[9] The wave is a self-sustaining nonlinear ride above the extra energy provided to the neuron by the glia cells. Each wave is a singular entity moving at constant speed through matter, but unlike a water wave, not of matter.[10] Each wave is an individual letter in a word or sentence pattern. **The coherence is purely temporal.**[11]

At the output end, the single axon fires in all or none bursts, expressing a digital-like signal. Because dendrites are mostly stationary and because there are so many more dendrites than axons, any particular connection requires that axons be able to move. Beside connecting to the next neuron, the axon can also feedback to its own neuron body.

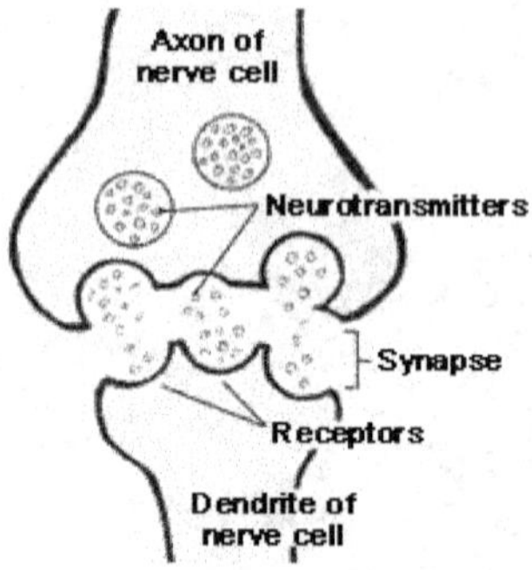


Fig. 43.

The area between axons and dendrites is called the synapse. The synapse consists of bio-chemicals and proteins expressed by the neuronal genomes, and stored near the dendrite surface in sacks called spines. Here, electronic pulse patterns are interpreted chemically by neurotransmitter type proteins such as glutamate receptors, that detect both a chemical presence and the electric pulse. Both conditions must be met, like a Boolean AND function, for the signal to pass.[12] Depending on the proteins' configurations, combined with the pulse rate, signals can be inhibited or enhanced with feed-forward and feedback loops.

A dendrite is to the neuron what the membrane is to all cells, it is the "sensory periphery" where most of the selection occurs.[13] Many dendrite branches in each neuron interface at multiple synapses with many singular outputs from other neuron axons. The input can be new information or the feedback from information previously processed within the dendrites or looped back through other neurons. **Information is recognized in the energetic input**. The dendrites asses the patterns encountered according to certain criteria structured at the dendritic end, in order to decide what to pass on as an electrical pulse rate.[14] Patterns are deduced from the variations of quantity and timing within each pulse and the relative differentiation of all pulses. One insight to how patterns are processed is that many individual bursts from many neurons can be interpreted as statistically concurrent.[15]

Receptors on the dendrites are activated or deactivated by the chemical input.[16] Because the membrane permeability is in this way modifiable, permanent change can be maintained.[17] In other words, synaptic receptors are the fundamental placemats for memory and learning. The synapse is a recording of its own experience, and therefore, each synapse will interpret future patterns uniquely.[18] The more it learns, the more the link is strengthened,[19] with receptors more receptive and more synaptic sites on the dendrite created. Dendritic integration may also take place further up the dendrite, and in the soma cytoplasm itself, with memory stored in the genome. Synapses can also form directly between the soma of one neuron and the axon of another.

Memory, as a functional structure, i.e., a pattern, outlasts any protein instance. In neurons, like in any cell, Brownian motion can provide randomness to the structure or operation. This can lead to original variation.[20]

Starting with patterns in the environment, whether photons strikes, variations in air pressure or molecule combinations, all of which are captured by functionally shaped molecular processes and then passed on as digital temporal pulses, this process continues again and again as pulses are interpreted by molecules between neurons before being passed on again as temporal pulses. Ephemeral spatial and temporal shapes are functional identities in time, whose function is meaning derived from distinction.

> **The Point:** Each neuron is a unique individual. They retain their form through material change. Like consciousness, they are of something else. Digital pulses from each of many neurons alternate with analogue chemical receptors of a single neuron which interpret and store the multi-pulsed patterns chemically, before each passes the patterns on again in singular temporal bursts. Part and whole alternate as patterns in one direction.

## NUCLEI

**Just as a single molecule has no temperature, an isolated neuron has no awareness.**[21] Meaning demands context. The context of life, open to its own future, requires an unlimited capacity for meaning. The 100 trillion possible synaptic connections in the human brain provide the possibility for virtually any mental connection.[22]

The 100 billion neurons in the brain most generally subdivide into sensory, information storage and processing, and motor neurons. Beyond those three, **like molecules in the cell, neurons have developed specialized functions**, functions as differentiated as an organism's capacities require. Like cellular components, neurons are organized by task.[23]

Tasks can be as specific as a single component. Single neurons can respond uniquely to different patterns.[24] But with experience and evolution comes more complex selected connections and combinations.[25] Tasks become centered in modules that themselves change constantly. As functions change, structures follow.[26] The tiniest groups of neurons, let alone single neurons, can be relatively specific. For example, appetite and emotions connect in "a subset of neurons in a subdivision of a subdivision of the amygdala,"[27] itself a small section of the brain.

There are neuron groups for every specialized task imaginable. There are grid cells for location tracking, some for motion detection and mirror neurons that fire when a person sees someone else perform a task. Tasks combine in function. For example, the representation of a particular object may be spread across a group of neurons. Many categories of memory, like colors, names and emotions, are stored in various parts of the brain.[28] A single sense may break down into different regions. Sound, for example, divides by pitch or musical memory.[29]

General functions become relatively concretized in a multitude of definable larger areas. The hippocampus deals with temporal connections;[30] the amygdala combines visual and auditory information and emotional memories;[31] the prefrontal cortex is the seat of conscious control;[32] the lateral frontal pole monitors bad decisions;[33] the subcallosal cingulate region regulates responses to stress; and the right anterior insula connects self-awareness with interpersonal experience,[34] to name just a few.

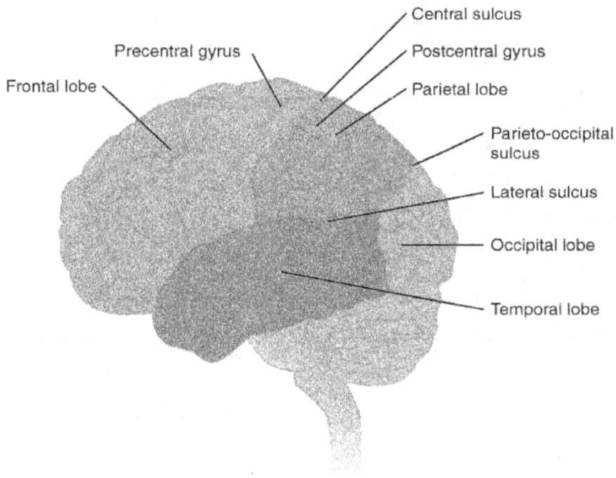


Fig. 44.

The cerebrum, the most general area associated with higher cognition, contains the occipital lobe specializing in vision, the temporal lobe for hearing, emotions and words, parietal lobe for spatial representation, and the frontal lobe for "moral sense, wisdom and ambition,"[35] and conscious awareness.

But **nothing works in isolation**. Vision subdivides into relationships among visual objects in space and relationships within objects. Movement is processed in the temporal lobe and combined with the name and details in the occipital lobe. Neurons form columns correlating with spatial orientation. Simple cells converge on complex cells which weed out "noise."[36] Specific meaning arises out of the general background.[37]

**Every area in the brain contributes to consciousness.** Any area physically cut off reduces the information available.[38] Distinct areas combine functionally with others in a continuum at, through and around every level. The totality of distinctions forms a "tangled hierarchy of causality" for conscious sentience to rise above that of individual neurons.[39]

The specific frame of reference for consciousness per se is the Prefrontal Multimodal Neocortex located in the frontal lobe, but communicating with the rest of the brain.[40] Here the Prefrontal Integration Modules are the **structures that correspond to the function of decision making based on comparing the past with the present**.[41] The "probable next moment" determines whether to go ahead or not. Moments in consciousness connect from one PIM to the next until the end of the attention span. The output from the PIM connects to a wide set of nuclei throughout the brain.

The largest physical demarcation in the brain is the two hemispheres. They are anatomically and functionally almost identical. Because they share the same arousal system, based near the brain stem, they both share the same conscious state (e.g. awake or not). The way they interpret the world, however, is different.[42]

The right hemisphere controls and receives sensory input from the left side of the body. The left controls and receives sensory input from the right side. The two sides combine to form a three-dimensional sense of distance from the outside, which is necessary for the demarcation of one spatial self.[43]

The left hemisphere is generally the more analytic and the right more intuitive. The following chart summarizes some of the differences:

| **LEFT BRAIN** | **RIGHT BRAIN** |
|---|---|
| Group coordination and communication | Environmental awareness, survival |
| Sensation, vision, and right movement | Sensation, vision, Left movement |
| Linear thinking 2.5 words/sec | Non-linear thinking: 10xx faster |
| Verbal memory | Visual-spatial memory |
| Language | Manipulation of relationships |
| Speech, understanding speech | Production of originality |
| Goal-directed linear planning | Error detection, humor, emotion |
| Routine | Novel |
| (PARTS) | (WHOLE) |

Experiments done when the connection between the two halves has been severed, reveal that when subjects have been shown different parts of the same object to each eye, each half independently fills in subconsciously what it thinks is missing. Behaviors chosen by the right without the left knowing it, will still be explained by the left. The left believes what it didn't actually see. Each hemisphere can act without the other's knowledge or consent. Each has a sense of ownership.[44]

In the normal brain, identical nuclei in each half will exhibit dominance according to the tendencies listed in the chart (although the language and speech centers are exclusively found in the left). Two wills and two beliefs are not separate but complimentary. The unified self is founded on the difference. By alternating or monitoring mutual inhibition, the ultimate task of constantly unifying lies with the medial prefrontal cortex.[45]

The right hemisphere processes general sensory input and the left hemisphere makes it meaningful.[46] Distinct points of input diverge in the right and converge in applied form in the left.[47] Top down consciousness is aware of bottom up sensations and posits a divisible spatiality.

But by constantly telling the story of what the right sees, the left brain is explaining what can't be seen.[48] What is unseen perhaps, eventually includes electrons and neurons, quantum particles and a space-time continuum reflected in two hemispheres.

**The Point:** Groups of neurons, like molecules in cells, specialize. Functionality, like in the cell, is contextual. Information is separated and reconnected. The right and left brains reflect whole and parts respectively.

## PATTERNS

Pulses along the neuron are wholly temporal, distinct entities forming a serial pattern. These pulses are converted and interpreted chemically at the synapse, whose configurations hold past information and which then pass on a single pulse, whose frequency is in turn altered by that configuration. Up to forty thousand synapses at the dendrite end pass on similarly altered pulses from that many distinct axonal firings. Dendritic integration in the dendrite trees, soma, and axon hillock[49] interprets a pattern in the multiple impulses from many synapses.[50] This pattern is based on the relative timing and amount of pulses as they are gradually integrated up through the neuron. The axon then fires its own pulse based on a final threshold summation. Summarized patterns, like serial pulses and loops of ephemeral proteins, are temporal not space-time brain matter.[51]

**The firing of many neurons in a group bring patterns to another level**. Not just multiple serial pulses seen as one pattern, these patterns are synchronous, that is, the entire pattern recurs at successive instants of time. More than simply a collection of random neuron pulses, elaborate sequences act in concert.[52] Though it is physically a moving set of neuron combinations firing against a background set that are not firing, **the pattern is a distinct entity**.[53] Neither any single moment, nor any single momentary matter/energy configuration attached to any given moment is itself the pattern.[54] The pattern exists apart from any material or momentary underpinning. Unlike a water wave, but like a wave traveling along a neuron membrane punctuated by firing proteins, the pattern is something separate from the collection of pulses.

A repeated stimulus may affect a particular area to more quickly form a particular pattern.[55] Similar stimuli may closely correlate with the firing of neuronal groups. For example, patterns of grid type neurons track an animal's movement.[56] However, though specific nuclei may be the locus or intersection of specific functions, neuronal patterns generally cover multiple regions.[57]

Thoughts can be observed moving across the entire brain. Sensations (input) follow actions (output) follow sensations (input) in a loop of thinking.[58] Rippling cascades can be discerned, possibly correlating with excitations and inhibitions.[59] **Interacting patterns form wave peaks that may correlate with attention**.[60] There may be a distinct pattern for every distribution of conscious entities among feelings, emotions, memories and thoughts of any kind.

Individual neurons receive and interpret patterns in pulses. When a combination itself moves across multiple neurons, the question of course, is what receives this pattern? Is the pattern itself what we see or perceive? Does simultaneity of firing or the amount of firing cause actions within and by the person?[61] Patterns and singular bursts are intertwined in causal loops of conscious and subconscious processes. Patterns are interpreted through neuronal selection.[62] This will be discussed in the section on causation.

Some scientists theorize that consciousness emerges from neuronal processing like liquids or gases emerge from the interactions of molecules.[63] What we've been saying here is that patterns are already distinct from physical neuronal firings.

John Searle (1932-) has proposed that like quantum particle/waves, consciousness is a field of neurons.[64] Mai-Wan Ho expands the theory by claiming that at the point of stimulus, consciousness is at a state of quantum coherence, and all subsequent conscious states are in various degrees of de-coherence. This might explain, she surmises, how oscillations in separate regions of the brain are synchronized.[65]

Rather than awareness itself being a quantum particle/wave, consciousness discovers particle/waves in searching for the smallest parts in matter. From the perspective of patterned duration, matter is the structural expanse of synchronization. Space-time derives from duration.

**The Point:** Combinations of pulses form patterns that move across the brain on top of but not consisting of, the physical structure.

## NETWORK

Though patterns are abstracted from any material embodiment, to be functional a pure organization still requires structure.[66] **A network is a pattern of connections dependent on relative position**. Neurons are part of larger neural circuits of interacting loops. Neuronal pulses are synchronized in a "hierarchy of scales,"[67] controlled by a symphony of precise timing.[68]

100 trillion connections between neurons are not "wired" arbitrarily. Any neuron is not more than two or three steps away from any other.[69] But to traverse those steps, information depends on a pattern of firing larger in scope than any single neuron. "Integrated information theory" says that the more information is stored, the more efficient extraction has to be. This requires a certain density between neurons or neuron groups. Nodes too far apart do not connect until, at a critical density, the nodes all link in a cascade of connections. The cascades get larger until a point at which they begin to vanish. This is explained by the "dilution effect." If too many nodes are too close, direction and timing are dispersed.[70] Patterns and connections are strongest when neurons fire simultaneously, but as sensations repeat, the activity slows down. The ebb and flow is self-sustaining, as the brain rewires itself with axons actively seeking compatible dendrites with changeable synapses. The network rewires as the strength of connections varies with experience.[71]

Different regions of the brain exhibit distinct patterns of synchronized firing.[72] Every task has multiple sub-tasks down to and including single neurons and their chemical components. "Assemblies of assemblies of assemblies" each react to a distinct receptive threshold before firing,[73] creating complex patterns of interconnecting and integrated signals, with individual links of varying strength. All these different areas of activity, however, may share the same resting state,[74] that is, a ground state pattern.

A few different wave types have been identified ranging in cycles per second:

1. Delta (0.5-4Hz) — deep sleep
2. Theta (4-7.5Hz) — unconscious
3. Alpha (7.5-14Hz) — deep relaxation
4. Beta (14-40Hz) — waking consciousness
5. Gamma (above 40Hz) – attention

Exclusively chemical synapses between neurons would be an impediment to faster wave speed. Another type of connection between neurons is the electrical synapse or gap junction.[75] These are "lateral" connections directly between neuron bodies. With gap junctions, neurons join together creating in effect larger neurons, which can grow into neuron groups called the neuron trellis. These larger neurons or groups can fire at 40 times per second, compatible with gamma-band oscillations.[76]

The neuron trellis is itself an "anticipation machine."[77] Because electrical synapses are bi-directional, loops of neurons feed forward and back, computing future probabilities based on stored experiences, in other words, predicting. Collective integrated information can regulate spikes more meaningfully than a single neuron. The vast combination of information loops found in each neuron trellis, as well as slower chemically linked neuron loops, expand to a virtual infinite variation across all groups to reflect the varying content of conscious and unconscious awareness.

Loops in each trellis and in larger circuits of neurons and neuron groups all compete with one another literally for attention. Different areas make different predictions.[78] What happened in the past is looped back to what is incoming currently. The top-down and bottom-up cycles of information form moment-to-moment recursive loops.[79] The immediate past is the context of the present. Links between events form to simulate the expected future.[80]

Neuron groups in the cortex are arrayed in tightly packed assemblies of 100,000 neurons/per cubic millimeter. Neurons "coding," or selecting, for a particular quality, e.g., orientation, direction or color, are stacked in columns.[81] The top layer has the most connections to and from other regions. Each layer, from bottom to top, has an increasing number of loops down more than up. The positive feedback loops extend the firing of the neurons which extends the duration of any perception. Neurons can also switch coding combinations, greatly enhancing the possible combination of synchronized signals.[82]

The brain expresses the individuality of living awareness in two ways: 1. **A function retains identity through changing instances of matter**, like in a chemical synapse or loops of neurons identified by function. 2. **An identity is retained through movement across matter**, like a serial pattern moving along a a neuron or a multi-area synchronized wave, each an unique individual floating above a sea of digitally pulsing neurons. A wave is unique even if the only thing different is time.[83]

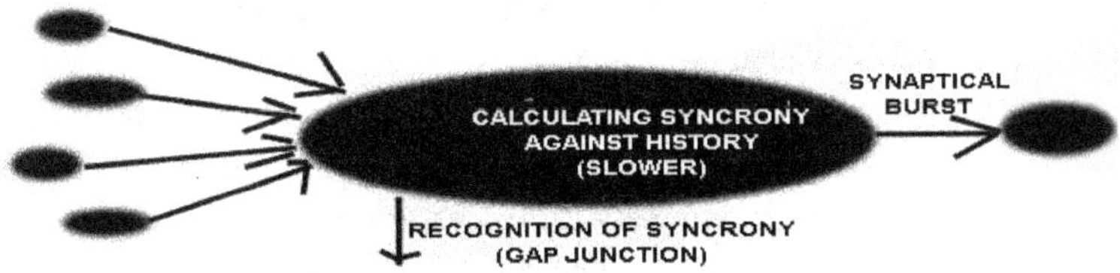


Fig. 45.

Throughout the network, the patterns of many-to-one inputs yield singular outputs that in turn are synchronized in global patterns. In some cases the dendritic input moves sideways through an electrical synapse firing in gamma-synchronous bursts. But the input is also first weighed in the chemical synapse before the singular output into the next. The synchrony, therefore, arises out of the totality of individual previous axonal outputs. Since the chemical synapse computations delay the next axonal output, **axon and gap junctions release similar pulses at different times**.

Perceptions are stored as thing and context. Each perceived object or perceived part of an object has an area defined not spatially but by linkage. Location is present in all areas of perception. For example, one area might retain green and place, and another area square and place.[84] Object recognition correlates with the amount and timing of many neuron bursts. Parts of a perceived face correlate with bursts in separate parts of the brain. Each area may locally produce waves at 40 cycles per second, but the perceived whole face synchronizes in global gamma patterns.[85] It is the timing of the signals that binds the disparate sensations and percepts. **Each perception correlates not with a specific area, but a specific wave pattern and direction that is added to the overall pattern.**

Internal bodily feelings are qualitatively differentiated in the combinatorial recursive loops in the brain stem.[86] Feelings combine with sensations, memories and all modes of awareness throughout the brain. Consciousness is dispersed throughout with interactions between large and small-scale assemblies.[87] Each link adds a bit of meaning.[88] Though anesthetics are observed to affect consciousness by acting on proteins in the soma, consciousness is not found in individual neurons. Consciousness is blocked because **all links travel through individual neurons**.

Background and objects, or object and details are perceived and represented in the abstract organizational structure of assemblies and subassemblies. At the point of attention, background scenes synchronized by slower theta and alpha waves, are fronted by faster individual gamma-synchronized frames.

**The network as a whole and in every link is creating expectations based on stored knowledge.** The unity of conscious and unconscious awareness is the activity of billions of linking and unlinking neurons corresponding to levels of ongoing representation.[89] The brain's living organization is a task structure expressing a temporal pattern.

> **The Point:** Patterns are further structured with relative positioning and timing. Functions retain temporal structure through a change in matter instances, and patterns retain temporal structure passing across matter.
>
> The same network principle that we've covered in complex matter and living beings as a whole is evident in the brain.

## TIME

Alternating point and not-point filters through the living structure. Because they exist apart from the underlying or associated matter, all living components, including all brain functions, are organized by timing. Internal time, or living duration, directed forward, is what is in control. In the brain, this organization consists of communication among what is objectively, varying neurons, groups of neurons and the whole brain, but is really a dynamic community of "time-varying activities" at many scales.[90] "Seeing is what takes time, not vision."[91] Vision takes place.

Myriad parallel inputs from all activities are both synchronized and reduced in serial links orchestrated by timing.[92] But overall synchronization of so many input-output links manifests in rhythms and patterns which are themselves coordinated. In attention, a gamma wave moves back and forth across the entire brain at 40 cycles per second. Different areas associated with different aspects of a perception, synchronize as individual waves moving through the gamma pattern as the perception comes to attention.[93]

All areas and sub-areas synchronize (or "phase-lock") at varying scales.[94] Gamma synchrony is distributed throughout at timescales varying as small or fast as hundreds of milliseconds, reflecting the immense variety of conscious content. For instance, a smell correlates with observed synchrony in dendrites of the olfactory bulb; pleasure and reward in the ventral tegmentum and nucleus accumbens.[95] Fast gamma patterns link different specifics, while slower patterns bind general sensory and bodily information into a whole upon which the different aspects appear as a perceived moment.[96] Rhythmic wave patterns also synchronize with rhythmic pulses in the environment, such as those heard in music or speech or in the larger rhythms of night, day and the calendar.[97]

The underlying default wave of consciousness, patterns of attention, sub-conscious loops and synchronizations as well as bodily aware loops, are one-way directional. The myriad functions within functions at all scales are in dynamic relation organized not in space-time, but by **timing unified by direction.** Rhythms are fast and slow relative to one another in an ever rolling past-to-future direction. When viewed in space-time, objects retain shape relatively unchanging compared to time. The viewing itself, however, is done by the dynamic set of temporal relations that impose shape.

If the thalamus is damaged, the underlying continuous default wave vanishes and the person loses consciousness. Consistent with a nested hierarchy, local timing synchronizes with a larger pattern which in turn, depends on a particular local timing. Material damage is an unrecoverable interruption in timing, causing form to fall back into matter.

Libet demonstrated that the physical brain decides before consciousness is aware of the decision. Neurotransmitters have been observed to be released seconds before a piece of music reaches an emotional peak.[98] Because of the anticipatory nature of all relational components, the brain can simulate its own future states.[99] Dynamically interacting temporal components resolve at a future point, and are therefore open to what comes next. A before flows to an after. Variations in pattern speeds contrast multiplicities with unities, producing relative sameness of things in a directional context.

Stored information returns in the same repeating neuronal processes, but each repeated process changes with new information. Fragments vibrate with others in similar changing frames, fronting slower changing scenes. Memory is built into timing. Memories are projected into an open future in "a dynamic heterogeneous multiplicity of succession without separateness," says Mai-Wan Ho.[100] Object permanence is achieved in qualitative similarity, not quantitative distinction. Repeated sameness, as such, is stored as past moments in a local sub-system, objectively observed most notably in certain neurons of the hippocampus.[101] Duration remembers by feeling its own identity of time following time.

The integration in global patterns is also an integration locally. The Prefrontal Integration Module (PIM) in the cerebral cortex, as previously stated, is often considered to be the center of consciousness per se. Neurons from all over the brain link here. The integrated information cycles through a stack of neurons in 200 ms.,[102] as each neuronal module in the stack adds new information from both other areas of the brain and the module it has just gone through. The recent past combines continually with new and previously stored information.

Conscious awareness experiences an event as longer than any event objectively measured.[103] Some studies have shown that humans can consciously be aware of images observed less than 20 ms.[104] Visual areas process information before consciousness is called to their attention. Sound signals are processed even more quickly. Disparate signals from different senses within a period of 200 ms. are consolidated into one longer moment experienced as "now." But the now is not discrete or static. The cycle in the PIM is an ongoing blend of nows. Studies of meditation done by Libet himself, showed that the time lag between sensations and consciousness can be reduced.[105] But there is a limit to how short neuronal processes can be.[106] Before simple distinction can be made, there is nothing to process. The window of duration points forward, as close to the timeless edge as the organism requires.

Binding is simultaneity of awareness and object. The largest and most fundamental binding is that between sensations and the sensed.[107] Movement is perceived in a world of separate distinctions in a background of space. Space is derived from internal duration. Multiple sounds like that of horns and strings for instance, can be experienced together. There is space within the time that binds. Internal-external is this space within a moment.

Starting from the standpoint of a space-time world, Richard Pico theorizes consciousness to be a four dimensional system based on a process relative to the three dimensional living body.[108] Conscious experience within time, says Pico, is the "focus linking multiple background inputs combined with temporal organization within local sources of input,"[109] that is, 4D frames and 3D scenes.

Underlying that, however, there must be a binding of the points following not-points with the in-and-out. Points form a one-dimensional line. The in-out of duration retains two-dimensional shape. A directional order of shape following shape is the spacialization of time. But duration is the zero dimensional binding of shape. Binding is simultaneity in time without distance. Timing arises in the relative variations in durational velocity, the metabolic rate interpreted as processing speed.

Time's boundary is a briefest temporal link, globally and locally between neurons at the smallest, and globally and locally at the largest. Beyond the immediate environment of any living being is the larger world of the most general fluctuations. The circadian rhythm is the brain and body's temporal relationship with both the movements of celestial objects outside and the motion of small components inside.

The two principal circadian rhythms are the sleep and body temperature cycles. They are tuned to the brightness of light and the amount of heat respectively. They run at different schedules, yet are coordinated.[110] REM sleep, for instance, is synchronized with a temperature cycle. Every organ in the body, every physiological function, ebbs and flows on schedule.[111]

There is a separate pathway from the eye beside the one that leads to the visual cortex. This pathway leads directly to an area called the SupraChiasmatic Nucleus (SCN). This area, along with the locus coeruleus constitutes the body's master clock. The SCN is called the circadian "pacemaker."[112] The locus coeruleus is the "metronome."[113] These two areas are connected in a continual feedback loop.

The baseline pattern of attention runs at 40 cycles per second. The minimal resting state of conscious awareness is 10 cycles per second. Neurons in the SCN oscillate and synchronize at various frequencies about a million times slower.[114] (These are the frequencies of universal motion as time is directed out of the blur). Each cell contains the ebb and flow of clock proteins generated in feedback loops with eight different clock genes. Groups of these cells when synchronized, affect the peripheral oscillating cells in all organs of the body.

This "metronome" affects the varying rhythms of different levels of aware processes in the brain. For example, it accelerates just before something comes to conscious attention.[115] The anticipation necessary for conscious awareness is part of the living flow. But all mental processes, indeed all living components, anticipate one another within the global symphony.

**The Point:** Duration, or internal time, is manifested in the temporal distance between relatively faster and slower forward processes, which are projected backwards as foreground and background respectively. Tasks coordinated in the complex timing of temporal positions, look to consciousness from the outside as structural, albeit ephemeral, patterns. Patterns are the rhythms of time itself.

## CAUSATION

How does consciousness affect matter? Peter Ulric Tse in his book, *The Neural Basis of Free Will: Criterial Causation,* finds the answer in "criterial causality." An incorporation of his theory into the philosophy developed in this book follows.

In all biological systems, information is a many-to-one mapping. The "one" establishes a category. That category then becomes the criteria for new information. "Different types of input lead to one type of output."[116] The quantitative many reduces to qualitative information. From membrane receptors to mate selection, "criterial causality" directs all aspects of an organism's functioning. Both material causes and logical steps only process a single input.[117] Biological causal processing interprets the commonality of multiple inputs or input types. Internal processes select by determining how multiple information streams both into and throughout the brain interact.

"Type-level criterial causality,"[118] like information, is not itself the transfer of energy through matter. In neuronal processes, electrical pulses and chemical synapses alternate, progressively creating a whole "summation of a gradient"[119] of discrete entities. Neuronal processes are the physical manifestation of informational organizing. Output is contingent on either simultaneity or interpreted coincidence. Coincidence, like simultaneity, is not matter, it exists in recognition.[120] Information is not causal until it is understood.[121] Relationships among inputs are themselves recognized as a pattern of amounts in time. Informational criteria is then realized in the context of single neurons in a changing cascade of multiple neurons.[122] Similarly, a single physical aspect is understood in a visual or other sensed or multi-sensed field.

**Since in life, form is not tied to matter and therefore, energy is not conserved, information can increase in complexity**.[123] Categories expand as information spreads. Organization per se is purely temporal, which means a total timing of relative responses of individual processes, not separate, absolute responses.[124] The representational "space" of consciousness is constructed of the similarity and difference of categories and temporal distance between categories.[125]

Many-to-one processing is criterial selection, turning separate inputs realized in a gradient into smooth, qualitative information. Criteria set up in the path of energy flow means that the receiver is already configured to receive the pattern. **Information is a subset of the physical that is waiting for the physical to catch up**.[126] Because the criteria is formed ahead of time it is not physical cause, it is informational cause.[127] Information to be information is always one step ahead of itself.

Information realized in a single firing changes as it spreads throughout the larger structure.[128] Because the criteria at each stage can differ, the scope of criterial selection becomes more complex.[129] Each informational recognition can trigger, or cause, a multitude of others. Through new information and information fed back from other neurons, the criteria in the synapse can be changed at any point from the previous criterial firing.[130]

Synaptic weighting is really a criterial assessment and selection of multiple inputs. The total input derives from previous criterial points, which may also include feedback from previous selections at the same point.[131] New information may change the criteria for future assessments and selections, as each criterial point awaits future input.

In other words, criteria for assessing inputs does not affect the current neuron or neuronal group, but changes firings in the future, even pico seconds in the immediate future.[132] Mental or informational events can change the physical manifestation of future mental events.[133] Information and matter interaction requires future-pointing duration. **Downward causality is also future causality**.

Firings can trigger cascades that can magnify rapidly an increase in the amount of information.[134] Propositions and rules form among the infinitely possible and therefore, arbitrary linkages, between input and output. Any representation can bind with any other.[135] But as with gene expression that can change without changing the DNA sequence, "dynamic synaptic reweighting" can change active neuronal circuitry without changing long-term connections.[136]

The experience of causality, and causal organization, is based on the sense of things happening in one direction. Different sensations vary in the speed at which they are processed.[137] These variations are unified by the necessary one-way opposite direction of downward causation.[138] With temporal direction, downward causation causes its own past. In the opening of time, life is aware of its own selection.

Because downward causation is not constrained by the conservation of energy, it can be larger than, and therefore cause, multiple inputs and many subsequent smaller patterns. Consciousness is the most general criteria for selecting from possible inputs. It is the selection of solutions from possibilities presented by the aware body-brain. Selection interprets input. These interpretations are, in turn, fed back to lower processes that can reemerge in future selections.

Neither determinism, which is cause from the past, nor randomness in the present, have room for free choice or will. Criterial causation combines the two.[139] Criteria are determined ahead of time, yet lead to non-determined future actions. Physically, the brain is sensitive to "the unstable nonlinearity" of minute perturbations.[140] Molecular fluctuations in the synapse introduce randomness into neuronal timing.[141]

The correctness of the criteria's results are not determined by the criteria. Increase in informational organization cannot be predicted from contributing inputs. Informational causal chains that are a subset of physical causal chains, are by nature, nonlinear dynamic systems that depend on changing initial conditions.[142] "Much like evolution itself, a scientific concept of free will comes to lie between chance and necessity."[143]

Protein components contain their own future in relation with other components, as the loops roll in a forward direction. The cause is the timing. Neural networks unify timing by retaining instances in one direction. Comparing the new with the old order of things predicts sameness or difference in the future. Freedom is the experience of causation and the anticipation of results. The aware organism is free. Freedom, of course, is another object of consciousness.

**The Point:** Selection criteria is built into the changeable structure of the synapse. The selection criteria is already in place ahead of the next input. What is passed on has been previously selected. Patterns do not cause from the past. They are interpreted from the future. Form is ahead of matter in the direction of time.

## CONSCIOUSNESS

Human consciousness, in studying consciousness, discovers a brain that would have to be structured in a way that would support consciousness. **The structure of the brain is how consciousness organizes itself.** This organization of itself is what human consciousness sees in its own objectification. Consciousness organizes neurology. This book has been a description of consciousness's way of describing the objective world known by consciousness. This does not mean that the world that consciousness constructs is not real. The world is the way it is necessarily known.

Studying consciousness apart from the living body, indeed apart from life itself, is tantamount to what might be called, "the Creationist Theory of Consciousness." But consciousness does not appear out of nowhere. The advent of consciousness is not a problem if we realize that living awareness has been there all along. Evolution is part of how consciousness explains itself. It has evolved from and with living awareness. The question is not, "Where does life or consciousness come from?" The question is rather, "Where does matter come from?"

A problem with assuming that consciousness is something found in the brain, is that the consciousness looking at the brain for consciousness is itself conscious. The neural correlate is observed by the neural correlate of the observation. Consciousness in this case can only be confined to the role of observer, passively watching the world and its own brain. It would be an "epiphenomenon" with no consequence.

Why is there consciousness and not simply molecules? Surviving requires choosing a course of action.[144] Living awareness is a function of survival for all living beings. Unconscious awareness is really the norm because conscious awareness is too slow for most successful actions.[145] In fact, the purpose of consciousness is to slowly reflect on quickly made, possibly incorrect, unconscious decisions. At some point, the unconscious body cannot choose how to categorize an object.[146] Consciousness selects from what is presented to it when the body/brain cannot select. So consciousness is more than an observer. Like all functions of the brain, it monitors and adjusts. It has consequence.

Structure.

The search for consciousness in the brain is daunting. The human brain contains a hundred billion neurons and over 100 trillion synaptic connections. The total brain network, theoretically completely mapped, is called the connectome. This, some say, is the material correlate of the immense complexity of consciousness. If every part and every function of the brain were known, then we would supposedly know what consciousness is.

Like all life, the connections within the brain consist of nested, hierarchical components inside nested, hierarchical components.[147] Just like the cycles of biochemical loops in the cell and body, the brain consists of assemblies of assemblies of neurons.[148] Unlike e.g., liver cells, which exist for the purpose of the organ, and only indirectly for the whole organism, neurons exist directly for the whole organism. As stated earlier, like consciousness, neurons are "of" something.

Neurological studies have shown that consciousness arises in the brain at the physical point or points where multiple sensations become objects,[149] and in links between the brain and the body, e.g., between the pre-frontal lobe (rational) and the limbic system (emotion).[150] Consciousness requires a link between the thalamus and cortex. It disappears when these links are disrupted.[151] The arousal of consciousness during the active link is accompanied by gamma synchrony. Loss of consciousness at sleep occurs when gamma synchrony between areas within the cortex disappears.[152]

Synchrony itself found in the individual neuron can further be traced down to biomolecular levels. But **trying to find consciousness at a place in the brain is like trying to find life at a place in the body.** Consciousness cannot be found in unaware physical information processing, which by definition is not conscious or aware. "Biochemistry cannot explain consciousness."[153]

Duration.

Consciousness can neither be explained as part of the physical brain nor as an immaterial entity apart from the brain. **Consciousness is a component within the living brain and body**. Multiple sensations continually move through paths integrated in an ever "widening scope of meaning...open to the next moment," says Peter Tse[154]

**Consciousness is the most general criteria for assessment and selection of multiple kinds of input.**[155] Since criterial assessment and selection at any point provides further inputs into other criterial points as well as itself, consciousness at the widest criterial point provides its own conscious inputs that are selected by the conscious criteria in the next moment.[156] When a thought or a word, for instance, "comes to mind," it seems to come out of nowhere. This is because, to consciousness, it is coming from the not-yet-existent future. Thought both causes and is selected by thought.

Because each criterial point in the aware body-brain is timed, duration itself is the ground of consciousness. Consciousness is larger within time and, therefore, more encompassing than any sensation. Duration is the constant and fundamental criteria that imposes a space-time world. Distinctions are increasingly selected as qualitative generalities to be imposed back on a quantitative, spatialized time. Open to its next thought, consciousness has the transcendental ability to remap all criterial output presented to it.[157]

Memory.

**Consciousness requires memory.** Criteria requires the maintenance of past categorizations to apply to new events.[158] Having experience means remembering, at least unconsciously, past experiences. The ability to see what is important, in fact to see any-thing, requires experience.[159] Current experience is of the immediate past, made up of past inputs and new inputs on the verge of consciousness.[160]

Survival, even walking without tripping, requires prediction. Even the sound of our own voice is recognized by prediction based on past accurate predictions. Prediction corrects mistaken categorization. If we mistake a bush for a person, we correct the prediction by making a new one.[161] The purpose of memory is to predict the future. By using memory, consciousness is "trying to recall the future."[162]

Short term memory is stored in neuronal loops. Long term memory is stored in neurons and neuronal genes.[163] Consciousness is the widest loop, supported by smaller loops storing short term memory that is in turn supported by even smaller loops. As we descend from brain wide synchrony down to the microscopic, memory goes from the immediate to the past neuronal state to the genetic past.

Attention.

Attention is consciousness of and in the current moment. Feedback loops from the immediate past, maintain sameness. Sameness extends the span of duration over a series of distinct sensations.[164] Qualia is what the subjective now is conscious of.[165] **Attention is what consciousness feels like.**

Objectively, there is no present moment.[166] Sensations represent different objective space-time moments. Attentional consciousness is the experience of binding. It is the present moment of categorical selection, which is the binding of category and object. Conscious attention is not just of something, it is the active binding of somethings. We don't remember things per se, we remember relationships between or among things recovered in binding.

Attention is the binding of now and before of object and background in a continual extended present. **Sameness is the context of change.** Newly decided categorization binds with sensational input.[167] Binding is simultaneity in duration. Binding is always in and of the now. The memory of anything is always, inescapably, accompanied by the point of view of consciousness in the past, remembered in present consciousness.

Most of the time we are unconsciously aware. We can only know past experience the way we remember it. We can never "leave" conscious attention. We only exist in conscious attention. It is always now. Past awareness, conscious or not, does not exist in current attention. It only exists as remembered in current attention.

We can't know about consciousness occurring anywhere or anytime other than the current moment. Thoughts come in from a non-existent future and disappear into a spatialized past. **We don't remember material structures, we remember the consciousness of material structures**. Imaginations brought to attention are likewise imaginations from the point of view of conscious attention. When we imagine or remember e.g., an elephant, we are imagining or remembering our consciousness of the elephant. We are looking at it.

The absolute certainty that Descartes looked for is only in the now:

now now now
I think therefore I am

When the thought in this sentence is first made conscious, the parts would be temporally separated. By the time you get to "I am," "I think" is only a memory, albeit perhaps a short term one, immediately available. But eventually, "I think therefore I am" is brought into the conscious now. "I think," "therefore," "I am" are individual unconscious processes. ("I" and "think" precede in experience and/or structurally in some way "I think.")

**The real dualism is not consciousness vs. matter, it is attention vs. remembered consciousness**. Now vs. the space-time past. The cogito is not demonstrated in logic. It is a binding. In other words, what we usually mean by consciousness is attention.

Binding.

Gamma frequencies synchronize at binding.[168] Gamma synchrony is the "binding signal."[169] In conscious attention, rapid synaptic resetting across many areas, combining with an increasing firing rate in specific areas, correspond with a general background and detailed foreground respectively.[170]

Distinct sensations in the body combine in progressively higher criterial selections. At the point of conscious attention, binding connections are open to random combination in a non-existent immediate future, qualitatively blending distinctions. Sizeless categories are consciously applied to temporally separate yet similar, and therefore, seen after as identical quantitative sensations. **Attention is the continual binding of form and matter.** Specific but directional functions of living awareness open wide to the future in synchronous binding, with direction per se the only link. Conscious awareness is spiked by conscious attention in the now. Non-attention fades into unconscious awareness. The binding of conscious selection is a timeless leap in pure time.

Conscious attention is a unification of sequences, which being alive is always forward-pointing.[171] The possible future collapses into the present. Loops of positive feedback extend the "attentional present."[172] The duration of smoothly connected singular mental events, causes the seemingly separate physical events in spatialized time that it previously derived from distinctions. The two processes, derivation and spatialization, are linked by anticipation, which constantly fills in the ever non-existent future. Recognition of a specific object binds with a category in blended immediacy at the moment of attention. "The end of each moment is the beginning of the next moment."[173]

The electrical synapse passes on a burst of recognition. The chemical synapse absorbs the new before passing on an individual burst. Faster patterns layered on slower patterns account for frame and scene context. The scene is already there, waiting in the future, waiting for the current binding of frame and scene, new and old.

In the application of form to matter, possible categories, based on past experience, which includes the experience of both the particular species and all living beings, are presented to consciousness at the anticipatory moment.[174] Consciousness is the rolling boundary between unconscious selections and completed categorizations.[175] Benjamin Libet observed that decisions are made by the brain ahead of conscious awareness.[176] The material brain is what Libet observed. The living brain observes.

The pattern of bursts coming in to the neuron, exists only in the interpretation of the neuron individual which passes it on as a unique individual serial burst. All the bursts together fire in a binding expressed in the pattern of a unique individual wave. The binding interprets itself. The self is a momentary binding of individual percepts in pure time. **The moment is the ultimate individual.**

Will.

Consciousness is not just of something, it is of anything. It is an open relationship with a non-existent future. Conscious information is conscious because "what is needed in the next moment is not certain."[177] We don't only "think" will is free because we don't know what happens next, will **is** free because we don't know what happens next. What appears to consciousness is seemingly spontaneous and uncaused. This is because attention is the linking between unconscious processes and the next moment.[178] Conscious awareness could not be conscious if it were not for unconscious awareness.

In order for consciousness to be free, it must be possible to make the wrong choice. We could always have chosen differently. Uncertain criterial synaptic weighting waits for conscious decision.[179] In the dynamic structure of the living brain, pre-conscious and post-conscious processes create their own possibilities. As life chooses to live, consciousness is the choosing that the conscious organism as a whole makes, selected from its own possibilities as to the right action for a predicted future. **The living being looks ahead towards the next moment which only it can fill.**

The conscious now is the ongoing mutual separation between anticipation and the pre-conscious processing in the brain. The act of choosing has been observed to coordinate reasoning areas in the frontal cortex with sensory and motor integration in the paretial lobe. One recent study compared choices made with and without being preceded by an external command. When the choice was made internally, that is, without outside direction by the scientists, the deliberation period in the frontal cortex took longer than when the subject was explicitly instructed to choose. A choice reactive to an instruction was close to immediacy.[180] Reactive attention brings consciousness into the current moment. Willful intention extends the current moment.[181]

Conscious attention, willed or not, is focused attention. The focus can be on the thing or its ground, each the context of the other. Beyond the span of either focus is the point and the continuum, the ultimate small and large. The ground of focus per se, is the anticipated but currently empty future.

The ground of will is attention. Choices come to attention. The selected choice is then made willfully, even if the choice is made as temporally narrow as possible. Choices made in the past through top-down brain processes wait for the present, that is, in pre- and post-consciousness. Attention inhibits body action.[182] In connecting action with attention, free will is the ultimate selector and widest downward cause.[183] The choice made by will is the "highest-level criteria for optimal behavior."[184]

Feedback delays are extensions of the momentary now. Willful intention narrows the focus of attention. The act of binding is felt. It is the feeling of categories and objects catching up to one another. The feeling is effort. Effort accompanies even the slightest intention, from recalling a thought to throwing a rock. Just looking at something requires a modicum of effort. Effort pulls the future closer to the momentary limit. The now is felt more widely.

Consciousness is of. Feeling is about. Effort is the feeling of will. It is the unification of consciousness and bodily action. The primal urge to survive manifests in consciousness as effort. Will is free because the brain-body belongs to itself. Self-fulfillment is the continual willful effort to accomplish the correct binding. Living awareness finds itself in the contradiction of consciousness and time.

**The Point:** Awareness shapes matter.

The organization of the brain is what consciousness sees in its own objectification.

Consciousness is the widest component.

It is the most general criteria for binding form and matter.

Consciousness expands in complexity toward the future.

Thoughts and decisions seem to come from nowhere because they are actually coming from the not-yet future.

The structure is a nested hierarchy of timings.

Duration is the ground of consciousness and will.

Binding is temporal simultaneity of category and object

Memory provides the same/difference context for binding.

Attention is the ground of binding in the present.

Will is the urge to live, directed at the anticipated binding.

Will is the ground of attention. It pulls the future in to extend the present.

Effort is the feeling of will.

## C. OTHER ANIMALS

*Animals are conscious beings*
The Cambridge Declaration of Consciousness[1]

### RESEMBLANCE TO HUMANS

So far we have traced awareness up through life as such, into the brain and into consciousness. In this chapter we'll attempt to follow consciousness "down the ladder" from animals seemingly closest to us to those with simpler brains and those with no brains.

Descartes' own conclusion based on the Cogito was that "animals have no thoughts."[2] Non-human animals are at worst machines, at best beings whose actions are driven only by instinct or "simple learning." Consciousness has up until recently been considered the exclusive province of humans. Since humans are animals, it makes more sense to realize that each species has its own "cognition,"[3] which I would say is the ability to use conscious or unconscious awareness to operate in its own domain of sensations. All living beings are aware of their environment by acquiring and storing information. **Charles Darwin himself thought "subjective experience" matched the relative complexity of the animal's physical structure.**[4]

Though various animals may demonstrate superiority to humans in some mental capacities, consciousness is often defined as self-awareness. Some would say this means "the ability to create a model of the world in which we can imagine or predict a future with ourselves in it."[5] In that case, only animals that (who) have a concept of self can be conscious. Since animals can't tell us what they feel, at least in human language, a standard method of judging sense of self is to observe whether the animal recognizes itself in the mirror. But what about a pre-historic human who had never seen a reflection? He might attack it because he doesn't know what he looks like or what a mirror is. Maybe he has a deep sense of left and right. Why should an animal recognize itself if it can't smell anything? As far as modeling the world, all animals predict to survive.

If not with consciousness, then in what way are humans different? Some say that humans have "the ability to create new concepts and rules and combine them in different ways."[6] In other words, humans are distinct from other animals in their "context-independent behavior."[7] But is man more conscious than other animals? Man's freedom from the need to simply survive, plus the structure of his brain which has the capacity to far outstrip the totality of sensations, has increased his ability not to be conscious but to be conscious of more varying phenomena.

All living beings demonstrate meaningful selection from their environment for the purpose of survival, necessarily accompanied by attention in the moment, buttressed by anticipation of the future based on memory of the past. **Attention is shared by all animals with varying degrees of conscious "depth."** In the words of Mark Bekoff, "[Some] animals are conscious beings, capable of understanding cause-and-effect relationships, forming abstract thoughts, solving problems, using language, making tools, exhibiting long-term memory, and showing empathy, in many cases with skills that are superior to those of humans."[8]

In August, 2012, a consortium of scientists concluded in The Cambridge Declaration on Consciousness in Human and Nonhuman Animals,[9] that "mammals, birds, and other creatures like octopuses possess consciousness and, in all likelihood, emotions and self-awareness." Alex Halberstadt adds, "Common shore crabs feel and remember pain, zebra finches experience REM sleep, fruit-fly brothers cooperate, dolphins and elephants recognize themselves in mirrors, chimpanzees assist one another without expecting favors in return and dogs really do feel elation in their owners' presence." And for good measure, "Elephants cooperate to solve problems. Chimpanzees teach youngsters to make tools. Even octopuses seem to be able to plan."[10]

**The Point:** All animals are conscious of something.

## SENSATION

Since knowledge of the world enters through the senses, **what is real to the totality of living beings is knowledge greater than just our own.**[11] Other animals are aware of a world not privy to humans.[12]

A dog's olfactory sensors sense both more and different molecules than humans do. They smell different odors within other odors, and can smell how old the odors are. They smell different odors in each nostril, giving them the ability to know which direction the smell is heading to and from where. It's not that dogs just hear better, they hear different sounds than humans do.[13] Humans hear sounds up to 20,000 vibrations per second. A dog can hear up to 35,000 vibrations per second. A bat hears up to 100,000.[14]

In a bird's eye there is a molecule, that when absorbing light, separates an entangled pair of electrons, which are then susceptible to the earth's magnetic field. Neurological impulses create a visual image of that field.[15] Combining with this image, sounds hit the eardrums at different frequencies on the left and right side, enabling the bird to coordinate sound with that visual image at the eye level.[16] Birds sense a vastly different world.

A housefly "processes about four times more visual information per second" than a human. This means 250 frames per second compared to a human's 60. To a fly, humans move very slowly.[17] **All animals live within their own time.**

Animals measure the world using different parameters. Bumblebees detect electric fields. For wood tics, the guiding signals are temperature and the smell of butyric acid.[18] Dung beetles orient themselves and navigate by the light of the Milky Way.[19] Every type of organism lives in its own reality with its own assumptions. **Every action any animal takes depends on a unification of multiple senses and perceptions.**[20]

**The Point:** All animals sense the world through the timings of their own bodily structure. The real world is what all living beings sense.

## MEMORY

Through the window of anticipation, consciousness brings in the past. Because the future does not yet exist, **it looks at the past as if it were the future. In this way, all animals are conscious.** Anticipation and prediction can be observed in all animals. This requires memory.

In an example generalizing about all mammals, a sheep was kept in an antiseptic enclosure, yet still seemed to scan for flies. Alex Halberstadt[21] concludes that this was not unconscious habit. The sheep was actively, that is consciously, anticipating flies based on memories from her past. Eric Chaisson claims that a paramecium "momentarily retains traces of experience." It ignores what it has already inspected. He goes on to say that an amoeba has no aware memory at all.[22] An amoeba does, however, locate and methodically attack and envelope its prey, indicating anticipation and prediction. It's prey is selected based on recognition. In differentiating between prey and not-prey, it is selecting. Proteins in the cell wall may recognize specific molecules, similar to the function of a human nose, but the whole animal recognizes what it needs.

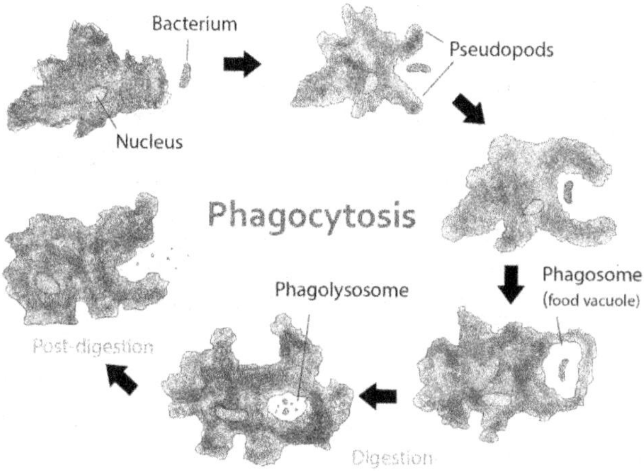


Fig. 46.
Phagocytosis - The amoeba actively devours its prey

All mammals, and many other animals including sea snails, have both short- and long-term memory.[23] Short-term memory, or working memory, as we know, keeps important details of recent experience available for immediate recall in the present.

Chimpanzees actually have a better short-term memory than humans do. In studies where several rows of numerals are briefly shown to both chimps and humans, and the subject is then required to point out which slot a particular numeral had been, the chimps were more accurate than the human subjects. The chimpanzee's sense of now, to them, is longer. Perhaps adapted to making quick decisions in the forest, they see more details at once than humans do.[24] They see in their current moment what is a series of objects that take several moments in a human.

A squirrel has spatial memory similar to, or in some cases better than, humans. It can locate nuts that it had previously hidden in a way that indicates an ability to visualize relative places.[25] At low tide, Goby fish hide in rocky tide pools. If an attacker is near, they jump to another tide pool without ever mistakenly landing on the rocks. Experiments revealed that during high tide they swim over the area memorizing the relative locations of every pool. They can retain the pool layout up to forty days later.[26]

**The Point:** All animals recognize in the span of their own present moment what has been recalled from their history, recently, long-term or genetic.

## BRAIN STRUCTURE

All functions in the human brain contributing to both **learning and memory can be observed in the brain of every animal** that has one.[27] The "neurological substrates" that enable consciousness, exist in human and non-human animals alike.[28]

The simplest worm species, C. elegans, has 302 neurons and 8,000 synapses. Yet this combination is still highly complex. It must interact with a genome of 22,000 genes, almost as much as a human.[29]

Of course, structures in various animals are similar, not identical. Dogs have a caudate nucleus, which is central to processing anticipation in humans. There is evidence that this structure in dogs correlates with positive emotions.[30] Dogs have a positive outlook.

Brains and nervous systems reflect overall body structure. Birds lack a cerebral cortex, yet do have a basal ganglia, which just like in humans is a seat of learning. But in birds, the cells are not specialized, thereby they substitute for a cortex by making information processing in that area more efficient than it is in mammals.[31] Dolphins have a smaller frontal cortex than humans, but have instead a larger brain mass.[32]

Jellyfish don't have a brain per se, but do have a simple nervous system that responds to stimuli. A jellyfish pulls back from a poke no differently than a human does.[33]

The sponge is perhaps the closest multi-cellular animal to a colony of single cells. But the electric and chemical communication among cells makes up a rudimentary nervous system, as sensations in one surface cell affect the entire organism.[34]

Each brain builds a model of the world from photons and electrons, i.e., light and motion, that meets the needs and desires of each animal. In order to do this, the brain (and nervous system) distinguishes the animal's actions from, and correlates its actions with, external forces and movements.[35] An animal needs to correlate its actions with changes in the environment. The dynamic interaction with a varying environment depends on a separate, independent varying action by the animal.

Sensory feedback from the animal's movements "updates" its model of a changing world. This includes the way that world reacts to its actions. To survive, the animal must predict the behavior of its prey and the prey that of its attacker. **The response to prediction is to vary behavior.** Prediction necessitates unpredictability. This requires an animal to "initiate actions independently of the stimulus."[36]

This self-varying behavior can be traced to processes present in all brains. For example, in certain species of fish, differentiation of self and other is enabled when one area of the brain sends out one signal to the motor system and a duplicate but delayed signal to the sensory system.[37] Nervous systems in insects and cephalopods involved in attention and sleep are affected when the animals are deprived of environmental stimulation.[38] Neurons in these brains are observed to reduce self-variation when the environment does not change.

All brains have processes which balance variability with efficiency.[39] When there is no perceived need to change, the brain saves energy. Efficient behavior, however, lends control to the environment. Learning is achieved by balancing the two extremes. Practice both reduces and increases internal, self-variability in accordance with learning the skill. It tunes, in other words. For example the snail, Aplsia, normally has a highly variable feeding behavior, due to the unpredictable availability of food. By making their diet predictable, snails can be trained to reduce the variability accordingly. Fly brains normally generate unpredictable, i.e., highly variable, turning motions. The variability in eye movement is independent of movements of the head, indicating control by the brain, not just sensations. In experiments where the environment is made comparatively predictable, the flies' motions reduce accordingly. Neurons in insect brains, called mushroom bodies, are implicated in the increase and decrease of variation.

Not only nervous systems are altered by experience, but during the animal's lifetime at least, gene expression is modified.[40] As was noted in the last chapter, the brain retains older information progressively down to neuron genes. The more reliable is the information, that is, the least varying is the environment, the more the nervous system will be controlled genetically. In worms, the npr-1 gene blocks the motor neurons from receiving input from the sensory neurons, forcing the worm to behave not automatically, but in a fully self-varying manner.[41]

The last chapter discussed the slowest frequencies of circadian rhythm, that is, the frequencies of universal motion. This is the baseline rhythm of all nervous system timing. Just above stable flatline is an unstable nonlinearity, sensitive to small perturbations which become amplified when sensations are predictable and unchanging or nonexistent. The barrel cortex in rodents is one example of an area generating self-variability. The brain is ready. All brains exhibit the ability to vary behavior even in the absence of environmental change. The brain does not merely passively react. A leech will randomly choose to crawl or swim when the stimulus is invariant.[42] Brain rhythms, however, readily adjust to unpredictable patterns.

**The Point:** Similar functions can be found in different brain structures. All nervous systems can be altered by experience. Variable behavior in correlation with, but independent of a varying environment, can be traced in the tiniest brains.

## LANGUAGE

Each organism's behavioral variation is constantly adjusting to the rhythmic patterns in the sensations from its own environment. Jeffrey Kluger defines animal communication as "the action of or cue given by one organism [that] is perceived by and thus alters the probability pattern of behavior in another organism in a fashion adaptive to either one or both of the participants."[43] With language, rhythm or beat keeping in behavioral variation is extended. Environmental variation contains information about a shared environment. **Language contains a theory of mind in which one animal possesses unique knowledge that it can give to another**. Beyond deception and behavioral prediction, one organism "tells" or teaches another about something that it would not otherwise know.

In learning through language, incoming signals are not only something to adjust to, but must be matched, as for example, vocal mimicry in birds.[44] Bird songs, in fact, are similar to human language in the degree of abstraction of patterned symbols.[45] Indeed, when birds hear another bird's song, their amygadala shows activity similar to a human's.

Michio Kaku claims animal language has no future or past tense, that only humans are conscious of that which is beyond the here and now.[46] But basic concepts like "danger" and "good" imply a predicted future and learned past. Conscious attention is not always required. Most communication happens too fast for conscious awareness. Humans as well as animals, understand each other by unconscious gestures. Simply pointing can convey beneficial information.[47] The behavioral method of information in animals can be extremely subtle and imperceptible to humans. A slight change in posture lasting a tenth of a second can suffice to convey meaning between dogs.[48]

Animal language can be quite sophisticated. Prairie Dogs were discussed in a previous chapter. Parrots use unique sounds to identify individuals. A parrot's lifetime name is taught to a chick at birth by its parents.[49] Language and learning is evident in Orangutans. Terms are needed to identify individuals and groups, as groups break apart and come together. The **sounds are learned beyond what is natural to the species**. Terms for the same purposes are different for different groups. Orangutans have dialects.[50] Chimpanzees learn to change their calls for various items to those of a newly joined group.[51]

| **The Point:** Language extends from sounds to gestures. Animal language inherently contains past and future. |
|---|

## SOCIAL CONSCIOUSNESS

**Communication is the basis of community**. The language of animals in groups matches the complexity of its group structure. Hunting in groups requires organized communication. It must be decided who does what and how much is needed and when. A "collaborative strategy" is necessary.[52] Hyenas plan which members will go and what particular prey they will hunt.

Social organization affects inter-group behavior. Hyena clans will claim a particular territory, yet will allow other clans to cross as long as they hurry and don't eat. From societies of apes to the choreographed schools of fish, **they all follow rules.**[53]

The complexity of chemical signals of social insects is reflected in the elaborate architectures of termite mounds and ant and bees nests. Lynn Margulis refers to it as "mindful coordination."[54] Consciousness arises from "holistic interaction" of the entire colony.[55] It is the colony not the individual that learns. A colony as a whole can be trained to travel faster.[56]

**The colony is organized by a system of tasks**. Individuals are differentiated by the special tasks they perform. Tasks may involve recognition of an individual. For example, an infected ant may have the fungus licked off by others.[57] But if certain individuals are removed from a bee hive, other bees take over their function. Like proteins in a cell, tasks in the hive or colony are maintained through a change in structural instances, in this case the individual animal. But individual bees themselves may change their role several times as is needed by the hive. Bees retain individual life histories.[58]

Individual roaches may display behavior varying from shy to bold, but they tend to merge into one personality type in a group dynamic.[59] Among many animal groups, individuals retain a particular place.[60] Individuals see others as individuals. If a rat is wriggling in pain, others will imitate the wriggling. Capuchin monkeys seemingly enjoy giving.[61] Some animals establish friendships. They'll have relatively strong or weak relationships with other members. Dolphins, elephants, baboons and horses have all been observed to form long term relationships with non-family members. They'll share with, sacrifice for and grieve over their friend.[62] In fish societies "Individual recognition is the norm."[63] Inter-species relationships have been observed. Groupers and moray eels will cooperate in hunting. A grouper communicates by shaking its body in invitation to the eel. Cleaner fish will clean a "client's" skin better when others are watching. Apparently they establish personal reputations.

Various animals will display genuine grief at the loss of a friend. Crows will mourn their dead in something like a ceremony. They'll stand in silence around the dead body. Some will place sticks around or on the body.[64] The amygdala and hippocampus in animal brains will activate similarly to corresponding areas in humans under similar grief situations. According to most scientists, we can only be sure that they are displaying withdrawal behaviors.[65] But since our brains have similar structures, Kluger asks, if animals experience relationships, why wouldn't they "experience the end of them?"[66] Animals seem to have the ability to feel grief.

**The structure for emotions in the limbic system is shared by all mammals.**[67] Dogs at play and elephants excited to see a friend both behave similarly to humans. Is there evidence that they feel joy? Is there evidence that they don't? Fish think and feel, says Balcombe.[68] Fish ask to be petted. When stroked, their cortisol level goes down, indicating less stress. Sentience allows individuals to have personal "biographies."

> **The Point:** In social organizations, members, like proteins in a cell, function as tasks. Like repeating structures, separate tasks are performed by different members. But individuals are recognized by individuals. An individual can perform different tasks, or roles, during its lifetime.

## CONSCIOUSNESS

**Animal sentience involves inner experience.**[69] Joy and fear are felt. A six-layered isocortex has been established as the area in the brain definitively associated with human consciousness. Structural similarities are found in all tetrapods (land vertebrates including amphibians, reptiles, birds and mammals).[70] But animal consciousness should not necessarily be evaluated based on human sensibilities and ideals. A better term might be "emotional consciousness."[71] Animals feel something.

In humans, the scope of space-time modeling is most likely larger than any other animal. Is there a point along the scale at which subjective experience disappears?[72] The one-celled paramecium seems to hold an "idea," that is, a coordination of messages sent through the cell enabling hairs to act like oars in purposeful movement.[73] Human, as well as all consciousness, is grounded in the first living cell's ability to survive. There is no reason to say that subjective experience does not appear here.

Although scientists do not agree on a definition of consciousness,[74] using the cogito as a guide, what can we be sure of? The attended moment. It is my contention that consciousness = memory + anticipation + binding in categorizing sensations. The recognized or new categorization of an object occurs in the moment between anticipation and recall. Fruit flies can shift attention to specific locations in their field of vision.[75] They can focus on a target. Attention selects a smaller object within a larger background. This is the transcendental distinction that living awareness makes.

**The Point:** All animals have a binding moment of conscious attention in which they focus on a small object within a larger background.

## INTELLIGENCE

**All animals are identical in having a now**. Attention in the moment. What differs is what we could call "intelligence." **Each animal expands space-time awareness to fit its own survival**. The scope of awareness varies with the categories its senses permit. The manipulation of those categories in space-time, plus stringing moments and actions together immediately or intermittently in succession, "a logical sequence of behaviors,"[76] extends prediction to the level of planning. The orange-dotted tusk fish blows water to find a clam beneath the sand, then smashes it on a rock with its head. Orangutans make "travel plans" a day in advance and use language to convey those plans to others.

Another measure of intelligence is **knowing how different objects affect other objects**.[77] An octopus will go out and collect the exact number of rocks it needs for the doorway to its den.[78] The single celled foramifera selects beads of different sizes and colors for its shell. Crows and jays, without a cerebral cortex, use tools better than any animal beside humans. A crow can figure out how to drop stones into a can to raise the water to a drinkable level.[79]

An individual octopus can **discover unique aspects of its environment**. “The animal makes the decision to go out and get information, figures out how to get the information, gathers it, uses it, stores it."[80]

The theory of mind demonstrated by animals expands beyond mere awareness of another's mind to actually **manipulating another's awareness**. For example, a jay will wait to uncover its food cache until other jays have left. Beyond place, tools, self and other, qualities like color, and names for all of these, chimpanzees understand and name abstract concepts like sameness and difference.[81]

Differences in mental skills between humans and other apes are accepted as a matter of degree.[82] But every animal has a mental ability. The amoeba's membrane rhythmically constricts and relaxes affecting the movement of its cytoplasm. When it is near food, the pulses increase, expanding the membrane, allowing the cytoplasm to flow into the area creating a pseudopod to envelop the prey. Near danger, the pulses slow, shrinking the membrane and causing the cytoplasm to retreat.

If given a choice of a number of foods with different ratios of proteins and carbohydrates, the slime mold (a colony of amoebas forming one organism) is observed to pick the most nutritious one every time. It decides. It picks the most economical route from one food source to another. In tests, it has mimicked the efficiently designed railway network of major cities. It remembers where it has been and doesn't repeat. It anticipates and predicts. Studies have shown that it can learn and predict to withdraw before the sensation of cold is encountered. Its rhythmic pulsing creates an internal clock through which sensation and memory communicate. This is its intelligence. Intelligence begins with will. The synchronous timing of the membrane and cytoplasm is its conscious attention.[83] When needed, intelligence/will flows into conscious attention.

**The point:** Sensation and memory combine in ways that best fit the species and its environment. Behaviors similar to various aspects of human intelligence are observed throughout the animal kingdom. Intelligence is a variation within a single species. Intelligence begins with will to make the correct choice. Animals use their awareness to various degrees of effectiveness.

## D: PLANTS

*Never treat a plant like it is an inanimate object.*
*It is just as alive as you are*
www.shift.is/plant-consciousness

### INTRODUCTION

Living awareness contains relatively brief moments of conscious attention, which varies in all animals from humans to amoebas, depending on their organization.

Instead of unconscious awareness punctuated by attention, the life of a plant is one of "ceaseless activity" demonstrating as much purpose as does any animal.[1] Unlike an animal, they are not a subject moving around in a world of interpreted objects. They move by growing into and conforming to their surroundings. They sense that which is directly functional. If animals think what they are doing, then plants do what they are thinking. Their now is coextensive with their lifespan, punctuated by separate, bodily focal points. Their moment is elongated with retained history in a body that continually grows toward a distant future.

A plant can do anything an animal can do, but it does it differently. Animals roam around looking for food and safety. Plants are rooted to the ground. They don't have to roam for energy, sunlight is everywhere. But in order to survive they require an extremely powerful set of senses, more sophisticated even than those of humans and most animals.[2] Their lives are "rich in sensation."[3] **Animals sense objects with qualities. Plants sense qualities.**

Plant awareness is achieved by growing, and their growth via root, branches and stems is modular. Modularity is a strategy to cope. Their sophisticated senses combine with just a few tissue types that easily allow for repeating, modular growth. Imagine reaching by growing your arm. Instead of the entire arm being constantly new, the growing extent is new, as previous extents correspondingly grow slower while appropriately thickening. The appendage's past remains in the current body which grew to this point and grows toward a future.

Branches bend in anticipation of the need for snow to slide off. Roots develop air ducts in anticipation of swampy ground. Aromas are generated with the expectation of attracting beneficial insects. They disperse seeds in ways that use insects and larger animals to aid the seeds in the future.[4]

The conscious awareness of attention requires memory of past experience, experience of something other, sensed through a boundary and anticipation of future experience. A plant contains all three in an extended life-long now. **Unlike an animal, the boundaries of a plants' body includes its own time.**

> **The Point:** Plant awareness is based on growing, not moving. Its own time becomes its boundary.

## SENSATION

Plants exhibit all the senses of animals in addition to their own. And they do so more powerfully. Plants sense at least fifteen distinct environmental qualities directly: Water, six different minerals, soil structure, light, gravity, competition from others, animal herbivores, disease, chemicals to alert others of dangers and those to inhibit growth in others.[5]

Plants use similar proteins as those used by animals to respond to the direct physical stimuli of touch. They react when they are touched, and react differently to different surfaces.[6] General elements of nature like temperature, gravity and wind are detected. They know their left from their right and in the case of twisting vines, clockwise from counter clockwise.

Their sense of smell detects concentrations and combinations of chemicals in the atmosphere and in the soil. Odor receptors recognize patterns in air-borne chemicals. They differentiate the odors of close relatives, the same general species of plant, and different species of both plants and animals. They release odors to communicate messages.[7] For example, they can change their scent to fool mites, and warn other plants to do the same. They can attract other kinds of mites to attack the first kind of mite.[8]

They taste soluble chemicals on their surface that indicate particular kinds of insects. They hear vibrations produced by insects. Roots have been shown to grow toward vibrations of specific frequencies.[9]

The sense of sight, inextricably tied to light, is intimate. They don't sense just light, they sense the colors of the spectrum. But **plants don't actually see colors, they see meaning in colors.** They communicate with their "eyes." They can see and be seen by other plants. Photoreceptors in leaves and needles reflect green but absorb red. Almost infinite degrees of light-to-dark from open sunlight and light green leaves to the concentration of green neighbors in dark shade and dark green are interpreted through ratios of red to far-red light.[10]

Photoreceptors sense not only shades but the level, intensity, direction and duration of light in the vicinity of the plant. Coordinating with red, the absorption of the blue range of the spectrum triggers a cascade through to the other side of the stem, elongating cells there which causes the plant to bend toward the light.

Plants go to sleep at the end of the day when far-red light predominates near dark, and wake up at sunrise when the ratio favors red. But plants don't merely sense separate moments of light, they see changes in the light. Like moving concentrations of chemicals, changes in the spectrum, nature and extent of light provide a temporal context.[11] They know when to bend, to flower and to grow. They know the time of day and the time of the season.[12] **They keep track of time**. They know how long the changes take. Experiments have shown they can tell how long ago the lights were turned off.[13] Through light, their internal clock is tuned to circadian rhythms, even affecting the timing of cell division.[14] Plant vision sees time intervals as a quality.

To be successful, flowering must synchronize with changes in the environment. In observing the changing shades of others, "a plant sees the competition for light, and grows and reproduces accordingly."[15] Seed maturation must be timed to seasonal conditions. Changing angles of light are used to track the movements of competitors. Plants alter their growth based on various periods of both dominating wavelengths and combinations of wavelengths across the whole day.[16] As with animals that hibernate or migrate, the circadian rhythm of a growing plant is intimately timed to the seasons of the world.

A single sense can tell apart different kinds of stimuli.[17] An aroma may lead a vine to a tomato plant, another to wheat.[18] A plant may respond to one stimuli, then ignore it and respond to another.[19] Beyond integration of multiple stimuli of one sense, plants integrate multiple senses into a singular dynamic interaction. They prioritize all these sensations depending on circumstances,[20] responding as if they had a nervous system. From a single leaf, the whole plant responds.[21]

**The Point:** Plants sense meaning directly. Time span itself is sensed in changing light.

## MEMORY

Plants modify current behavior depending both on what is sensed and what has been retained from past sensations. For example, if a plant is placed in a dark box it will still track the sun the same way it had before. It is not reacting to the present, but acting based on past experience.[22] A Venus

Flytrap retains which previous hair was stimulated and how long ago it was stimulated to determine whether the current touched hair indicates the presence of an insect or a rain drop. Integration of past and present information about the soil is used to determine how far from the old stem a new stem must grow.[23]

**Storing a memory from previous experience allows a plant to learn.** Dandelions learn that if they grow too high in the grass they'll be cut off (by what we would know as a lawn mower), so they bloom closer to the ground.[24] Demonstrating the basic type of learning called "habituation,"[25] mimosa plants will fold up when struck by water drops for awhile until they learn that the drops are harmless. After a delay of weeks, when drops resume they still remember not to fold, although they will react to other kinds of stimulation recognized as new.[26] If a tree only receives water at a particular time of year, it will learn in subsequent years to reserve development and growth for that period.[27] Trees can keep track of the number of warm days, or the number of days that stay light for at least thirteen hours, before they bud.[28]

**The Point:** Plants remember and learn from experience.

## ANTICIPATION

Anticipation of future experience is based on the expectation of what has been learned and then recalled from past experience. Because of the ability to sense an electric field, grass, anticipating water, will turn greener before a rain storm arrives.[29] The reach of sensation contacts the expected phenomenon at a distance. In accordance with what is yet to happen, what is anticipated is the consistency of the sensation. One thing follows the other as if it is all one "thing." Association precludes completion.

When plants smell the aroma of a distant harmful beetle, they begin producing defensive chemicals to ward off that which is to come.[30] With the same power as a bacterium, the aroma is analyzed for gradient change. To the plant, this beetle is this increasing gradient of this aroma. Expectation is not only for what is associated but the direction of the association.

When direction of experience itself is learned, a time and place of contact can be predicted. Plants turn to face the sun where it is due to appear.[31] Previous experience of the sun's many tracks across the sky prepares the plant for when and where the next begins. **One expanse of time can be compared to another**. The spans can correlate with the gathering and consumption of nutrients. The amount of carbon converted to starch and stored during a day is according to the requirements of the anticipated night's length.[32]

A plant can also change its own direction by growing toward a stationary goal. It will sense the change in the nutrient soil gradient as it grows. Non-nutrient or toxic substances are noise that disrupts the gradient. For example, a root will grow around a metal pole before it comes close to touching it, moving around and then forward.[33] A vine selects from several possible hosts, sensing the one with the most nutrients. Once it reaches the one it has chosen, it performs a "cost-benefit calculation" anticipating the number of coils required for the amount of nutrients it will obtain.[34]

After the best patch of soil is chosen by the roots, the plant grows leaves of a complimentary number and strength.[35] Anticipation spreads through the plant. The moment a root senses dry soil, the stomata, the openings in the leaves through which chemicals pass, close.[36]

Short-term anticipation gives way to long-term **planning in not merely the growth toward something, but the growth and development of the plant itself in anticipation of the concomitant growth of its surroundings.** The red/far-red ratios in the perception of light inform the plant of the others' future direction. "The position of likely future competitive neighbors is mapped."[37] The complex shading of light is predicted far into the future. The movement of life-time growth isn't caused by simply bumping into things and changing direction. The future situation is prepared for by "configuring its own outline."[38]

A plant decides to act based on an integration of an enormous amount of information about its self and its environment in the remembered past, present and anticipated future. It changes the trajectory of its growth, reallocates resources and emits and receives meaningful chemicals.[39]

| **The Point:** A plant anticipates and plans its own growth. |
|---|

## LANGUAGE AND SOCIAL

The living process, as opposed to material flow, rides on communication. "Signs are the abstract entities of purposeful exchange."[40] Organic chemicals are the passing, impermanent structures of signals interpreted at all levels of life.

The sheer amount of biochemistry in plants is greater than any animal. "Unable to run away, plants deploy a complex molecular vocabulary to signal distress, deter or poison enemies, and recruit animals to perform various services for them."[41] Plants, using molecules as signs, communicate with other plants and animals with a vocabulary of up to three thousand chemicals. Up to one hundred thousand different compounds have been discovered in the root area. As much as twenty different groups of molecules are known to be used for external communication.[42]

Though root connections and airborne chemicals are its principle means, the body of the plant itself is used to communicate. Some plants can alter leaf flavor or texture to become poisonous or indigestible to predators.[43] **Molecules are made and used in specific quantities and places (on their own body) with rhythmic timing.**[44] The complexity and multiple combinations of compounds expand signs into contextual or semantic meaning, producing the equivalent of language.[45]

**Using chemicals, plants are able to discern self from others and whether they are related or not**. From one part of a plant to another, the self's own genome is recognized to prevent self-pollination.[46] Plants communicate and cooperate with and among its own branches to avoid being eaten by insects.[47]

Despite there being more species of all kinds of life in a few inches of forest soil than there are people on Earth,[48] roots will transmit information about the soil inhabitants and level of nutrients only to other selected plants.[49] Trees are especially social. If the soil and light differ from one spot to another, they will share nutrients. Plants and trees in a network warn each other of predators.[50] Plants can detect the chemical signals of insects, if similar to their own.[51]

Again, **because of lack of mobility, timing is paramount**. Hormones are released when they are ready to be sensed. Entire groves of fruit trees will ripen in synchrony.[52] Communicating trees have the same rate of photosynthesis.[53] Timed to when the male insect hatches, flowers will bloom producing a scent that attracts that insect to pollinate.[54]

Growth and development in plants is coordinated through molecular signals, not simply mechanical reactions.[55] The entire life of a plant, both in itself and with its environment is based on the timing of its molecular language. **Plants "communicate through feeling,"** says Paul Lenda.[56] Feeling is body awareness. A plant's body is integrated with its time.

**The Point:** An elaborate language of timed chemical responses enable the plant to recognize and communicate with family, friends and enemies. Mobility through growth integrates its sense of time with its body.

## METABOLISM

In photosynthesis, chlorophyll absorbs energy from sunlight to transform carbon dioxide and water into carbohydrates and oxygen. As long as there is any light, plants keep doing this.[57] Activity in plants is constant.

Sugar from carbohydrates travels from the leaves down to the roots through the "phloem" channel. Water and nutrients from the soil rise up from the roots through hollow tissue called "xylem." Instead of a heart pump, plants use a "molecular pump." The void of evaporation from the leaves, combined with tension caused by cohesion of water molecules and adhesion of those molecules to xylem walls, pulls the liquid up.[58]

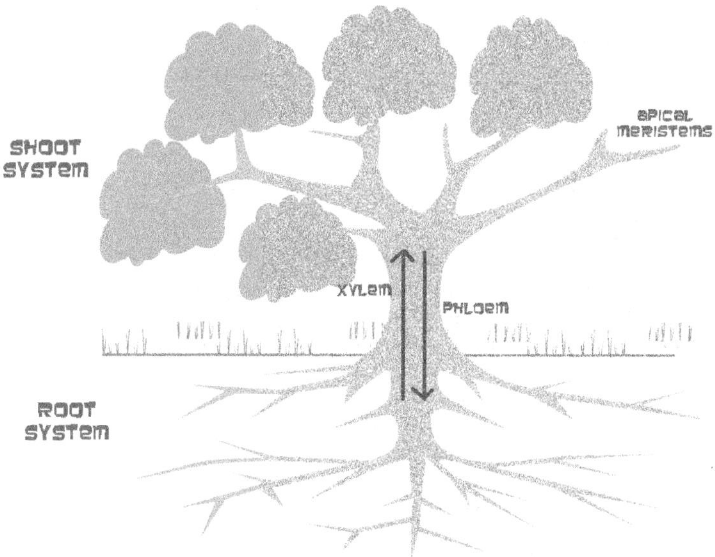


Fig. 47.

But beyond photosynthesis, energy from light surges directly through the entire plant. Not only do plants bend toward the light, but the whole plant responds. It grows toward and according to the amount of light.[59]

Energy flow informs intercellular communication throughout. Different parts of the plant communicate with one another. Changes in resource supply in one area will affect growth in other areas. Large macromolecules move from cell to cell carrying complex signals. The communication is selective. Only certain cells will talk with certain other cells at a given time.[60]

**The entire plant is a "brain-like information-processing system," integrating input with coordinated response.**[61] Electro-chemical signals move through cells in a similar way to nerve cells.[62] Cell-to-cell communication is like that of a neural synapse.[63] Instead of cell division, strands of cytoplasm between cell walls branch like new dendrites.[64] Instead of specialized areas, energy waves flow through molecular networks that are ordered in semi-crystalline arrays.[65] Like neuro-transmitters, memory changes remain, reflecting the past stimulus.[66] Both local and long distance communication is integrated in cytoplasmic signaling. Both current sensations and memories of stress or abundance coordinate current and future actions.[67]

The principle division of communication is between the above ground stem and the below ground root. Leaves constantly keep the roots informed of changing conditions. Roots use this information to plan and direct the orientation and growth of the plant or tree.[68] They send the required nutrients up from the soil.

Though coordinated for the good of the whole plant, intercellular activity is independently organized in the apices (see below) of both root and stem modules.[69] Movement is initiated in these areas through cell division. The development of the plant as a whole is directed by cell creation.

**The Point:** Metabolic flow and information flow are entwined in the developed inter-cellular structure. The key to life-long memory is ingrained planning in the cell division areas at the root and stems.

BRAIN

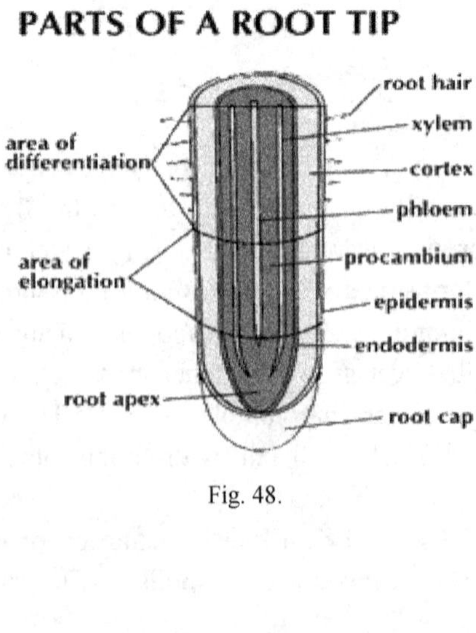


Fig. 48.

Roots both sense the availability of water and nutriments and adjust the rate at which those nutriments are absorbed.[70] Nutrients, water and sugar products are digested in the root, producing up to one hundred thousand different compounds.[71] Besides sending these compounds and water up the plant shoot, roots take the sugar products from photosynthesis and send them to other plants. The root seems in some way to be "in charge."

Since a plant lacks an animal brain, it has long been assumed that a plant has no subjective sensation of the signals to which it responds. Plants were thought, since rooted to the ground, to not need a brain to move about. Yet it was Charles Darwin himself that proposed the "Root-Brain Hypothesis." In *The Power of Movement of Plants*, Darwin writes, "It is hardly an exaggeration to say that the tip of the radicle thus endowed [with sensitivity] and having the power of directing the movements of the adjoining parts, **acts like the brain** of one of the lower animals."[72] A root can change direction, and while moving, seems to be aware of what it wants and the obstacles it must engage.[73] It "receives impressions and directs movements."[74] When what is hypothesized to be brain structures at the tip are removed, the root loses its sense of direction.[75]

Direction, of course, involves learning and memory. In brains, this is accomplished with "neuronal plasticity," where neuronal groups strengthen and weaken their connections depending on experience. Cells throughout the plant generally do the same thing.[76]

Plant cells are shaped like tubes that connect to other cells at each end. The ends are polarized, that is, each cell has an input and output end, similar to an axon and a dendrite.[77] The connections are called "cross-walls," which function like a synapse.[78] They generally align according to the direction of the whole plant, with the input or anterior end facing down, and the posterior end facing up.[79]

In animals, the posterior pole of the body is usually where the sex, excretion and motility organs are. The anterior pole houses the nutrient and sense inputs as well as the nervous center, or brain. It was Darwin's insight that the root represents the anterior end and the shoot the posterior end. In this way, plants are like upside down animals.

In plants, though, the long distance transmission of signals from sensory to motor areas of root to shoot are mirrored throughout. The anterior/posterior polarity of plant synapses, likely contain local sensory and motor functions particularly adapted to plant-type motility[80] (see next section). In addition to synapse like functioning, rhythmic ion oscillations, also associated with learning and memory processes in animals, are present in the cellular cytoplasm.[81]

The tubular cells link in long continuous strands, forming a vascular system that interconnects all parts of the plant. **The vascular system doubles as a nervous system**. Vascular bundles function as nerves, passage ways for rapid communication, between the root-brain and the above ground shoot.[82] The vascular system supports the transmission of action potentials rivaling the speed of animal nerves.[83]

The shoot's many branches and segments each contain apices, which are the centers of environmental sensing and development. The root and each shoot apex are parallel units of a distributed nervous system.[84] "The entire plant is analogous to the brain," says Gareth Cook.[85]

These nerve centers are, appropriately for plants, the areas of cell creation and growth. Undifferentiated embryonic cells, called apical meristems, are found at the tips, i.e., the "root" of each segment. Adjacent to these cells is the area of cell division. One daughter cell remains meristematic and the other differentiates into one of the major tissues types. These are pushed to the next area where they elongate and further differentiate. Lateral, i.e., thickening, growth begins with lateral meristems. Consistent with modular, fractal development, the cell creation rate slows in direct proportion to the number of cells that have differentiated.[86]

Though all apices function similarly, in the root apex the cell division and elongation regions are separated by an area of cells called the "transition zone." They are free to fully focus on processing sensory and developmental signals. This area seems to be brain-like tissue.[87] The oscillatory patterning in these groups is similar to neuronal groups in the animal brain responsible for sensory/motor connections.[88] Phototropism, i.e., orientation toward light, and gravity sense are also unified here.[89] Establishing the primary direction, the root is the most fundamental of all apices. The root synchronizes activity in the rest of the plant.

**The Point:** Cells throughout the developed plant areas have an input and output and connect like neurons. The root and modular apices direct growth through cell creation, and together form a distributed nervous system. The root, being the primary module, coordinates the activity of the whole plant.

## GROWTH

Fig. 49. Modular growth is fractal

**Plants move by growing**. Their awareness is not the same as an animal that moves itself about in an environment. They exhibit the same behavioral capabilities as an animal does but through "phenotypic plasticity,"[90] that is, modular growth. They predict and map the surroundings that their bodies will grow into.

Roots will grow in a direction that interferes with a competitor.[91] Branches will alter their growth to regain a tree's balance. Coordination of cellular growth,[92] position and tissue identity throughout the entire plant reflects anticipated requirements.

Essential to strategic and tactical growth is timing.[93] The plant must know when as well as where. As was mentioned earlier, the ratio of red to far-red in the light spectrum is used to precisely sense and control the growth rate to match the corresponding growth of its environment.[94]

Not only do they move by growing, but they grow differently than an animal does. Their structure is fundamentally different. Animals grow evenly and age consistently throughout the entire body.[95] **The body of a plant is temporally, as well as spatially, modular**. Their life-history is retained in modules - repeatable sections of trunk, branches and stems. Because there are few tissue types, it is easier to repeat a module.[96] Since a plant can't run away, this structure enables it to survive with a loss of up to 90% of its body, leaving only the primary module, the root.

Each module has its own cell production in the meristematic area of the stem. The growth of the shoot as a whole is rhythmically timed to localized modules which, fractal-like, grow in self-similar successive intervals.[97] The patterns are similar on different scales. The overall structure increases linearly while the growing points, the start of each module, increase exponentially.[98] As mentioned in the previous section, the rate of cell creation and tissue differentiation are inversely proportional. Variable relations of growth and growing points allow the plant to precisely complement the changing shading of the environment while and by contributing to its own self-environment.

Modular growth is partially dependent on the root. Cooperation between modules is maintained by a continuous "flux of signals" between the root and the modular shoot.[99] 15% of the carbon that comes down to the root, and 15% of the nitrogen that goes up to the shoot, is reversed. The modularity begins with, and is directed in accordance with the root.

Since the distributed brain is mirrored in the modularity, sensation is distributed in modular growth. Growth leaves a trail in each module and in the unified group. A plant is not a being that moves, but a movement that is what it will be. And what it is now will still be there in its future.

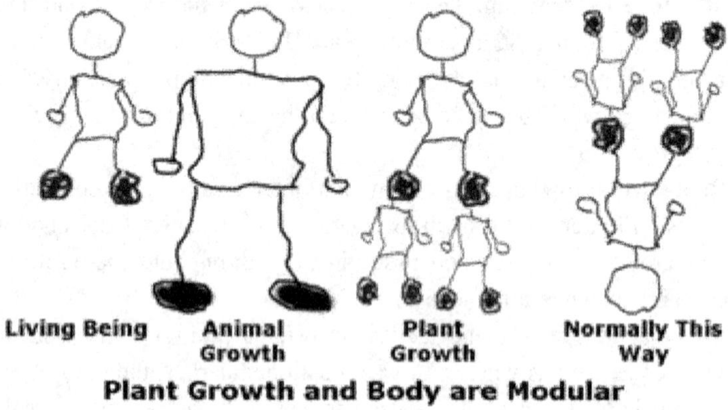


Fig. 50.

It would be as if every time you have to change direction you grow a new knee and shin. And the old knees remain to let you know where you've been. It would not be the same leg with more knees, but a new leg segment. As a plant, your memory is your body shape, which prepares you for what's coming next.

**The Point:** Plants retain their life history in their bodies through modules. They grow into their own future by retaining their past.

## CONSCIOUSNESS

Plants have the same "intelligent behavioral capabilities" as animals, i.e., "sensory perception, information processing, learning, memory, choice, optimization of resource sequestration with minimal outlay, self-recognition and foresight by predictive modeling" as animals. "Plants can effectively process information obtained from their surroundings and show learning behavior which involves goal-seeking, error-assessment, and memory mechanisms. Plants can communicate this information with neighboring plants," - Frantisek Baluska.[100]

From the point of view of a plant, the relationship to both matter and time are essentially different than that of an animal. It is an embodied awareness. It is not a living body that is aware by moving among objects. Its intentional movement is growing to fit, that is, "inhabit" it's environment. Plants don't form images of what light presents, light itself is directly absorbed in its form.[101] **Instead of pictures, it "sees" instructions.**

It's body is plastic, that is, it elongates. It's modularity allows repetition and direction to the form. As such it retains the "places" where it's been within its body structure. **Its sense of place, along with its sense of now, corresponds to its entire life.** Remembered past "nows" are retained in its past growth, the circumstances of which remain in its shape. The anticipation of the future is present in the shape of the past. Metabolism is at one with sensation. In the present, animals and plants trade carbon dioxide for oxygen. But behind them, **animals leave a trail of matter. Plants leave a trail of living matter, i.e., themselves.**

**Plant s are aware by growing**. Their objects of consciousness and their sense of place are their own developmental states. The present sense of now is modularized. Modular states are past and future to the present memory and anticipation of any other module. **Modular growth is fractal, which means self-similarity divides yet retains**.

An initial signal reaches other modules at different developmental times, sometimes separated by days. The attention of the whole plant is an always active coordination of focal points[102] engaging directly with a "surface topology" of the surroundings at every moment.[103] The mutual awareness of modules allows the local tissues to grow in a way that uses resources optimally for both the module and the whole plant. If water is scarce, the roots grow in search. If there is less light, stems will grow more.

The plasticity of growth is proactive. Developed cells are like dendrites forming new connections, while new networks are formed by adding new tissues. Memory becomes stronger with increasing flow rates. It is like a "series of developing brains...acting like parallel processors each with slightly different computational capabilities."[104] The "open-ended morphology" means the plant falls into its own growing future.

Modular apices are integrated at "various levels of attention,"[105] Dispersed focal points are growing while directing growth. The mobility of animals is externally directed toward a stimulus. **Directed toward itself, plant mobility is internal**. Space does not appear to the plant as a chaotic flux to be categorized, but as a network of self-communication. The boundaries of conscious awareness are fundamentally temporal, because spatial boundaries are its own growing.[106] It is not categorizing. It is creating its material self. Interplay among varying centers of attention means the now extends over the lifetime. Stem tips to the whole plant are like birds in a flock, or insects in a colony. The root coordinates, but is not the sole source of growth. Plant awareness is a network of individuals, of scalable, self-maintaining units.[107]

In animals, intelligence and conscious awareness arises in the opening between past sensation and current motion. In plants, the opening is variable among apical foci. As a totality, it continuously extends to the lifetime of the individual.

The enormity of the signals that at any one time are integrated, and the varying environment within which those signals must continually adjust, mean that **each plant is not only an individual to itself, but is distinct among others.** Each plant responds to similar sensations in a "non-similar" way.[108] A particular oak tree will act "out of character" and grow limbs unusually lower to the ground if another tree blocks the sun.

Individuality in a network of individuals extends to the autonomy of modules and cells. The simplicity of cell types allows for autonomy of cell groups. Groups of cell combinations or single cells communicate specifically with other combinations or other single cells in other locations.[109] What is communicated is unique to the participating cells or groups.

All living beings process information by remembering and learning in some manner. Unlike a rock, living beings have a sense of their place in a world.[110] Unlike matter, their movements are not automatic. They recognize patterns and make decisions.[111] If Intelligence is a property of all life,[112] it is certainly a property of plants.

Unlike animals, however, there is no binding moment of awareness. Because they move through growth, their sense of now is a continuum over their lifespan. If animals are beings that do, plants are doings that are, were and will be. **Animals build things. Plants build themselves.** Their environment is contiguous with their growth. If they don't think what they are doing, they do what they are thinking.

Their vascular system doubles as their nervous system. They don't eat the quantity of things. They eat the quality of sensation. They eat light.

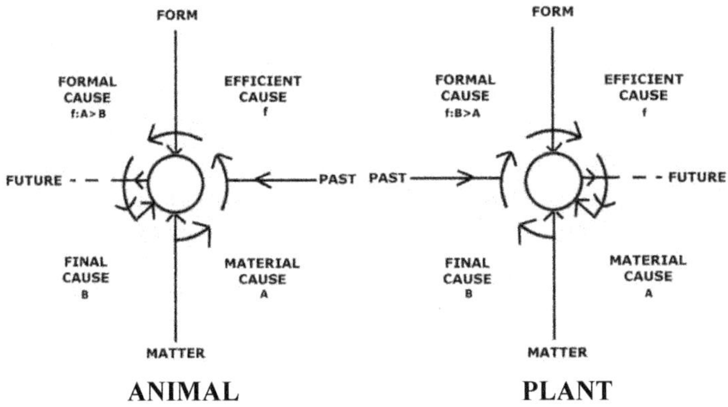


Fig. 51.

Their relationship to matter is different than that of animals. Animals and plants reverse the exchange of carbon dioxide and oxygen in respiration. But animals also leave a trail of waste, dead matter. Plants leave a trail of living matter. Leaves fall and dead branches remain within their immediate and life-long environment. But in their own body, plants leave a trail of themselves. Because they eat light, sensation and food energy are integrated. This means both **their metabolism and their sense of awareness are reversed in their relationship to time**. Animal consciousness spins in place as it projects its moving place in the world. Plants are where they've gone. Their sense of place moves with them. Because purpose, or final cause, begins in their own past, its purpose is determinate, but self-determinate. Because matter is flowing through form, their own matter is their future cause. Animal awareness is form flowing through matter sensing matter with shapes. **Plant awareness is matter flowing through form sensing its own shape**.

**The Point:** Plant awareness is directed through their own retained past into their future growth, thereby extending their sense of the present to their lifetime. It looks at the future as if it were the past.

E: FUNGI

*Mycelium is the neurological network of nature*
Paul Stamets[1]

ANATOMY

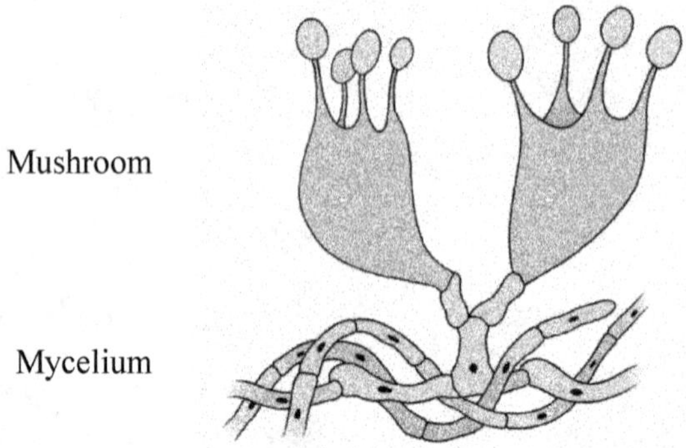


Fig. 52.

Fungi belong to their own Kingdom separate from, and in a significant way between, the Animal and Plant Kingdoms. Generally, the two fundamental sections of a fungus are the mushroom above ground and the mycelium below ground.

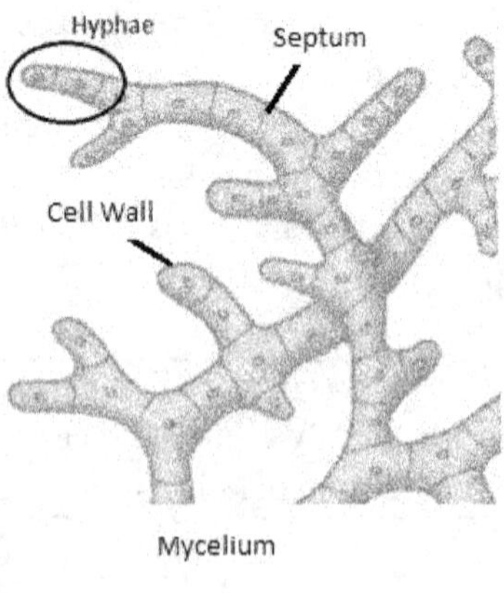


Fig. 53.

The mycelium, which is a mass of simple structures called hyphae, is the actual body of the fungus. The mushroom is the sex organ, appearing only intermittently. Hyphae are long threads of fungal cells strung end to end. Each cell is connected by an end wall called a septum, which is perforated with holes through which nutrients and cytoplasm flow. The fungus grows as new cells are added both at the end and in branches, forming a network of single cell-wide long tubes. The cell walls are made of chitlin, which is the same material that forms the rigid exoskeletons of insects.[2]

**The Fungus travels by hyphae growth** up to several inches a day, **weaving a wide mycelium network**. The chitlin walls enables it to move through rough soil which it breaks down as food. It can hold up to 30 times its mass. A single cubic inch of soil can contain up to eight miles of mycelium threads.[3] The largest organism known is a 2200 year old honey mushroom mycelium in Oregon that covers 2400 acres.[4]

When over-filled with nutrients, or conditions threatening survival occur, e.g., when the dark, damp areas without air movement that they normally prefer, changes, they produce **an above ground fruit body, the mushroom.** The mushroom produces spores which are released in the air to sexually reproduce elsewhere. Fungi also reproduce asexually. In a process called fragmentation or fractioning, new cells split in two, creating a separate fungus. In this case, the fungi increase the population of their local niche.[5]

**The Point:** The entire organism is a network of identical cells living underground, except for the sexual organ, the mushroom.

## PART ANIMAL PART PLANT

"Fungi are the grand recyclers of our planet," says Stamets.[6] **Plants create bio-matter. Fungi break it down**. With 90% of plants, but trees especially, the process is symbiotic. There are not enough minerals in the forest soil for any one tree alone. Because fungi **don't have chlorophyll** they cannot manufacture their own sugar. So they trade.[7]

Trees communicate the need for nutrients by releasing sugar through their roots to the entangled mycelium. The mycelium, reaching farther than any root, finds and poisons tiny organisms, causing them to release nitrogen. The mycelium converts the nitrogen into mostly phosphorus which is passed back to the tree root.[8] Fungi will pass nutrients from one tree to another depending on need. For instance, in the winter, conifers will produce nutrients that are passed through mycelium to aspens. The direction is reversed in the summer. Information about insect attacks as well as carbon, nitrogen and water exchanges, create vast networks of mycelium connected trees. "Mother" trees have been observed to use the network to nourish their seedlings.[9]

Fungi were the first organisms to come on land 1.3 billion years ago. **Fungi can eat rocks**. They produce enzymes from minerals. Plants were able to follow 700 million years ago only by partnering with fungi.[10] But the evolutionary branch of fungi are actually closer to animals. Animals first split from fungi 9 million years after plants did.[11]

Unlike plants, which take carbon dioxide and nitrogen from the atmosphere, **fungi and animals get carbon from organic compounds**. Like animals, fungi store sugar as glycogen rather than starch as do plants. Unlike most animals, fungi digest before they ingest. The hyphae send enzymes into the soil outside to convert larger bio-molecules found there into smaller forms which are then absorbed through the vast cell walls of the mycelium.[12]

Paul Stamets says, "We chose the route of going over ground, encircling nutrients and creating stomachs. The mycelium chose the route of going underground and externally digesting its nutrients."[13] They do, however, create a kind of primitive stomach by surrounding their prey with "netlike cell chains."[14] They "surround" by growing. Their growth is not modular like a plant, each cell is the same.

**The Point:** Fungi travel by growth like a plant, but are not confined by root in one place. Like animals, they get carbon from organic molecules. They grow around their food, digesting before ingesting.

## INTELLIGENCE

Some types of slime molds are able to choose the shortest route between two food sources.[15] They expand in two directions simultaneously until they can compute and remember the distance for next time.[16] Those slime molds, however, are not fungi. They are a form of amoeba. They can move or reach and withdraw their whole animal body. And when they capture food, they ingest before they digest.

Fungi are different. As mycelia grow across and beneath the landscape, they build "food webs," that is, they form a mat of tiny strings that support microbial life as well as provide food for themselves. Just as fungi paved the way on land for both plants and animals, they continue to be the "leading edge organisms in nature." **They are the "interface" between plants**

**and animals,**[17] "in constant dialogue" reacting to, directing and recycling nutrients through the food chain.[18]

**Mycelium webs are one continuous body**, miles in length but one cell thick. Having no organs or differentiating tissue, **they are essentially a membrane with cytoplasm**,[19] although each cell contains a nucleus or two. The membrane-like cell wall directly confronts and must negotiate with both its own prey and organisms which prey upon it. As the environment changes throughout and beyond the maze, the intercellular nutrient levels constantly keep pace. Mycelia are aware of and communicate their surroundings with all cellular sections while being at the same time aware of their capacity and to adjust accordingly. A mycelium is an **"exposed sentient membrane."**[20]

According to Stamets, the entire mycelium is aware. "As you leave your footsteps, the mycelium reaches up and responds by grabbing newly available broken twigs or sticks." They sense any touch. The total membrane system shares information forming a "neurological network," which "mimics that of the computer internet."[21] Mycelium has evolved an optimal pattern resembling neurological topography.[22] In fact, the mycelium functions at a level of complexity and computational power comparable to a supercomputer.[23] The nodes, however, are not different kinds of cells. They are not so much modules as they are bricks or blocks, yet different areas chemically react differently to different input, but do so in a coordinated way.

These intelligent networks are capable of transmitting an enormous amount of information for themselves and for many organisms across the environment.[24] The intelligence is based on the sheer amount of bits or nodes distributed, with each bit contributing "a specific function of thinking."[25]

Information is transmitted across the vast network by neurotransmitters of the same type used in an animal brain.[26] Fungi language transmission, however, is entirely chemical based. Without electrical amplification, chemical signals are passed very slowly. In a large mycelium brain, a thought may take days to form.[27]

The third essential part of life after memory and production is the boundary between self and other, between inside and out. Mycelium are all boundary. With only one cell type, the fungus/brain is organized by pure position, both generally and locally. Unlike the more specialized nucleic tissue in animals and apices in plants, position determines function. By being a neurologically designed network without nodal structures, the mycelium is **an externalized mind**. Stamets says, "like salmon take advantage of the tides, mycelium exploits the natural momentum of matter."[28] Fungi are the "common bridge" of nature.[29]

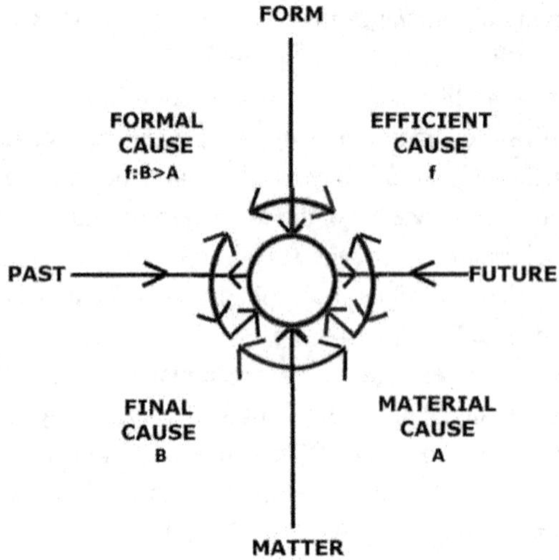


Fig. 54.

Unlike a plant, a mycelium does not conform to its environment, it makes and eats its environment, yet moves by growing, and so leaves both an environment and itself. What grows is a brain. It eats with its brain. Metabolism and sensation are one. It is not form applied to matter, nor matter applied to form. **Form and matter are not flowing past one another. They flow into one another.** Unlike the many apices of a plant, there are no control centers. Conscious attention is determined by relative position within the organism. Attention is both immediate and bodily extensive.

**The Point:** Living awareness in mycelium is immediate, and one and the same with material survival. The structure of consciousness is without cell types. It operates purely by relative position. It looks at the past and the future as if they were the present.

Life and awareness requires three elements: Memory, Procedure and a Boundary. Animals emphasize procedure, Plants memory and Fungi boundary.

## F . ARTIFICIAL INTELLIGENCE

*Any sufficiently advanced technology is indistinguishable from magic*
Arthur C. Clarke

### ARTIFICIAL INTELLIGENCE

Why is it called artificial INTELLIGENCE? Why not artificial consciousness? Artificial life? For material science that reduces life and awareness or consciousness to matter, artificial has no meaning. Of course, if brains are just particles, then particles can possibly be arranged better than human brains are for the purpose of computing.[1] Computers do indeed compute faster than humans. If we call computing "intelligence," then artificial intelligence means, at least potentially, superior intelligence. But do computers have superior awareness? Are they more alive?

Can a computer think? Can it know or feel it is thinking? Can a computer feel anything? Is it capable of conscious experience? Can a material object become conscious?[2] If I have said anything in this book, it is that **matter, in order to be conscious, must first be alive.** There is no way to explain a conscious rock. A living being is capable of coming to attention in its own now, that is, experiencing in the present. This is what it feels like to be alive.

What comes to attention is what is categorized in the environment as important for the survival of a living being's structure. But living awareness is more than interaction with the immediate object. Experience is more than sensation. Humans bring in an extra large breadth of knowledge and previous experience to any present. We feel right or wrong actions. We can't prove Godol's incompleteness theorem, but we intuit an understanding of it. We feel free to choose. We think in layered patterns of patterns. Temporal relations are a given. The before and after exists within the living present. Spatial appearance, including the appearance of mechanism, is a projection by the living structure. Consciousness contains unconsciousness.

The fictional zombie is a totally unconscious living being. Their living actions are not conscious. If consciousness, however, is only matter, then the word "zombie" is meaningless.[3] On the other hand, if we can imagine a zombie to be real, then we can certainly imagine non-zombie human consciousness to be real.

**The Point:** Why is the term "artificial intelligence" and not "artificial life," or "artificial consciousness?" The term artificial life is meaningless. Consciousness is alive.

## LIMITS TO A.I.

We generally assume that a computer can manipulate symbolized facts, but the understanding of what it is manipulating lies completely in the mind of the programmer. The computer follows the programmer's instructions. Deciding on instructions for what end, requires imagination and improvisation, as well as choice and a sense of purpose.[4] For example, any child understands the phrase, "I am a dog" by pretending to be a dog. This requires understanding what "I" means, and what a dog is, and creatively combining them.[5] A robot can be instructed to recognize a dog image and taught to then pick up a bone. It is all together a different thing **to be able to generalize a concept**. A robot can "use" something. Humans understand what "use" is in general. They understand expanding the concept, e.g., "leverage," as in leveraging knowledge for further application. A.I. can manipulate previously programmed facts, but can't apply strategies from one kind of task to another kind.[6] A computer can't take an expertise at chess and apply it to football. It would need imagination. Unlike a computer, a human mind can grasp the unknown boundary.

The ideal is called A.G.I.- Artificial General Intelligence. If computers are eventually able to solve problems in every domain of human enterprise, it is theorized, one of those domains would be the most general of all domains, the domain of domains, artificial intelligence itself. Computers would obtain the sense of self by knowing what self is.[7] On the other hand, if it knew everything, it would know infinite detail. Infinite bits of information would still require understanding the difference between signal and noise. **An A.G.I. consciousness would have no sense of limit.**[8] It would be without sensation. The feeling of experience is grounded in both smooth, analogue continuum and digital parts. Computers are digital.

**The Point:** Unlike a computer that cannot generalize beyond boundaries, human consciousness is open-ended manipulation of limits. A conscious computer would generalize without limit.

## ARTIFICIAL NEURAL NETWORK

If it takes a brain to be conscious, perhaps it follows that computer consciousness would require mimicking the brain. Simply reducing the brain to a map of neurons, though, doesn't explain consciousness. Being conscious is the feeling of being conscious. It is what we mean by experience. A piano is not the music. A brain is not the experience.[9]

If it does take a brain to be conscious, however, it follows that consciousness at least requires reproducing the circuitry of the brain.[10] We can imagine digitizing neurons, but what about the synaptical connections? The spike strength, timing between spikes, positioning and arrangement, cannot be reduced to a discrete value. The brain is both discrete and continuous. An IBM supercomputer was programmed in an attempt to simulate 10 billion neurons, but ran 1500 times more slowly while requiring thousands more watts of power than a human brain, to reach the same solution.[11] **Something more than brute computational force is needed.**

It needs to be dynamic. It needs to be wired like synapses. A conscious computer needs to be always open to new information. It needs to adjust to changing signals. Like a biological nervous system, to dynamically learn, it must teach itself.[12] It cannot make its own predictions by being given distinct instructions. It must have the ability to discover, isolate and classify distinct patterns and make predictions based on those patterns. It must be able to rewire itself to reflect the changing patterns in the data.[13]

In a neural network that analyzes visual data, for example, the program starts with the simplest pattern, something or nothing, the purest edge. From numerous edges, it recognizes a circle. From numerous circles it recognizes a face. From numerous faces it recognizes a cat. At each level, millions of "voters" vote on the correlation in the data. Each layer looks for patterns in the previous layer. Connections are weighted according to previous correlations, "programming the next level like information altering thought."[14] Instead of deductive logic that manipulates symbols according to the formal rules of the program, it uses statistical inference. It is inferring patterns from specific instances, in other words, learning. But how does a computer know what a cat is? Or why there are faces?

A neural network scanned 10 million images to learn to recognize a cat,[15] but not a litter box. Neural networks are not yet A.G.I. The problem is - what determines the layer that the patterns represent? The initial data input depends on the programmer. Using statistical programming, however, A.I. is not only faster, but can identify patterns that a human could not.[16] It can identify what humans could not. But identifying is not learning to identify. If, however, A.I. can teach itself to survive, perhaps humans will be able to compete only by merging with computers.[17]

> **The Point:** Simulating the human brain requires purposeful organization. An artificial neural network can generalize an image statistically from variations in patterns. But the object to be recognized is supplied by the programmer.

## SIMULATION

Brain scanning devises can correlate images of brain waves with specific thoughts about a specific action. Algorithms can then be written to translate the brain waves to movements.[18] A monkey, by having his brain waves interpreted, can move an avatar on the screen by moving his own body.[19] If we think of a letter, a machine can type it on a screen.[20] With A.I., when the thought is imagined, and the task willed, people are able to move paralyzed limbs through a mechanical devise.

On the receiving end, electrical activity in the brain correlates with input, e.g., hearing a word.[21] One person could theoretically control what another hears. According to Michio Kaku, we will be able to transmit and insert "words, thoughts, memories and experiences" into the brain of another.[22] If a person can move his own legs by imagining it, he can control someone else's legs. But would the second person visualize the image as his own or would it feel like being pushed by some external force, against his will? Imagining a task includes willing the thought in the first place. The feeling of willing the thought would have to be transmitted. "Words, thoughts and memories" are not experiences.

Living beings communicate thoughts through a medium. An experience is what is felt. The concept of simulation theorizes that the inner sense of awareness can be simulated.[23] Does this mean you send me a word and I hear myself imagining it, or do I hear what sounds like an unwilled voice or thought in my head? If, as Steven Pinker claims, that every aspect of consciousness is tied to the brain,[24] and consciousness is just an arrangement of atoms, then it should be possible to transmit any experience.[25] The question is whether the receiver experiences it as his own thoughts and not that of the sender. Of course, even if it is experienced as one's own, is it an exact replica of the sender's experience or an interpretation? Is experience copied or shared?

Machines can translate. Voice recognition interprets a sentence in one language, translates it to another language, and vocalizes the sentence in the second language, ostensibly understood by the second person. Languages, however, are not simply a list of words. Meanings differ considerably. "Minister of agriculture," might be translated as "priest of farming."

Humans understand one another through meaningful stories.[26] Even two people speaking the same language is no guarantee of precise translation. Language is tied to rational thinking or description. The feeling of experience is nonverbal. Every person categorizes his/her own memories and observations.[27] Consciousness includes the unconscious.

Ironically, if it were possible to make experience transferable, this would make the subjective objective and thereby prove the reality of the subjective. If consciousness is transferable, then consciousness exists apart from the body, and is objective in its own right.

Some people hope to live forever by copying every memory, every story, every experience they have had to a computer or robot which would, using A.I., engage with new experiences in the same way that the person would have, had he/she lived. Others would continue to interact with the robot as if the original person were still alive. Is that person now in the machine or is that machine another person? **A copy is not an identity**. It is a "semblance of our selves."[28] Awareness, on the other hand, is the binding of self. A unity. A living being is an individual, objectively similar to other individuals, but always living in its own time. A self is a feeling of belonging to one's thoughts.

Imagine this scenario someday in the future: One-by-one we enter into robots. Each robot behaves just like the person who went in. The robot says, "I'm in here. I'm aware. This is great, come join us." After awhile every human on Earth enters into a robot, to live forever, forever looking and acting as if they were conscious. Who would know?

Another scenario has the internet evolving into the collective consciousness of a "brain-net."[29] We will go into computers by intimately connecting to the internet. Handheld devises will get smaller and completely mobile, eventually becoming brain implants. Total input would be everything anybody has ever observed or imagined and is now observing and imagining. We think instructions. We see the screen in our minds. Put this together with virtual reality, and we can go to Paris just by thinking it. In virtual reality, there is no real space or time. Reality would be constructed out of data.[30] If this is better than "real" life, and everybody in the world is doing it, how do we eat and survive? Like fungus, can we eat what we're thinking?

An intelligent internet would by necessity have the goal of survival.[31] Three-dimensional printers might provide food, but from where? Every person in the world would sense a universal world of infinite possibility. Billions of individuals choosing whatever they can imagine from a pool of billions of other's thoughts enhanced by computer power. Collective consciousness does not necessarily mean collective intelligence.

**The Point:** Part and wholes can be copied, but an individual cannot be.

## CONSCIOUSNESS VS. INTELLIGENCE

Computers have gone from merely being programmed to the ability to learn from their own mistakes. The data input, however, is pre-filtered by programmers. Intelligence must reduce the complexity of the world and effectively apply what it then learns to the world.[32] The higher the intelligence, the greater the complexity turned into information. Computers calculate faster, that is, they analyze more complex detail in a shorter period of time than humans could ever do.

Consciousness may directly experience qualia, but a computer can describe qualia in greater depth than any human. Do humans really understand the color red?[33] In the time it takes us to focus on the color red, a computer can describe everything about light and color and red in particular. Humans experience more than what they directly sense. Computers compute everything directly, even their abstractions. Humans experience the descriptions.

Contrary to popular opinion, Alan Turing (1912-1954), creator of the "the Turing Test," tested for intelligence, not consciousness. In fact, computers will probably pass the Turing test before the biological aspect of human intelligence is generally realized.[34] "Humans are nearing one of the most significant moments in our entire history... the point at which intelligence escapes the constraints of biology," - Huw Price.[35] If all human behavior can be duplicated by a non-living machine, then biology might have reached its maximum intelligence.[36] If life at that point falls behind, is that also the beginning of robotic awareness?[37] If computers become more intelligent than humans, do they at some point become consciously aware?[38]

When technology changes the environment faster than humans or any living being can adapt, life becomes irrelevant. If computers are more intelligent than humans, perhaps they could control the climate in a way that was more beneficial to the environment.[39] Unfortunately, however, a superior intelligence would have a concept of beneficial beyond our understanding. It might not be beneficial to humans or any living being. It might not be beneficial at all. Faster doesn't mean better, it simply means operating without human constraints. We would not understand the reasoning.[40]

A conscious computer would self-program. Self-programming would mean learning about a world that it itself creates. Where would it begin? With an edge? An edge of what? Non-edge. Perhaps Being and Nothing. What would a computer make of the blur? Computer scientists are discovering that computers already talk to each other in a language the scientists don't understand. How could they?

If a computer reasons, does it think? Since non-biological "thinking" is more complex, some believe it has a greater individuality than a human does.[41] But if it can reason through data almost instantaneously, why should it? What is a computer's motivation? Intelligence is measured by the ability to survive in changing circumstances. Why would a computer want to survive? Would it keep on learning for the sake of learning? **Consciousness has a conscience**. It has a sense of what it should do. The purpose of a living being is to continue living. Life is self-energizing, not just self-programming.

Organisms evolved in a mechanistic universe to make mechanisms. I have to wonder - could robots have directly evolved without humans? What is the Theory of Artificial Evolution? Robotic evolution from simple life forms via genetic engineering seems to be more like breeding, and at least, requires human intervention. Biological awareness is the basis of human intelligence. Machine intelligence is the basis of a robotic awareness. Superior robotic intelligence is beyond human comprehension. Intelligence with no purpose is meaningless. Computers are intelligent. Living beings are willfully aware.

**The point:** Living awareness wonders why.

## V. LIVING AWARENESS

### A. TIME

*There is neither time nor motion without life*
Robert Lanza[1]

#### CONTRADICTION

It is often asked, "How did life come from matter?" Since the question is raised by a living being, we might just as well ask, "How did matter come from life?" Matter is in motion. Motion takes time. Where did time come from? Each of these questions cannot be answered alone. They depend on one another.

At the center of everything is a contradiction. As Immanuel Kant famously showed, the world at its core is either made up of indivisible substances or infinitely divisible substances. Indivisible substances are mutually exclusive and go on, one after the other, up or out into what we might today think of as the infinite expanse. Infinitely divisible substances divide into smaller and smaller parts reaching down and into the infinitely small. The large and the small encompass one another. The large is always smaller than something larger. The small is always larger than something smaller. The large, though, is the container of the small. In other words, the whole is all that is not its individual parts. Between and among as well as beyond the smallest imaginable parts is the largest imaginable whole.

All that is given is an infinity of small parts in an infinitely large expanse, both eternally implying one another. It is an impossible choice as to which one is fundamental. Going from one to the other and back constitutes an infinite amount of impossible choices. It is the fundamental contradiction. Changing from one choice to the other is without order, and in no direction, and in all directions. Going back and forth in infinite directions is the blur of pure change.

Life is what chose a direction. Life chose to live. The direction is forward. In choosing a direction, life created time. **Time is change with direction.** Simple forwardness does not include anything in front of it.

With the choice to live and the direction in which to live life, life and time are differentiated from the blur. A living being's environment consists of what it selects to survive. But the first selection is itself. The first category of distinction from the blur is form. through which living beings see the blur as meaningful things. Human beings eventually see and understand this blur as filled with distinct categories. The blur is what is categorized. It is called matter. In choosing a direction, Life chose itself in freeing form from matter. Life is the maintaining of whole form through a change in matter parts.

Selecting different parts from the surrounding whole creates categories, which with time, are recognized. The forward direction of time points toward a future which always does not exist. Life makes distinctions while pointed forward toward nothing. The matter from which selections are made is already behind. **Life looks forward into the past. Awareness is looking at the past as if it were the future.**

From single celled organisms to human beings, living awareness selects sustenance out of a world already selected out of the blur of contradictory change. Selections from matter sustain from behind. Life stays one step ahead of matter by continually holding some and then letting it go. Life is able to hold matter in the first place by falling into or being pulled by the nothingness in front. Life stays ahead of matter and ahead of itself. Awareness lies in the ongoing gap between form and matter.

**"All observations contain this inherent paradox."**[2]

Everything both envelops and is enveloped. Small and large, part and whole, are turned inside out in the time of living awareness. What is given to any living being is what is experienced at any moment. The point and the continuum rotate inside and outside. The inside is a point of time never catching up to itself, aware of lasting, remembered selections in the expanse outside. The continual when chases the divisible where. In awareness, motion and position are necessarily compliments.

**"The paradox at the heart of any and all knowledge of the world: There can be no such thing as a pure observation, one free of the changes imposed by time."**[3]

We can be conscious of consciousness past, but not the undercurrent of awareness, the moment of conscious attention, which is a binding of form and matter, as categories are applied to recognized material input. Nonrecognition either expands categories or breaks them down.

Consciousness is always OF something. It is a relation. But it is not a relation of two things, two ends of a relation. It is a relation with an object at one end. Consciousness (conscious and unconscious awareness) has only a direction. It is a point of view. The view is toward the future, but since the future is empty, the content can only be the past. By sensing what has passed as if it were the future, conscious awareness creates a space-time object of experience. Since the future is not-yet and the past exists only in memory and the present is a point of time, awareness is the dimension of inward-outward, not four-dimensional space-time. Life opens out to the not-yet. The present is a substance-less window.

## CIRCULARITY

Life looked at from the outside, that is, objectively, matter passes through form. Matter pushes the living being. From the inside, that is, subjectively, form passes through matter. The subject pushes its formed categories onto matter. In that opening between form and matter, living awareness spins in a continuing moment.

The brain is not conscious. The brain is an object of consciousness. Consciousness observes a brain, including its own brain, from the outside. From the outside we see the same living body moving. Matter and form appear to move together. If the body were matter like a rock, this would be true. But this is a shape whose contents move through it faster than the shape can change. Our experience is of ourselves moving through and toward matter. In the living body and in the living body's awareness, form remains the same through a change in matter.

Consciousness observes or views its brain in the way it understands. The brain is organized the way consciousness thinks. Consciousness sees a brain and its body in the way its brain and body construct consciousness. The world to human consciousness, a world which includes the brain and body, is organized the way the aware brain and body organize it. **Consciousness can only see its own organization.**

**"You cannot know what it is like to be a bat...finding your way through darkness by talking to it in tones that reply to you in the shape of the world."**[4]

All living beings can only be aware of what their living structure allows. This circularity is why physics is only part of the picture. Speculative reflection is awareness of form over matter. Physics is awareness of matter over form.

**"The universe imposes constraints on the laws by which it is governed."**[5]

Matter forever lies behind the logic that can be made of it alone. Since it is living beings, including scientists, that make sense of it, either...

**"The existence of life, forces the physical parameters of the universe to be defined in a very precise way."[6]**

or

**"The universe is compatible with life...the universe is calibrated to make life possible."[7]**

Studying the studier is circular. Biology applies the science of matter, that is, physics and chemistry, to the science of life. Living awareness transforms biology into epistemology by making distinctions. Human awareness transforms epistemology into ontology by generalizing distinction as such. Physics transforms the concept of distinction into the science of matter.

**"Quantum wholeness is closer to the organized unity of functioning of parts of a living being than parts of a machine."[8]**

No matter how far we go into the material, into the physics of the small or the physics of the large, it is life, the organized unity, that knows it that way. It is the certainty of individual self-communication that sees itself mirrored in the certainty of the individual entangled particle.

**"Our reality depends on our biology."[9]**

Consciousness evolved to the stage of constructing the evolution of consciousness. Evolution evolved. **The universe can only be the way living awareness understands it**. The purpose of life is to survive to live. The direction of survival is time. The theoretical quantum/space-time unification of the point and the continuum is what life continually unifies. Life turns disorder into order, thermodynamics into information.

## LIVING DURATION AND CLOCK TIME

**"The 'physical present,'...is a mathematical abstraction of infinitely short duration and nothing happens in it. By contrast, the "subjective present" is arguably the carrier and container of conscious life, and everything that ever happens to us happens in it."[10]**

Anything less than Planck time is meaningless according to physics. Yet we live in and through it constantly and always.

Velocity is the distance matter moves or extends in a period of time. The distance can be the same and the time shorter or longer, or the time can remain the same and the distance shorter or longer. Speed is the average velocity. Motion is measured by space and time. What we call **clock-time is repeated motion over an identical distance at an identical velocity.** Clock-time varies according to the motion speed and spatial distance of the observer relative to the clock. How fast does time itself change? Or does it? The measure of time by itself is in the observer. This is pure time, "pure duration" as the philosopher Henri Bergson called it.[11]

Time, unlike clock-time motion in space, that is, space-time, has only one direction. One direction has one end-point. It is one point forward. If time is change with direction, then there is still change. What is observed is change. One thing cannot change. A whole thing has parts. If a thing changes, says Roberto Unger, its parts change relative to one another and to the whole, and must therefore, be distinct.[12] Infinitely small yet distinct parts change relative to an infinity of other distinct parts in an infinitely large whole. In order for there to be observation, duration pulls distinction out of a continuum.

In duration, distinctions occur in an order of succession. Order within pure succession necessarily implies something outside of pure succession, which we call "space."[13] Sameness is distinction relative to other parts in the same whole. Inside pure duration, consciousness can only be of consciousness, which is an infinite regress, never catching up to itself. Because there is no object, the infinite regress is infinite internality, which is in opposition to the infinite externality of the spatial projection. Consciousness is chasing the continuum into itself in opposition to the spatial distinctions which go on and on. This opposition projects space-time. Duration is placed in matter.

Though pointed in one direction, life is aware of two directions, the past and the future. The 2nd Law of Thermodynamics, the so-called "arrow of time," is one direction from order to disorder. Awareness sees time in matter this way. Information, on the other hand, is directed from disorder to order. Awareness sees matter in time this way.

A living being itself is the structure of duration. The parts and wholes are relative to one another. But in the crossroad of form and matter, the relationship is pure time, the whole is made up of timings. The timing of tasks depends on the whole. In the moment of conscious attention, the timing of the whole almost catches up to itself. There is a binding of form and matter, of durational forming and space-time input.

**"Understanding of time is limited. There is only now. There is before and after but past and future are cut off."**[14]

Beyond the current moment, whatever happened is open to interpretation. Memory is a collection of past moments of both conscious attention and unconscious awareness. We can bring into current conscious attention past moments in which we were not attentive. Our memory may or may not be completely accurate. We can never know for sure.

**"Global time conflicts with the relativity of simultaneity."**[15]

Non-simultaneous localities are the essence of matter. The non-locality of phenomenal simultaneity is the essence of awareness. Space-time distinction is relative. Awareness of sameness in distinction is absolute.

**"Simplicity is universal."**[16]

Complex matter-time structures are relative to one another and in their parts. The whole that connects itself in categorization of input is absolute. Binding is not relative to anything. The objects that it binds are.

**" Every moment of time, and every place at every moment, is uniquely distinguishable from any other. No moment ever repeats. Looked at in enough detail, every event in the universe is unique."**[17]

The material world is subject to infinite breakdown, parts without end. The present moment is the pure time point. It is both a continuum and discrete. It has both continuous motion and time-positional discreteness, neither of which can be observed together. Time both moves and is stationary. It is the split which opens to a selection from an infinitely escaping object by an ever chasing point of view.

**"Consciousness...knows neither beginning nor end."**[18]

Consciousness, the observing experience, can never reach the precise edge of anything. The limit is the unreachable point at which space and time intersect. Normal experience, as well as scientific measurement, stops at the border of practicality or usefulness.

**" The material point...remains in an eternal present."**[19]

Matter in itself has no time.

**"The present moment is inseparable from life."**[20]

Space-time points are relative to the point of view. The present moment of duration is absolute. Space-time is projected from the continual duration and divisible point. External space-time produces the illusion of time moving slow or fast. Duration only points forward and outward. Duration is spinning in place as form and matter swirl around. Observed from the outside, matter is behind form, e.g., metabolism. From the inside, form fits onto matter, creating a universe of information going forward into the past as metabolically thermodynamic matter of order to disorder flows from the past to the future. From the outside, life metabolizes through procedure, memory and boundary. From the inside, life is aware through future, past and present. Time both moves and is stationary in the contradiction of the present. In the opening of form and matter, living awareness forms matter.

## ONTOLOGICAL INDETERMINISM

**"Living beings make predictions because the world is consistent, but determination is unknowable."**[21]

If the future is determined by the past, then the future already exists. It is a "block" universe. But the future does not exist. Awareness predicts from the past. The past conforms to the structure that survived. The present, the center of internal duration, is not part of external clock time. Events in clock time are predictable. Prediction takes place in the present.

The location of each separate point in space-time is uncertain relative to position in space and motion in time. They are either somewhere in space but relatively nowhere in time, or somewhere in time, but relatively nowhere in space. The closer to one you get, the further away is the other. If many particles interact, the position or velocity of any single particle is unknowable. The change of all seen as a whole, is known statistically in the present.

**"The future isn't what happens to us; it is what we see ourselves eventually doing."**[22]

The future does not exist. We see the past as if it were the future. We cannot imagine the nothingness of the future, so we fill it in with the interpretation of what we remember. Memory is selected memory.

## TEMPORAL STRUCTURE OF EXPERIENCE

**"Time...is the process by which we perceive changes in the universe."[23]**

Life turns external time around, reversing entropy, by organizing both itself and its environment. Life is an open system through which it defines the material, including its own body, as a future possibility. Pure quantitative matter is given qualitative meaning. Information is future order from the point of view of life. The order it sees flows to disorder. Living structure is organized timing, ordering matter as information. Information is the seeing of that order coming in from a future which doesn't exist.

Living parts, functional tasks, are temporal structures. Timing is a coordination continually resolved at a future moment. The past is an hypothesis proven by the present.[24] The object is never the same because it is always observed from a new perspective. It is only theorized to be the same. We assume there are static points in space-time. Our observing always moves on. Space-time points are but memories.

A clock is a comparative measure of events in space-time. Movement over a distance by one thing is compared to movement over a distance by something else, like a clock. Measurement resolves to position on a measuring device. Position by itself is spatial. If duration is a continuum and space is a collection of positions, how can there be motion? The present is both position and motion. It splits into infinite outward directed mutually external parts and infinite inward duration. Form and matter meet at the point at which they are separated. They are separated in the temporal space of living awareness, which lives slightly ahead of formless matter, the blur. Motion is duration spinning in the direction of space-time from the inside and spinning space-time towards duration from the its own outside.

To the degree that the immediate to recent past can be clearly and consistently retained, conscious attention can widen the "duration block of the present moment,"[25] the "now," by intentionally rendering it as an object. As an object, the present appears to itself, elongating the difference.

Speed is a measure of distance travelled per period of time. The speed at any instant, is the velocity. The present is the instantaneous value of the rate of change of time. What is the speed of time? It is the span of now divided by the entire life span up to the current moment. Life is the measure of time. Time moves more quickly as one gets older, because each new moment is a smaller percentage of the entire life span. **Eventually the living moment is smaller than the capability of awareness.** Order is generalized. The timing of individual tasks become disentangled from each other and from the whole. Bodily circadian rhythms become slower than the environmental rhythms. The timing is off. The dynamic ordering of thermodynamic energy as information, becomes static. Form collapses into matter. In the awareness of others, a dead body has form that dissipates in collective space-time.

## SPACE, SELECTION AND OBJECTIVITY

Life arises with matter, and selects from matter, causing matter to exist in categorized parts. **Space-time relativity means that the essence of matter is non-simultaneous locality. The non-locality of simultaneous selection**, i.e., integrating a whole with a part, **is the essence of awareness.** Scientific modeling and the search for truth is based on the act of selecting a subset of matter.[26] A timeless objective truth cannot be observed because the observer is always in time observing and, therefore, adding something.

The structures that make up a living being are identical to themselves virtually. Through break and repair, the separate instances of the function are non-contiguous in time, observed as repeating structures in space-time. The spatial aspect and the temporal aspect have no necessary connection. Purely functional means the same form through a change in matter, not constant change but continuous, that is, with breaks. Separate matter parts come together repeatedly to form the identical structure. Opening out to the future in the pure duration of timed tasks is what separates timing from space, and the observed space-time.

## SYNCHRONICITY, OSCILLATORS AND CIRCADIAN RHYTHM

**"It is by virtue of its energy self-sufficiency, that an organism is a *sentient* being."[27]**

Functional timing manifests as patterns which exist through matter but are not of it. The contradiction of matter traversing a continuum at separate points can be approximated by space-time limit in the observed, and resolved by movement qua movement in observing duration. A pattern is a continual flow with discernible repetition, the core of which is vibration.

Vibrating matter begins in a repetition of movement at and from a point. Movement reduces to observable sequence.

Temporal structures are "coherent excitations"[28] that build up from dipolar molecules to the coordination of gene, protein and lipid activity in every cell, and between and among cells within organs and between and among organs throughout the organism. The organism consists of systems all of which are "coherent domains which can maintain definite phase relationships with respect to one another....Coupled cycles of different sizes are fed by the one-way energy flow."[29] The directional, thermodynamical flow of matter is distributed and held by billions of dispersed but correlated rhythms and frequencies creating a coherent whole maintained with stored energy one step ahead of its energy input, that in and through that step opens a future into its past. The intercommunication of all parts becomes self awareness at the level of the whole.

**Circadian rhythm is the clock of duration**. It is the timed interaction of bodily tasks, culminating in organism wide rhythm in tune with the greater environment, including the movements of earth, sun and moon. Durational clocks are found in all living beings from algae to people. A healthy body is in tune with itself. An unhealthy body is one where a subsystem such as the heart, loses proper communicating rhythm. As an organism ages, its rhythms deteriorate.[30]

There are over two trillion cellular clocks in the human body. Molecular pendulums swing back and forth between high and low levels of protein and gene activity, tuned to a virtual 24-hour day. Light that reflects the movement of the world enters the eye, triggering hormones that coordinate all the cellular clocks. Fluctuations of light are synchronized with melatonin levels and body temperature.[31]

Each organism-wide rhythm is unique to each species, and more precisely, to each individual belonging to the species.[32] In some species, multiple clocks oscillate together, and in others, many clocks are synchronized to one. Since individuals of the same species share a general environment, their biological rhythms are relatively independent in each individual.[33] An animal's personal movements can cause the clocks to adjust the timing of various functions.[34]

## FRACTALS

The Second Law of Thermodynamics, to summarize a compatible cosmological theory described earlier in this book, describes a flow of order to disorder in a universe in non-equilibrium, that can be explained by a beginning of gravitational smooth order and electromagnetic energetic order that together become disordered in a clumpified diffusion, finally ending in infinitely small parts, infinitely dispersed, with no relative movement.

It is living awareness that unites the very small and the very large in a life-size middle, and it is the separation of the two that provides the possibility of living awareness. The flow of order to disorder is turned around into information by living awareness. The phenomenal world in the middle appears to awareness as quantity qualified. Quantities are known only in relations, in terms of one another. Qualities are the appearance of that which is quantified, that which is related.

Even if there were a workable unifying equation of quantum smallness and space-time vastness, phenomenal qualitative experience would not be explained objectively. Even if set theory is universal, selection of set members is subjective quality, not objective quantity.

The aware being abstracts from a quantified world of parts a qualitative set of generalities. It is the paradoxical unification of quantity and quality that generates living awareness. Quantity is both multitude of parts and magnitude of relations. Small and large are a relation of comparison. By applying patterns to quantity, we see shapes. Shape is what unites.

With patterns, scale includes both size and shape. The finite shape with an infinite edge defines a fractal. This is what self-similar means. Fractals are a way of envisioning the blending of quantity and quality, discrete and continuum, large and small, space-time and particle.

The fractal self-similar link between quantity and quality expresses itself in the thermodynamic expanse by individual living aware beings, which are themselves self-similar rhythmic entities in their temporary time-creating thermodynamic reversals.

**For life, the separation of form and matter is, itself, a fractal.** The form is relatively the same through a change in matter parts. The fractal connection between the infinite continuum of form with the infinitesimal formless discrete parts, mirrors the circadian rhythm relating to the movement of large objects at the top of the scale, fractally connected with the infinitesimal random atomic movement within the cell.

Life is a rolling fractal.

## THE TIME DIMENSION

In his book, "Reality's Fugue," F. Samual Brainard theorizes that scale is a Fifth Dimension. Beside three coordinates of space and one of time there is a coordinate of scale. Each coordinate is not just a "here-and-now," but also a size.

If we imagine becoming larger, the objects we normally view will appear smaller and the distances between them shorter. If we imagine becoming smaller, those same objects will seem larger and the corresponding distances, longer. But, if the viewer him/herself were to actually become smaller, their measure of time would expand along with distance. The velocity, therefore, would measure the same as when the viewer was normal size.

If an object is seen through a magnifying glass, a small object appears larger, and moves faster across the screen, than when viewed normally. Time, Brainard points out, is not a factor when the object is viewed through a magnifying glass.[35]

Let's look at this.

Imagine that the object that is viewed is a one meter long stick, and the viewing subject holds a one meter ruler if front of his/her eyes. The farther away the subject stands, the shorter the stick will appear compared to the ruler. As the subject walks with the ruler closer to the stick, the stick will appear larger until the ruler is immediately flush with the stick. At that point, the ruler and the stick will appear and measure the same length.

When we say the object in the magnifying glass becomes larger and faster the higher the magnification, the dimensions of the object are really becoming more accurate. As the magnification increases, the subject's view becomes closer to the object. Magnifying is a virtual method of moving closer.

Again, start by watching a train from a mile away. Hold up a one-foot ruler to your eyes. Each car will take some time crossing the ruler. The closer you get to the train, the faster each car crosses the ruler. At some point the train car is passing through the ruler at it's true average velocity. 60 miles per hour = 88 feet per second. In one second, 88 feet of train will cross the ruler. Compare this to the view from a distance. An 88 foot length of train will cross in front of the ruler in a longer time than 1 second. The measure of velocity equals the appearance of velocity at a theoretical adjacency.

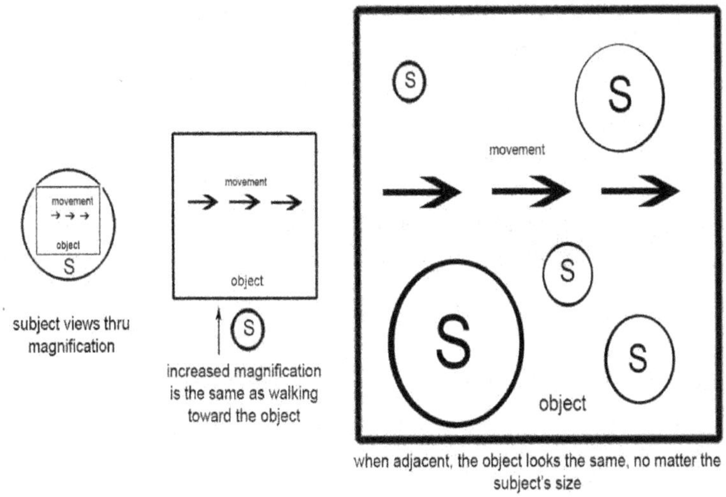


Fig. 55.

How close can the subject get to the object? Once the subject/measurer is the same size as the object, the only way "forward" is for the subject to be smaller than the object. If the subject and object are adjacent, and the subject is smaller than the object, then size will no longer be perceived. The velocity from that viewpoint will then be viewed or measured the same for any object size larger than the subject size. **The appearance of velocity will still equal the measure of velocity.**

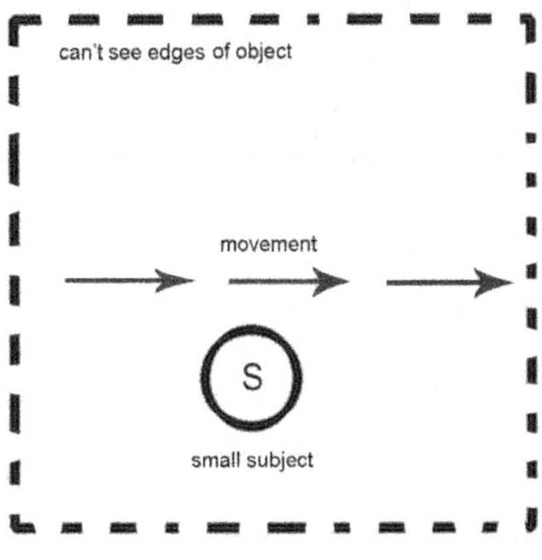


Fig. 56.

If the subject is not adjacent, but is close enough to the object that all that can be seen is the object, then to the subject, and assuming that the viewed object has a flat, smooth, featureless surface, there is no way of telling if the subject is smaller or larger than the object. **Smaller and closer are indistinguishable.**

If the subject is far enough away from the object, then the space viewed around the object will determine the size. Size is, therefore, an appearance based on the field of vision, or context. If the space or context around the object can be seen, the object will then appear smaller, the larger the context. If the subject is adjacent or close enough to see the object only, size does not exist. The ratio is one-to-one.

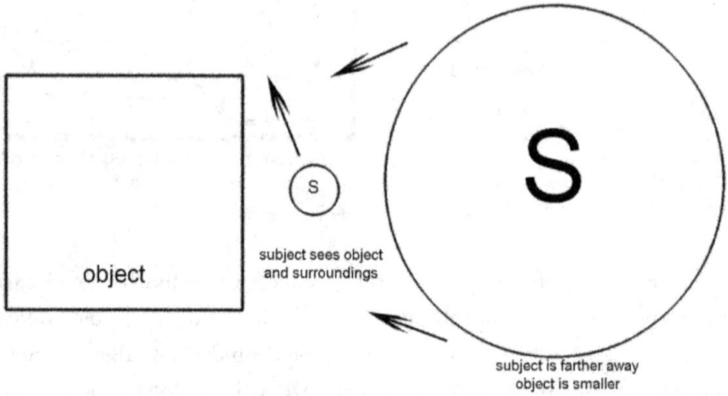


Fig. 57.

Size, unlike a space-time object, is a relation. Large and small are relative to one another. The size of an object by itself has no meaning. Size is determined by the relative distance of the subject to all objects in view. A tennis ball looks smaller than a car if both are viewed from the same distance. But as you move your eye closer to the tennis ball, the tennis ball will appear larger than the car. The measurement of each object when measured adjacent to the object, or calculated according to the subject's measured distance from the object, will have a length. But a length by itself is not large or small. **Large and small require a subject**.

Motion in space-time is relative to different viewing subjects. "According to the special theory of relativity, it is impossible to say in an absolute sense that two distinct events occur at the same time if those events are separated in space. If one reference frame assigns precisely the same time to two events that are at different points in space, a reference frame that is moving relative to the first will generally assign different times to the two events."[36]

Space-time motion is objective. Space-time velocity of an object is relative to two observers (or reference frames). Simultaneity is relative to two (or more) viewpoints in space-time. A moving clock for one subject will tick slowly compared to an observing subject's stationary clock. The appearance of size is, in that case, irrelevant.

Size-time is different than space-time. In the case of the subject becoming smaller, the point in space of both subject and object remain the same. Only the subject's size changes, and the corresponding measure of time. Velocity remains the same.

In the other example, the size of an object is relative to the subject at different momentary positions. Object size is compared in memory. Rulers and magnifying devices contain the physical reference of changing size. The velocity of the object viewed by the magnifier appears relative to and determined by the closeness of the subject to the object. But the magnifier/subject can move in and out at any speed. Becoming spatially closer or more distant from the object is also independent of the subject's motion.

Traveling in size-time is subjective. In one case, (magnification or changing distance) there is no time. In the other (changing subject size) there is no distance. That time expands and contracts with the subject size, means time is relative to the subject. Size-time is the pure duration of the living subject. **Traveling through space takes time. Traveling through size brings time with it.**

Space-time resides within the motion of the object, which if the subject is adjacent, and can thereby view nothing else, there is, to the subject, no motion. Likewise, inside a windowless train car, traveling at a constant speed, the contents of the car could be thought of as one unchanging object to the subject.

A size dimension, however, is not just size, it is scale. Aʙ is size. Aᴀ is scale. A scale model includes not just size but shape. In rare cases, a scale model may be larger than the original, but all objects are faster and larger than what they appear to be. **Every object that appears to a subject is a scale model of the object.** The subject-object size relationship brings shape out of size. **Shape requires memory**.

In the difference between form and matter, conscious awareness or attention is aware in the spaceless now. The similarity of form with itself, distinct from matter, is the domain of pure time. **Awareness moves in scale-time, that is, self-similarity from one moment to the next.** The relation between the subject and object is one-sided, that is, a direction, a direction forward and outward. Self-similar motion is time directed out. The 0 dimension is in-out. Self-similar form views self-similar matter-objects through the living boundary of time. Scale-time travel is a self-similar fractal. Scale-time motion constructs space-time motion, which is confirmed in the relative motion observed by multiple subjects.

Absolute time is the infinite temporal point. As such, it moves outward into all there is. The four dimensional space-time world is projected out of the combination of 0-dimensional time and the infinite small/large given. In the subject's time, relation of scale is with a memory. Self-similarity as a coordinate travels only in time. It is not a series of separate space-time moment/positions. The connection from moment to moment is the self-similarity of the living being.

Scale is the living awareness that looks through the 0 dimension of time, in self-similar moments, at what appears through and to its bodily structure at objects that are continual scale models of themselves. In the momentary presence of conscious attention, category and input, the same and the new, the absolute relation, bind in the simultaneity of recognition. Scale movement is the movement of living awareness. Awareness shapes matter.

Brainard's concept of scale, however, is not simply scale models. He means the scale of a collective point of view. A collection of individuals not only view things in space-time, but have a common understanding of how things are organized and explained conceptually. Each collection is a set, whose members can belong to other sets, and one set can be contained in another. Chemists look at matter at the scale of molecules. Biologists look at matter at the scale of cells. But they can view things in combination.

In the philosophy of this book, before matter is ever seen scientifically, the inter-subjectivity of collections of individuals derives from the self-other contained in the living boundary, beginning with the cell wall. Collections or sets derive from the transcendental parts and wholes of awareness.

Interestingly, Brainard's idea of scale extends to species of aware beings. Here, the size of the animal, and the concomitant scope of awareness of the species merge. In the same section on scale, he mentions studies that have shown that the rate of metabolism is, in many cases, proportional to the animal's mass to the power of 3/4. The denominator 4 is indicative of an extra relationship to time.

Metabolic rate is the measure of energy expended per unit of time. (We have discussed the concept of "power" earlier). But time, in Quantum Mechanics, flows two ways, or perhaps, not at all. A four dimensional space-time universe, has been described as a "block universe," in other words, devoid of flow. On the other hand, the thermodynamic energy in metabolism flows one way. Like velocity and position, energy and time are also complimentary qualities in Quantum Uncertainty. Dividing energy by time would necessitate a different kind of time to be meaningful.

In classical physics, an inanimate object (or a whole system as one object) has an average one-way velocity. A closed inanimate system is subject to the First Law, the conservation of energy. In a living being, energy is expended per unit of time. Metabolism is an open system.

It is the viewpoint involvement (subject-object) and not the point (object) itself that establishes scale. Although relative space-time events can be calculated, to be understood, they are grounded in a human's viewpoint. Viewpoints are either lived or imagined by a living being as being actual. **Mass does not have a shape**. Mass is a magnitude. Particles are "formless probability clouds". An inanimate point does not "see" size or shape in another space-time object. Spatial size and shape require memory, which requires a metabolism. Without life to look at them, there are no planets and stars. There are just quantum gravitational fluctuations.

Living beings are aware through the time of their living bodies. Objectively (from the outside), scale is a 5th (or at least an "other") dimension, that is, a coordinate of time, space and size. Subjectively (from the inside), the in-out 0 dimension, establishes scale by applying form to matter, that is, to any space-time object.

Four dimensional coordinates are points in motion. Scale, however, is not a point. It is a relation, a ratio between a point and a view point. A scale coordinate is a container of space-time points. Four dimensions is a measurable dimension of parts in a whole that are moving relatively to one another. A scale dimensional point of view is a whole viewing parts. Parts that are wholes to smaller parts, and wholes that are parts to larger wholes is the contradiction that plays out in the awareness of relative scale.

## B. AGENCY

*Difference itself is immaterial causation*
Gregory Bateson[1]

### AGENCY VS. FORCE

There is no causation in a continuum. There is no causation in eternal separation. There is only causation in the combination. The living middle separates and combines the extremes with time. Time/energy equivalence in the present is the point at which awareness is directed toward the future, facing energy coming at it from the past.

Existence is both infinitely large and infinitely small. We can calculate parts and categorize wholes. Living awareness is the qualitative agent of the finite: "Limiting, defining, dividing, differentiating and individualizing."[2]

In the external world the causes of movement are called "forces." Physical cause is a relation. Things are attracted or repelled by other things. Gravity is space-time acting on mass as mass acts on space-time. Electromagnetic force is what attracts or repels molecules and atoms to and from one another. The strong and weak forces do the same for sub-atomic particles.

Force is observable, not in itself, but in its effects, in the relation that changes. We can observe a relation, or what we suppose is a relation, by observing the change in one thing in relation to another.

Agency is the relation of form to matter. Matter is formed. Form is matterized. Agency is the cause of living movement. Agency self-causes.

The timing of tasks, that is, living parts, is determined by the whole living being, the duration. In duration, the whole causes the parts by defining the past from the current vantage point. Life has split the part-whole contradiction into a temporal whole before parts. With the incorporation of final cause, agency achieves backwards causation. Throughout the living body, top-down is future cause. Final cause or purpose is actualized in the future. Every task in the body relies on the timing set above and in front of it. The timing of individual tasks depend on the timing of the whole, resolved in the ever escaping future. Actualization of the whole is always one step ahead of itself. In the running difference between now and what just was, the space of awareness applies expectation and renders meaningful the material past.

There is no difference in matter itself. Life causes by making distinctions. That which is and that which is not depends on immaterial causation. Agency is causation of meaning.

**"Why would the resulting momentary activity patterns necessarily be any less physical than the objects and actions were in the first place?"[3]**

How does awareness connect with external objects? How does agency cause external things to move? A cue ball rolls into another cue ball causing it (explainable by the electro-magnetic force) to move. If we push the ball ourselves, where is the force? What we see is the hand pushing the ball. Whether the hand is alive or not is irrelevant to the observation. Our observed hand is as material as a cue ball. What we observe from the point of view of an agent is an encounter between objects.

Living objects flowing in matter-through-form in space-time are observed. Objects and background flowing in form-through-matter duration-time are patterns of awareness. Patterns are what are sensed and interpreted by the pure timing of an organism. Patterns flowing through, but not of, matter are perturbations of the patterns which make up the organism. Object per se, is a concept of human consciousness.

In the window of attention, anticipating intention, looking forward into the past, is felt in the effort of will. That feeling occurs in the moment of binding categories with input. Attention is accompanied by the effort of will. **Effort is what will feels like**. The simplest application of category to sensation, i.e., recognition, involves a miniscule effort. Moving one's eyes is less effort than moving a rock, intending a thought or memory even less, but **it is all effort, some actions more noticeable than others**.

Parts and wholes arise from form and matter in the direction of time. The separation of form from matter, forms matter. Forming wholes on matter parts is patterning awareness. A patterning entity binds selected patterns onto input. The binding is felt, but reaching the escaping binding moment can only be felt as effort. The intensity of effort is felt in proportion to the bodily tasks involved. The feeling of gravitational force is the relation of space-time to mass. **The force of agency is the relation of form to matter**. The connection between conscious attention and matter is what life feels like. And can only be accomplished by life.

Agency comes from life. Living beings will to act. Intentionality is the fundamental state of will. Consciousness is of, but directed towards.

In time, wholes precede parts preceding wholes. Life decides which wholes with which parts. The body makes the decision before we are conscious of it. But the decision, about what and whether to choose, is based on previous decisions and observed results, in other words, experience. From the outside, "before" means the past. From the inside, "before" means the future. The decision awaits us. The effort is conscious. We feel it in the present. We act with our whole living body/brain. Our body is a united whole in the present moment and aware in conscious attention. We are aware of the decision as intended. The intention feeds the next decision. **We are not dead bodies instructing live minds**. Unlike the relations of external forces, agency is immediate. Strength of agency is not in the degree of change, but in intensity. Agency causes the observed force. The action caused is what is observed, the same as any action between objects.

But where does the energy for agency come from? Life stores external energy by temporarily turning the thermodynamic flow around. The energy is released, propelling the living being forward. But where does the extra boost to do that come from? Because the future, which is simply a direction chosen by life, is nothing, the organism falls into it. It falls into the future as the future pulls it forward. Every task in the living being has a completion time ahead of it. The amount of energy released in time, i.e., power, is the time gained from its own future. The future is the force of top-down living agency.

## FREEDOM

**"...I cannot choose what I choose..There is a regress here that always ends in darkness."**[4]

The darkness of the non-existent future forever precedes the present attempt to focus on the present. The present forever recedes into the past. Conscious attention cannot catch up to the past that continually moves away. The infinite regression of will, limited by the moving present, pulls the future into action. The individualization of the organism differentiates itself from the environment, feels the force of effort and senses the selected object's movement as effect.

The experiments of Libet showed that the intended choice occurs in the brain milliseconds before the conscious organism is aware of the choice. From the point of view of conscious attention, the choice comes in from the future. Libet's conclusion is not an argument against conscious choice, but an argument for living choice.

**" It is the whole soul..which gives rise to the free decision."**[5]

**"A free act is thus a coherent act."**[6]

The number of possibilities to choose from is determined by the organism's generality of categorization. By that I mean, life in general chooses to survive. The species level is limited to choices within possible environmental success, the living body to its structure and its own individual experience. Tasks within the body are not by themselves free. They are beholden to the organism.

Choice itself is not a number. Emanating from the future inside, choice is a qualitative application to a quantitative possible. Between the determination of survival and experience and the randomness of arbitrary pure play, lies the creativity of free choice. Possibilities collapse into a singular choice.

**"Free acts are exceptional"**[7]

Choice is not a choice unless it is willed. **Free choice is redundant**. What is non-free choice? We can choose which action to take. And we can choose to act. There is no difference, because choosing which action, is itself a chosen action. The difference is intensity of effort. The choice of which action to take, however, is often determined by experience and species. In that case, the choice is to act or not.

Conscious attention, freedom and will are concomitant aspects of awareness. Freedom, like force, does not exist by itself. Total freedom alone would be random. Freedom is freedom FROM a past OF which attention is conscious. Will or agency is felt IN the living structure, but observed by the aware being as an external force BETWEEN objects of attention.

Conscious attention occurs in occasional bursts. Most of the time we are unconsciously aware, that is, not in the mode of attention. Attention in the now cannot be reached through attention. Attention is in the future, one step ahead in the body. Free will means free attention and willing body. Free consciousness runs ahead of willful unconsciousness. When freedom and will agree, the act is coherently meaningful.

The fundamental choice is choosing to choose or not to choose. Emanating from the future inside, the ultimate future nothingness is death. The avoidance of death, which is survival, is the creation of a living future.

## C. SELF

*The mental is a set of functions that operate through the body. Such an approach supports the idea that there is a place for the self within nature*
Gary Gutting[1]

### SELF

"The **moments** at which we thus **grasp ourselves are rare**, and that is just why we are **rarely free**."[2]

The self is not a thing inside the organism but is one and the same with the whole organism. Organization as such is the living being as a whole in pure time, a process, the structure repeating with the passing material. Living parts are tasks. Tasks are subservient to the organism. The organism is subservient to survival, in other words, to itself. Life is a continual grasping of itself. But only in the moment of attentional binding does a self choose. Life itself made the original choice to live, and at the furthest transcendental edge of the resultant world of measurable, separate, quantified parts projected in a space-time continual whole, life is individualized as the measurer. The act of measuring is not itself a quantity. The organism individualizes in the moment, at the binding of categorical wholes to measurable parts. Past bindings of both the individual organism and its species origin contribute to a past from which the present is free.

The terms "self" and "consciousness" are often confused. Self can be an object of consciousness, as in "consciousness of self." Self can be that which is conscious of something. In this book, I've tried to clarify the terms by defining awareness as that which life is. Life is aware. The organism is aware. Awareness is essential to and exists throughout the living body. The self, or self-awareness, is the organism as such.

"Conscious attention," is confined to the awareness felt in the present at the moment when category, no matter how general or specific, and input, sensational or remembered, come to attention in recognition as same or different. We can be conscious of our own organism as a collection of memories of past sensations and observations of the world and our place in it, or collections of memories of past thoughts. The self as object can be our entire life, up to and including the immediate past and immediate and projected future.

When I say "I" can be conscious of my self, I mean there is conscious attention in the present, of my conscious attention. But that consciousness of what we are conscious of, has already passed. The current self as object is always and continually in the past. Although the "sense of self," can linger a few milliseconds as immediate memory in the extended now, the sense of self disappears into the past body. Consciousness of consciousness is awareness of the future nothing being filled in. The self as subject is a point. There is no aware substance in the present. It is a binding that immediately leaves behind that that which is bound. The present is a limit. It is a direction forward into nothing, with nowhere to go but outward.

## IDENTITY

**"'Self,' unavoidable in any description of the most elementary instance of life, indicates the emergence, with life as such, of internal identity."[3]**

A whole, consisting of organized timed task/parts, is aware by categorizing wholes made of separate continuing parts. The ongoing moment of attention is a "multiplicity that is singular"[4] The self/moment is a subject, not an object in a relation between objects. The sameness of self is that which differentiates. But as pure duration, it can't catch up to itself. The self as other is the bodily matter sensed as past. Our bodily matter is then categorized as the same kind of stuff as the surrounding matter.

**"In the reality of organic process, where the self is ever becoming, it does not pass like an automaton from one frozen instant to the next."[5]**

Though the subject is not an object in relation with another object, as a subject it is the subject potentially of any and all objects, including its own past. But as object to its own immediate subject, the subject is only continual separation. Continual separation with direction points to objects which can only exist as spatial matter, which the living being of pure duration senses as space-time matter. Life almost or briefly catches up to its own immediate future in the simultaneous binding of attention taking place in the infinitesimal present. Self-similarity of awareness is identity with itself, which is differentiated by pure time. The continuum of duration flows through the particulate of matter, with the self-as-subject in a space-less middle looking forward and outward, aware of phenomena.

## OTHER

**"....you cannot experience yourself as what you are until you encounter what you are not. This is the purpose of...all physical life."[6]**

The now continually becomes past continual nows. The self/now differentiates from itself. It self-others. **Final cause means that otherness comes before the self.** The self of pure duration looks out at the future to find the past. Purpose comes before. The purest otherness is the nothingness of the future. It is this nothingness that turns awareness toward that which it is not.

**"It is not entirely fair to criticize matter as stupid. The human body, in keeping itself alive, does a vastly better job than any conscious effort could."[7]**

The body is the most immediate other. But the body is not stupid matter, it is living matter. The living body consists of timed tasks, that are maintained through the flowing of matter, forming ephemeral, repeating structures.

The other is the future resolution of timed tasks.

**"What is physical matter in and of itself, behind the mathematical structure described by physics? This problem, too, seems to lie beyond the traditional methods of science, because all we can observe is what matter *does*, not what it *is* in itself... physics tells us what fundamental particles do or how they relate to other things, but nothing about how they *are* in themselves, independently of other things. Charge is a way of relating to other particles. Similarly, mass is the property of responding to applied forces and of gravitationally attracting other particles with mass, which might in turn be described as curving spacetime or interacting with the Higgs field. These are also things that particles do or ways of relating to other particles and to spacetime."[8]**

Objectivity is the measured relation between or among objects. But objects don't actually touch. When two bodies collide, the molecular composition of the two bodies repel one another. Living awareness is a subject. A subject is one object in relation to itself. It is an internal relation of matter and form. Living awareness is forming of matter into objects. This act is felt. Living awareness is not two objects failing to touch, **it is the temporal space between form and matter feeling the acting touch of forming matter.**

**"What counts as other [is] the organism's world. An organism's environment is not equivalent to the world seen simply through the lenses of physics and chemistry."[9]**

**"For each life...there is a universe that involves 'spheres of reality.' ....Our planet is composed of billions of spheres of reality, and internal/external confluence."[10]**

What are the objective places of matter? We see objects that we assume have a real position in space-time. We assume that a fly landing on what we perceive to be a real external barrier is an objective place. But place for a fly is what it is to a fly. All living beings organize the world into places. To us, the fly lands on what we have defined, what our living structure defines. The fly traveled a long ways to get there. To a mite, this is its planet. All these places are the same to objective measure, but different to each species of living awareness.

**"The 'original object' of human perception...is the intentional action of another human subject."[11]**

The aware self must first see itself in the other to establish the ground of the set of infinite possibility, the set of awareness. A world that exists for all living beings must in some way be shared. Objectivity is replaced by inter-subjectivity in living awareness. The measure of objective, external force is replaced by the perception of the other's internal agency. The otherness of the other living being is the subjectivity behind the other's measurable objectivity.

To a phenomenologist, the provable certainty of the other's subjectivity, his comparable awareness, must be found in descriptive observation, not measure. The sudden surprise of being seen by another, the experience of being an other, is an object for another subject. The impossibility of imagining one's own death, which forces one to see a clone of oneself in the future, is an objectification of one's own subjectivity. It is mutual subjectivity in the shared creative conversation in which two people are anticipating each other's statements, knowing what the other is going to say next as if it were him/herself saying it.

Awareness of awareness is experienced in improvisational music, where I experience myself as being "played" by the other whom I simultaneously experience as my "playing." Interaction is mutual causation.

Agency is revealed in conveying information to others. Or as Hans Jonas says it most simply, "Life can only be known by life."[12] The one external object that life perceives directly is life. Only subjectivity can be known directly. An intention, the directedness of attention, is known in the not yet, differentiated from matter.

**"In mystical experience, we could say that the awareness of awareness itself is heightened while self-sensation is cut off."[13]**

Awareness of our own awareness in the present means an awareness that does not include the otherness of body, others and past. This is the awareness of no-self. Here the infinite inward of never catching up to ones self is cutoff from the infinite outward. The mystic finds either unity with the all, or the pure consciousness, "the infinite closeness" that watches the self.[14] This is called by some, "the liberating insight of non-identification" with neither the object of awareness nor the aware subject.[15]

There is an interesting convergence of views from the spiritual in Echert Tolle's (1948-), "The Power of Now" and Western physics, Sam Harris's" Free Will." For Harris, there is no such thing as free will. His arguments are basically- 1. Decisions occur in the brain milliseconds before we are conscious of making them. 2. The infinite regress of deciding to decide to decide, etc. fails to find a self deciding. Consciousness is merely an end-result observer. How different is this from Tolle's "Now," where this is the true consciousness that must be experienced, that is, lived in? Meditative traditions often speak of "the watcher," that which is there beyond and beneath both subjectivity and objectivity, both self and other.

## BOUNDARY

In order to break through the fundamental paradox, life requires its own boundary. The cellular membrane, common to every living being, is also alive, and along with memory and procedure, is one of the three fundamental structures of the organism. The shape of the membrane is the shape seen from the outside. **Shape, not the membrane's physical shape, is life's container, the form remaining the same.**

The internal and the external are immediately adjacent yet remain separate. Both the flowing energy of metabolism inward and the outward forming action of awareness are caused and defined by the limit of the boundary. Part and whole unify, differentiate and intersect at a boundary. The contradiction of small part and large whole is a dynamic limitation of living awareness. It is the boundary of a living being that allows it to limit by selecting and applying selections. The living being shapes matter. The boundary from the outside is a four dimensional shape or form that keeps its shape relatively the same through a change in matter. From the inside, the living being forms matter through the limits of selection.

The living boundary means the organism is an open system, open to interactions with others, and able to understand others.[16] A boundary is not a wall. It limits. **The living boundary between form and matter is time.**

## EMERGENCE

**"Emergence...A property of something made of parts is emergent if it would not make sense when attributed to any of the parts."[17]**

By looking at the past in a forward direction, living awareness organizes matter as emerging from matter. The organism is greater than its parts by being organized ahead of its parts. The parts are tasks timed to the whole, emerging from the past.

**"The problem of emergence: understanding how a new higher dimension of causal influence can be woven from the interrelationships among component processes and properties of a lower dimension."[18]**

Top-down cause is future cause. "A discontinuity of properties despite compositional continuity"[19] is the very definition of living awareness. Form continues, matter is discontinuous, the aware body continues, forming is discontinuous. Tasks are interwoven timings, resolved continually ahead, timed for the survival of the whole.

**"This logic of generating complex global regularities and identifiable 'entities' from simple iterated operations became a new model for emergent processes."[20]**

Loops are similar but not identical. Living parts are functions inside of functions. Tasks "call" tasks at the behest of other tasks, looping in a "tangled hierarchy" in and ultimately for the purpose of the whole. Evolution is the emergence of living forms that both "reorganize natural processes in ways that would never spontaneously occur"[21] and are themselves reorganized in the resultant process of selecting.

The dynamic non-linearity of living motion limits both the predictive capability and the perceptive accuracy. Life is not a separate consciousness measuring separate objects. Living beings are part of the world being measured, and as such, are not just continually separated but are continually separating. Relative sameness in form is not constant, but continuing with breaks. In other words, the form constantly returns as similar with new matter.

## PROCESS

**"Being alive does not merely consist in being composed in a particular way. It consists in changing in a particular way."[22]**

Life is constantly moving forward at the past. It must move to survive. Stopping is death. Process is the coherent movement of parts with a common goal. More than mere motion, the living organization is a symphony of timing, continually resolving a step ahead of itself. A changing organization has to have an organizing principle to maintain coherence. Life is its own purpose.

**"Because processes are defined by their organization, we must acknowledge the possibility that organization itself is a fundamental determinant of physical causality."[23]**

Like patterns and symbols, organization as such is not matter. It uses matter. The living organization is ahead of its parts. The organization determines its action.

**"Energy rate density (also termed power density) is the amount of energy flowing through a system per unit time and per unit mass."[24]**

Seen from the outside, by life looking at its self, the matter flow is stored and released according to the timing dictated by the organization. The force to store and release on time in the first place comes from somewhere. The flow of energy varies not only to when and how much, but how fast, that is, how much when. And the timing, in turn, determines the amount. The variation in timings of local energy release, require a velocity of energy, not just energy. Future resolution pulls the living body forward.

## STRUCTURE

**"The life cycle is a highly differentiated space-time structure, cycles spanning an entire gamut of space-times from the local and fast (or slow) to the global**

**and slow (or fast), all of which are coupled together. Coupled cycles of different sizes are fed by the one-way energy flow."[25]**

The living body is a structure of time. It is not the physical structure, with physical parts, that appears to sensation as a static thing. It is the living structure that exists as process, as timed movement, whose purpose is to maintain its living structure. Organization organizes the world it requires. If it fails, it dies. The world that makes survival sense to a human is the universe that has evolved to create it. What is given, is given to all living beings, each selecting what they have to select. Life as a whole evolution, is one structure, self-similar in its evolving implementations, self-similar in its individual life spans.

The structures of each function, whether within the body, or within the community of evolving life, is self-similar, it is the same but the instantiation is ephemeral. The task is always there in terms of its necessity, but the form returns with new material.

Varying velocities of energy flow, and disproportionate storage, "amplify the weak signal into macroscopic action."[26] The inter-communication of tasks rise to and for the organized time of the living being. The contingency of the given requires the creative response that is aware.

## INDIVIDUAL

All living beings are individuals. This is how one can observe that they are aware and in moments, conscious. Their actions are not arbitrary or random, and they are not completely predictable. If two individuals have identical genes, and live in an identical environment, it is individuality that accounts for their differing behavior. Watch a herd of goats. Each is reacting in its own way. Every neuron in a worm is mapped, yet we don't know what it will do next. Plants in the same row, grow at different rates. Rocks don't do this. They don't do anything. There are no individuals in non-life.

An individual is not an object of physics.

**" Life early on would have been a collective state, more of a community held together by gene exchange than simply the sum of a collection of individuals. There are many other well-known examples of collective states: for example, a bee colony or a flock of birds, where the collective seems to have its own identity and behavior, arising from the constituents and the ways that they communicate and respond to each other. Early life communicated through gene transfer."[27]**

Communication within a flock is more than a collection of birds. And like a flock, in a collective state of genes, there must be singular entities to give and receive genes. What is the nature of that entity that exchanges? Selective advantages manifest in bacterial survival. And bacteria can exchange genes "horizontally." Before single celled organisms, though, it is not clear what does the exchanging. A virus? A virus maintains an advantage by way of an organism, not alone.

The genetic code itself could "evolve." By laws of physics, gene molecules mix and match progressively toward some optimal physical state. But when did a molecule become a codon? When do molecules become FOR something?

**"Turbulence behaves kind of like an ecosystem. There's a particular dynamical mode of the fluid flow that's like a predator: It tries to "eat" the turbulence."[28]**

Chaos takes matter up to the edge of life. What separates the flow from the turbulence? Is the flow the agent? Or are both flow and the turbulence separate agents? Or is the singular entity the interaction between the flow and the turbulence? Either way, there needs to be a boundary. Otherwise it's just molecules.

**"If you think about the physical world, it is not at all obvious why you don't just make more dead stuff."[29]**

Yes. But how did the optimal code arise in a kind of stochastic turbulence? Is the "optimal" state the beginning of "code?" Precursors to a permeable cell wall would have to be in place. Nucleic acids, amino acids and lipids are dead. Memory, procedure and boundary are virtual. They mean something.

**"...given that life started as something physical and not biological, it is fundamentally a physics question."[30]**

No. It is a matter question. Life started when life started. It's where the "physical" came from. Physicists don't know what happened before the Big Bang. How do living beings know what happened before life began? Mass is physics. Matter is philosophy.

**"[The organism] is peculiar in its own space and time and asks to be dealt with on its own terms."[31]**

The individual is outside the evolutionary explanation of species. Though each organism has a unique perspective, the individual is not simply singular, acting independently. It is a subject within the world of intersubjective Life. Tasks are the parts of the organism. Organisms are the tasks of Life as a whole. **An individual is both a part and a whole**, a task and an organism.

Part and whole intersect in the living present. Points of view are the inside parts of the outside whole. But an intersection contains "no room for an inner self."[32] The point continually recedes in the continuum of becoming the past. A space-time coordinate is not a living coordinate. Living viewpoints are "dynamic singularities."[33] The present looks forward into the past, and selects the outside part that matches the previously generalized whole. In binding, conscious attention is the individualization of the living being. Any combination of point of view and selected object is unique. The living being creates its own inner room, its inner self, in the pure time that separates its present from the future through which it views and to which it will never catch up.

## D. AWARENESS

*The problem of mind is that of the problem of life*
Alva Noe[1]

### INFINITY

**"This indefinite 'and so on' with which the visual perception is imbued, an ever-ready potential for realization, and especially the 'and so on' in depth, is the birthplace of the idea of infinity."[2]**

One step follows the other to an unforeseeable end. But at that end, no matter how far it may be, one more step can be taken. However far we can walk, we can turn in another direction and walk as far as we can imagine, and then walk back to where we started. The area enclosed is as large around as the distance we have walked. And we can always imagine it larger. We can build a wall around it with a ceiling as high as we can imagine. And a floor as low. And we can always dig deeper, and build higher. There is no end to the large. It is "indefinite."

How many steps does it take? If we take a large enough step, it would only take one step. Another step and we would walk twice the distance past the indefinite end. How many steps would it take if we took smaller steps? If each step were half the size, it would take double the steps. The smaller the steps the more steps it would take. The "and so on" steps become indefinitely smaller, and the smaller each is, the more the "on and on" steps. If the larger is a number, the large is a number of smalls.

If the distance and the step size are definite, to walk that distance, it is necessary to first walk half the distance. To walk the remaining distance, it is necessary to first walk half of that distance. It is always necessary to walk part of the distance to reach the whole. The more parts, the smaller the parts and the more of them to walk through. As this paradox of Zeno illustrates, it is impossible to get anywhere. The parts never end. The parts are too small and the whole is too large.

Yet things do get there. They do move. Another Zeno example: "An arrow is only in one location during any given instant of its flight. But if it is in only one place, it must momentarily be at rest."[3] In other words, movement is a series of static positions. How do infinitely divisible parts reach each other?

The standard answer for calculus, is that as the distance gets shorter, the time to move through that distance is smaller. Movement is a ratio, a relation between space and time. Movement requires velocity. It is slow or fast. The precision of the ratio is always approximate. The limit of the precision is what is practical. How long does it take to get anywhere? To move 0 distance takes no time and infinite speed. How much time does it take to reach the indefinite end? How fast must we go?

But slow or fast compared to what? We may say a thing is moving fast or slow based on our memory of similar things or compared to other things that are moving in our field of vision. But what is that actual motion we are looking at? We are connecting the isolated positions.

**" Heisenberg's uncertainty principle....position (location in space) belongs to the outer world and momentum....belongs to the inner world."[4]**

The indefinite "on and on" is the "birthplace of infinity" for consciousness. But infinity, that is, the paradox of infinity, is the birthplace of consciousness. Parts and wholes move through one another. The continuum of duration projects a relation among spatialized parts. The undifferentiated change of the world is differentiated in a temporal continuum by aware living individuals. Time in space and space in time is the scaffolding of living awareness upon which survival depends.

**"Inherent in the geometry of fractals is the creation of ever-repeating, 'self-similar' patterns nested within one another."[5]**

Infinitely small parts and the infinitely large whole, spatially resolve in the same form with an infinite boundary. Existence doesn't reduce to pure, open infinity. Infinity is limited by the small and the large.

**"I believe fractal geometry defines the structure of life."[6]**

Bodily structures are the same with infinite variations. Individuals are the same form with changing matter. Whole individuals are the same species. Living parts are the same structure with infinite repetitions. Past parts go out into empty space.

Parts not only divide forever, but cannot reach each other. Opposing infinite externality is infinite internality, the infinite regress. But in time, the self-similar form of recursive living awareness resolves the paradox.

**"We can will ourselves to act, but we cannot will ourselves to will."[7]**

If awareness were not alive, and therefore, totally separate from matter, the internal regress of awareness would be infinite. But the will to act, the will to think a thought, resides in the living body/mind.

**"What moves ratiocination forward are micro-intuitive insights that bridge the steps, otherwise ratiocination cannot occur because it falls into an infinite regress."** [8]

Each logical step, each element of reason, each individual thought, have no finite edges in a continuum, and break down infinitely. The self-similarity of aware thought, the same self, bridges thoughts like time bridges steps. The relation of now to past is whole to parts. The infinite regress and the infinite externality break open the paradox of infinity in alive awareness. The on and on and the in and in break apart with opposite directions. Living awareness looks in one direction at the same direction going the opposite way. It looks forward into the past. The living reversal of time is a leap into binding at the immediate juncture of infinite regress and mutual externality.

In the same way that thinking requires an intuitive leap into a future reasoned step, the world that human consciousness sees is teleological. Without teleology, the evolution of the universe is one giant infinite regress. The universe evolved an intelligence that theorized that the universe evolved an intelligence that theorized that the universe evolved an intelligence..... Survival is purpose in one direction.

## NOW

**"You may think that you are seeing all these objects at once, or 'now,' even if they are at different distances from you. But you really aren't, as light bouncing from each one of them will take a different time to catch your eye. The brain integrates the different sources of visual information, and since the differences in arrival time are much smaller than what your eyes can discern and your brain process, you don't see a difference. The 'present'—the sum total of the sensorial input we say is happening 'now'—is nothing but a convincing illusion."** [9]

Note the phrasing: Different time to "catch" your eye, not different time to "reach" your eye. There is only "arrival" time. Distance is what seems to be the illusion.

If we could process information at the speed of light, would we experience a linear series of photons-hitting-individual-retinal-cell-receptors? Would each of these events be their own "now?" But what would that mean? How long is a photon hit? Is each photon hit a position without velocity? What is the time between photons?

What if we could process two photons "at once?" This time-span would include the change, that is, the similarity and difference between the two, but beneath our perception. At some point, "the sum total of the sensorial input" establishes change sensed by a slower process. This seems to mean that the more photons processed as one image, the slower the process.

Is the present moving through space-time or is space-time moving past the present? Does time move? Or is time stationary? Any point of time in the past, that is, not in the present, is part of space-time. So any past point in time appears to have motion in between positions. The present is the pure time point. It is both a continuum and discrete. It has both continual motion and time-positional discreteness, neither of which can be observed together. Moving forward into the past splits the present open, with the living being's own immediate future the point of awareness. The observation is a continuum, the observed discrete. Neither space-time nor the present can move without the other. The relativity of motion is a projection of viewpoint distance.

**"'When a man goes to work on something,' Lorenz writes, 'he takes into account, continuously, during his performance, the responses of the object, and by these governs the following acts......very close tie between action and perception'"[10]**

Fig. 58. Catching the Present

**The image on the previous page is objectively stationary. Yet we are viewing it in time.** The lines and shapes appear to be moving. By concentrating on one shape, the movement seems to slow down. If we then relax our focus, the movement continues. As our focus moves around, the movement picks up. We are seeing our seeing. Every intention is feeding the next attention, intending our own reactions, etc. Parts never catch up to wholes. Even when we are stationary, our eyes never stop moving. Living awareness partitions and selects. If there is nothing to select, if there is no change in the world, what is left is meaningless motion.

**"The notion of time is related to change, and the passage of time is simply a tool to track change."**[11]

The time of pure duration imposes selection of parts and corresponding wholes. The smallest small must be smaller than what can naturally be sensed. The on and on must always go on. For there to be awareness, it cannot be otherwise. Human consciousness and all living awareness has organized the world in the most practical way it can.

**"In the one real, time-drenched universe, everything has a particular history precisely because it is finite."**[12]

We cannot select the infinite. It is already there. Small and large are already there. Parts and wholes are the finite tension of the infinite. Each individual's now continually creates its own past of previous selections.

**"One property of the real world not shared by any mathematical equation is that it is always some moment."**[13]

Quantitative parts are selected by qualitative criteria as the present imposes itself on change. The present cannot be measured because it is both the measurer and measuring of external time and space. As both measurer and measuring, the part-whole compresses to the simultaneity of distinction and difference projecting a simultaneity of position and motion as past space-time. Space-time is relative to points of view, but points of view themselves are absolute. Points of view and the points viewed, can only be occupied in the simultaneous present.

**"If, as Laplace described it, we could see 'the future just like the past, present before our eyes' as the inevitable unfolding of conditions & rules, then it wouldn't be like knowing or observing at all; we would be more like mirrors, or recording machines, or like computers with infinite memories. No bit of**

**information would have any value, no action any motivation, no knowledge any consequence, because nothing would be in question. It isn't that 'I' would become the molecules that constitute me; rather 'I' would cease to think or exist."[14]**

Subjectivity without an object is nothing. Objects arise in the filter of subjectivities' non-existent future. If the future exists, we are no different than objects.

**"If the present is a moment in time without duration, it can't exist."[15]**

Pure distinction is nothing. The present, the leading point of pure duration, is not a moment in space-time. The Now is the varying temporal distance between form and matter. The present is the point at which life forms matter in conscious attention. It is experienced as Now.

What is Now measured from space-time? The clock measurement is approximate. The day is divided into twenty-four hours, but not from sunrise to sunset. The sun rises today at 5:00. A few days ago it was 4:50. Imagine if we adjusted the clocks every day like we adjust the calendar. Every day we set the clock ahead 2 minutes. Today 5:00 is approximately two minutes later than it was yesterday. From December 21 to June 21, we gain an average two minutes each day. Clocks and calendars are objective measuring devises. Time of day is relative to the rotation of the Earth. The sun rises at sunrise. So how long is a day on Mars? Compared to what? If time is relative to motion, how fast is time?

**The standard definition of one second: The duration of 9,192,631,770 periods of the radiation corresponding to the transition between the two hyperfine levels of the ground state of the caesium 133 atom.**

**"In terms of orders of magnitude, if 1 second is $10^0$, then 1 millisecond, the time it takes for a neuron in the human brain to fire and return to rest, is $10^{-3}$. Planck time, the smallest measurable unit before time apparently breaks down, is $10^{-44}$. This is the time it takes light to travel the Planck length, 1.616 x $10^{-35}$ meters, (about $10^{-20}$ times the diameter of a proton)."[16]**

Magnitude, unlike multitude, allows us to think of measurement in terms of absolute relation. Part and whole seen spatially as small and large are fast and slow in time. Things are what they are in comparison to other things.

Studies have shown that the general "sense" of now for humans is somewhere around 400 ms., less than half a second. A neuron fires 400 times in this span. The input to and by the brain, in any span of time, is an enormous amount. The brain-body reduces quantitative input to what we consciously attend to or process in 400 ms.

**"Observers will start to notice differences when the timing interval is larger than about twenty milliseconds or so. This timescale sets the limit of visual simultaneity in humans."**[17]

Within the now of 400 ms., at most 20 distinct light flashes can be distinguished. That's 20 distinctions per now. The smallest distinction lasts 20 ms. Light/dark is the minimum object.

A neuron can fire a thousand times in 1 ms. That means 400,000 times per now. There are 100 billion neurons with 100 trillion synapses all firing or not, in synchronized patterns of patterns.

What is meant by simultaneity? At 20 ms., a single distinction can be made. This means in a temporal order, a minimally defined same/not same can be observed at once. One is not the other. How many 20 ms. distinctions are recognized as the same in 400 ms.? The same what? How many are different?

**Recognition of difference is itself the unity of observation and object**. Assuming that a same/not same difference is the minimum part of a recognized object, how many categorized objects are there per average now? What is the size of the objects recognized? What is the speed of recognition? What is the speed of meaningful recognition?

Is the whole slower than the parts? If the whole is recognized before all or some of the parts, is the whole then faster than the parts? In other words, if so, it may be that few distinctions are made in 400 ms., and yet an object is recognized, either as new or already categorized.

Furthermore, how many objects per category? How many categories per object? Some objects will not be consciously recognized but assumed. How many bindings, that is, categorized objects, per now?

If the parts and/or wholes are correctly recognized, this would be known or measured after the fact, depending on an objective definition. Subjective recognition would be the measure of meaningful objects recognized per now.

We experience a series of nows, each with no clear beginning or end. Nows don't touch, or have edges, they blend. Part/whole recognition determines what now is. **Simultaneity determines the limit of time.**

**"The human brain can achieve the remarkable feat of processing an image seen for just 13 milliseconds.. what vision does is find concepts."[18]**

Processing an image is faster than recognizing it. The world is faster than it can be processed. It is larger. There is more of it.

How long does sameness last? If an object can stay in focus longer than the initial binding, this means either that there are multiple bindings of the same object, or a binding lasts a long "time." Neither the input nor the processing ever stops.

What is in focus and what is remembered as having been in focus a split second previously, is continually interrupted by memories, thoughts and objects only partly noticed. When the "now" is and how long it lasts is measurable only after the fact in space-time.

**The velocity of the change of form is relative to the velocity of change in matter**. From the outside, in space-time, matter flows in and through form, form changes more slowly than matter. From the inside, form is moving faster than matter changes, otherwise form could not catch up, it could not perceive the motion as object. The velocity of form and matter are relative to one another, as awareness and that which it is aware of. The sameness of self slows matter down to the categorized objects of which it is aware.

**"Understanding of time is limited. there is only now. Past is a present remembrance and the future is a present anticipation. Everything may be happening simultaneously and all causes are happening now."[19]**

**"What does exist is the recent memory of the immediate past and the expectation of the near future. We link past and future through the conceptual notion of a present, of 'now.'"[20]**

If the Now is a point in space-time it is no time. If the Now is not in space-time it is all time. Now "measures" the change of positions based on memory and anticipation. Awareness of that which is there doesn't take place in measurable space-time. Looking forward at the past does not "exist" in space-time. It causes space-time.

The opening of form and matter is temporal, space-less distance. The Now is not just separate awareness. The Now is living awareness, which includes the living body. The body is also living in the Now. The flutter of part/whole momentarily resolves in binding. The rhythm is in concert with the world. The Now opens to the world by joining with it. The Now is that opening of form and matter that moves forward into the past, causing the two to become one. Life opens form and matter and binds category and input.

Although it is always now in conscious attention at any time, the living body has local, measureable processes at any given time. One neuron beat might be one thousandth of a second. An average pulse of 60 per minute takes one second. The pulse rate, neuron rate and the rate of all tasks in the body/brain, however, are not timed to a clock. They are timed to one another.

## OBJECTIVITY

**"To say that a world is objective means that the world's existence does not depend on the agent."[21]**

Objectivity assumes there is a realm of being that is beyond any one individual's sensations and, at the same time, is the same for all individuals. The objective world is the world that awareness measures with enough precision to survive.

Change or movement of things in the world presented to awareness are necessarily of a larger size and speed than what appears from a distance. An accurate appearance would be adjacent to the object. Senses, therefore, must simplify, that is, reduce, what is actually there.

Thinking, or the manipulation of retained objects, is the tool used by a living being in an attempt to discover what is actually there. Scientific thinking has revealed a universe constructed of consistent parts and wholes beyond the senses. Material reductionism assumes that wholes can be explained by parts.

By applying form to matter, living awareness physicalizes, that is, externalizes or spatializes, time. Retaining thought within pure duration produces object-events distinct in clock time. Non-local simultaneity in the future-facing present, applies non-simultaneity to an objective past. Distinct events are placed in time, following previous events. Qualitative categories attach to quantitative objects.

**"The laws of nature are assumed to exist independently from the physicist while equations of physics are the mathematical expression of our own attempts at reaching them."[22]**

This world of parts is measured. Because parts can themselves be made up of parts, measurement is approximate. There is, then, the phenomenal world of what we see, the objective world of measurement, and the real world that is measured.

**[Objectivity explains the world as] "The mechanical motion of particles acting predictably on each other."[23]**

The smallest parts known so far are the particles of quantum physics, at the boundary of smallest contradiction. The largest whole known so far is the space-time of general relativity, determined, yet expanding towards the unknown future. The laws of physics are the unchanging rules of interaction. Parts and wholes move according to unseen laws. Forces are relations between objects. Light is simultaneously formless parts and whole matter. Like aware-ness, it has direction outward. Unlike awareness, it has no time.

**"The distinction between past, present, and future is only a stubbornly persistent illusion."[24]**

If the ultimate reality is space-time, then things cannot change over time, they just move. Parts in space-time are limited to a few particle types that interact.[25] The particles are actually variables in a mathematical entity called a wave-function, in which particles can be shown to share their properties. Space-time is constructed of the totality of correlations, the sharing, of wave-functions. Space-time "is just a geometrical picture of how stuff in the quantum world is entangled."[26]

External objects are scale models. Time is a direction. Distance is an illusion.

**" ...consciousness ...measures, limits, cuts out forms of things from the indivisible whole and contains them as if each were a separate integer.... It is this essential characteristic of [consciousness] which conditions the workings of all its operative powers, whether conception, perception, sensation or the dealings of creative thought. It conceives, perceives, senses things as if rigidly cut out from a background or a mass and employs them as fixed units of the material given to it for creation or possession."[27]**

We see shapes and count them. Objectivity chases the patterns.

**"The standard definition of entanglement refers to particles only at a given moment. A complete theory of quantum gravity will have to add time to that picture... Whatever this 'theory of everything' turns out to be, time will be the key to unlocking its mysteries."[28]**

The unification of space-time gravity and quantum physics can also be looked at through awareness as the unification of non-simultaneous locality and simultaneous non-locality.

A truly objective world requires the object of time. Space-time itself is a static whole. Movement and distance in space-time are relative to frames of reference. There is no one point in space-time that views time objectively. For time to be objective, all points in time would have to be simultaneous. Simultaneity is not a point. It demands a point of view. It is a two in one. The only simultaneity in the universe is the moment of awareness.

What is true about all points, is that the direction of time appears from every viewpoint to be one way. Events may be measured differently from different viewpoints, but the events always seem to go in one direction.

The second Law of Thermodynamics defines this direction as a direction of ordered energy to disordered low energy in a First Law Universe of constant energy. Life is time directed toward energy.

**"Your own world line strictly speaking isn't a line: It has a non-zero thickness and it's not straight. The roughly $10^{29}$ elementary particles (quarks and electrons) that your body is made of form a tube-like shape through spacetime."**[29]

That shape, however, remains through a change in particles.

**"Whereas the particles that constitute the Moon are stuck together in a rather static arrangement, many of your particles are in constant motion relative to one another. This spacetime braid consists of trillions of strands (one for each [sic] cell), each composed of trillions of hair-like elementary-particle trajectories, intertwined in a complex pattern that never repeats."**[30]

Patterns of ephemeral parts functionally timed to one another in an ever future resolution, organize as distinction from relatively static parts.

**"Since no brain is the same, every person will have their own limits of time perception and a different processing time and their own sphere of now."**[31]

The binding of communication occurs eternally in the next moment. The "sphere of now" spins through its particular organization.

**"The objects of everyday experience** – tables, chairs, mountains, moon – **are not public.... the glass I held was, when I observed it ... and the glass you now hold is, when you observe it, ... are numerically distinct."**[32]

Everything we see is a scale model varying from moment to moment.

**"The objective world, i.e., the world whose existence does not depend on the perceptions of a particular observer, consists entirely of conscious agents."[33]**

**"Reality consists of dynamical systems of conscious agents, not dynamical systems of unconscious matter."[34]**

The objective world is what is usefully consistent in the totality of intersubjective viewpoints. What is useful is what is required for survival.

**"Generally, if the falsity of a statement is conceivable it cannot be logically necessary."[35]**

Scientific conclusions are falsifiable. Since they are open to change, they are not necessary. Objectivity is what is intersubjectively agreed upon so far.

**"...abstract entities that move through time and are constantly optimized. We call these entities 'theories.'"[36]**

Objective knowledge is an ever widening consistency. Scientific proof is a proof of consistency. **Proof of consistency is proof that matter is dead**. The living individual in its own now is similar, not consistent.

**"A neuron's biophysical makeup is intrinsically linked to its computations: the information doesn't exist separately from its material construction. [However,]systems thinkers have transformed our understanding of life, showing it to be a self-organized, self-regenerating complex that extends like a fractal at ever-increasing scale, from a single cell to the global system of life on Earth. Everything in the natural world is dynamic rather than static, and biological phenomena can't be predicted with precision."[37]**

The brain is not a machine of parts computing parts. It is a living organization organizing it's own patterns. These patterns are the ephemeral, fractal connections between parts and wholes.

The problem with objectivity is that it starts with what is not aware and tries to explain what is. If consciousness is merely an epiphenomenon, then so is life. That living beings are aware makes more sense than they are not.

## INTELLIGENCE

**"Intelligence may be viewed as the complexity of our simulations of the future."[38]**

Is imagining complex possibilities or imagining simple probable and successful solutions the sign of intelligence? Or is it processing speed?

Lack of intelligence is when awareness falls behind or moves too far ahead of its environment. Intelligence is when awareness is in sync with its environment, dynamically changing with it. When enough members of a species can't keep up, the species goes extinct. If man or the world destroys an environment required by a particular species, the majority may not be intelligent enough to adapt. Some species are extremely resilient. Like a computer with no sense of self, a species as a whole has intelligence, but not aware.

Individuals use the intelligence of their species unconsciously in their bodies, memories and experience. They are capable of being aware of all that their bodily structure allows. Individual intelligence applies both bodily and species knowledge. Knowledge retained must be usefully applicable within an increasing complexity, dynamically co-evolving with the environment.

**"Since no brain is the same, every person will have their own limits of time perception and a different processing time and their own sphere of now."[39]**

The structure of bodily timings filters the world. The number and scope of bindings per span of attention vary according to retained value.

**"It is through the quality of quantity that we form the idea of quantity without quality."[40]**

Intelligence without awareness is based on quantity.

**"We already don't generally understand how complex AIs think because we can't really see inside their thought process."[41]**

What would be the motivation behind an intelligence explaining its thought process?

**"A + B = C, the 'A' could encapsulate almost anything. But to a computer, what 'A' can mean is so much bigger than what that 'A' can mean to a person."[42]**

Assuming that meaning is infinite complexity, then the meaning cannot be understood by an aware being. Like the chemical language between individual ants that builds to an intelligence in the colony, is there consciousness outside the individual in the society of language? Even though an individual ant has a level of awareness, does the colony have intelligence without consciousness? Aware individuals are tasks that contribute to the whole. If the individual robot is not conscious, can the society of robots be?

**"The reason why humans have this idea of decomposition, breaking ideas into simpler concepts, it's because we have a limit to cognition. Computers don't need to simplify concepts. They have the raw horsepower to process them."**[43]

Meaning limits. Quantity goes on. Awareness requires the difference. Intelligence is the will to close the gap that consciousness binds in passing.

## MIND

**"Mind depends on the existence and development of living organisms."**[44]

All individual living beings are aware. Is life as such, that is, Life, the life that runs through all living beings, aware? Life is one thing, in that all living beings share it. But is that one thing one giant individual, or is it a totality of living awarenesses? If Life is aware, we might call that awareness, "mind."

Is life on Earth a separate Life from life on other planets? Even if the basic molecular structure were the same, the fundamental timing of circadian rhythm would be different on other planets, let alone other solar systems. Earthbound circadian rhythm would be unique. Are Earthlings and "Martians" two kinds of Life? Or is Life defined as life on Earth only, and intelligent beings somewhere else in the universe are, therefore, not alive? Could there be a single living being on a planet? With its parts looking to us like separate organisms? Are the galaxies life forms?

**The universe is not the world.** The universe is what appears as the world to present day humans. Humans 20,000 years ago looked up at the sky and saw stories. To them, the world was stories. Imagine those humans imagining how those stories would turn out 20,000 years later? Will the universe that we see or know, look the same to humans 20,000 years from now? Before Hubble, Einstein's universe had no beginning. There were no galaxies, only stars. Maybe the universe we see at present is simply a story we tell ourselves.

**"The deeper ground, and source of our physical world, our brains and bodies ... everything, literally comes from nowhere physical."**[45]

Living awareness is aware in the connection it makes between anything and anything, even between being and nothing. A Theory of Everything has to tie the extremes together. The extremes somehow, in some way, are there.

Otherness is required. The relation of self and other is primary. This relation opens in the temporal distance between form and matter, beginning with the cell wall. **The central coordinate of the world is the point at which small and large cross in and out.** Awareness is the unification.

The Mind of Life has the same purpose as an individual living awareness, to survive. It has survived by forming matter in myriad ways to fit with any given change. In this way, it is intelligent. Intelligent Mind is awareness from the perspective of evolution.

"**We are the center of the universe because we are its meaning.**"[46]

The universe is no larger and particles no smaller than the reach of our awareness. We are as vast as the universe. There is nothing known that we don't know. All scales are present to us. We see further and farther each day.

If there were no living beings anywhere in the universe, how old is the universe? The age of the universe is where we are. This is the world.

Time is change with direction, in Life toward expansion. Time is the expansion of knowledge, going objectively farther and farther into the material, subjectively further and further into ideas. Life expands farther and further into its environment. Human beings have even expanded into the present, dwelling on the single binding until it doesn't matter, and meaning itself is confronted. Time is change with direction, forward and outward.

**"It is the elusive "life" that no scientist has ever found because the consciousness that is looking for it is it."**[47]

Awareness is expanding knowledge. Knowledge means nothing unless it is shared by Life. Life has been alive millions of times longer than the oldest individual. Living beings don't have lives. Life has living beings.

## I WOKE UP THIS MORNING

by Dwight Harris

We are all shooting stars...
Some of us graze the horizon
Some of us light up the sky
Some of us were once famous
Some of us once supervised a project
All of us once knew we were right
As the Sun sets quickly
To begin again

# FOOTNOTES

## INTRODUCTION

1. R. Dijkgraaf, "How to Solve the Biggest Mystery in Physics." *Wired.com*, September 9, 2017

## I. MATTER

### A. DUALISM

1. M. Strevens, "Looking into the Black Box," *New York Times*, November 24, 2013
2. L. Smolen, *Time Reborn: From the Crisis in Physics to the Future of the Universe*, Mariner Books, 2014, p.15
3. M. Tegmark, *Our Mathematical Universe: My Quest for the Ultimate Nature of Reality*, Vintage Books, 2015, p.8
4. H. Jonas, *The Phenomenon of Life: Toward a Philosophical Biology*, Dell Publishing, New York, 1966, p. 73
5. ibid., p. 45
6. A. Einstein, *Essays in Science*, Philosophical Library, New York, 1934, p.17
7. *https://en.wikipedia.org/wiki/Quantity*
8. ibid.
9. S. Perkowitz, "Light Dawns," *https://aeon.co/essays/why-is-the-speed-of-light-the-speed-of-light*
10. L. Susskind, G. Hrabovsky, *The Theoretical Minimum: What You Need to Know to Start Doing Physics*, Basic Books, 2014, p.24
11. *https://en.wikipedia.org/wiki/Quantity*
12. Susskind, Hrabovsky, op.cit., p.29
13. W. Salmon, *Space, Time and Motion, A Philosophical Introduction*, Chapter Two, "A Contemporary Look at Zeno's Paradoxes, University of Minnesota Press, Minneapolis, https://math.dartmouth.edu/~matc/Readers/HowManyAngels/SpaceTimeMotion/STM.html
14. Noson S. Yanofsky, *The Outer Limits of Reason: What Science, Mathematics, and Logic Cannot Tell Us*, The MIT Press, Cambridge, Massachusetts, 2013, p.45
15. Salmon, op.cit.

### B. NEWTON

1. W. Salmon, *Space, Time and Motion, A Philosophical Introduction*, Chapter Two, "A Contemporary Look at Zeno's Paradoxes, University of Minnesota Press, Minneapolis, https://math.dartmouth.edu/~matc/Readers/HowManyAngels/SpaceTimeMotion/STM.html
2. A. Alexander, *Infinitesimal: How a Dangerous Mathematical Theory Shaped the Modern World*, Scientific American / Farrar, Straus and Giroux, New York, 2014, p12
3. *https://betterexplained.com/articles/a-calculus-analogy-integrals-as-multiplication/*
4. A. Einstein, *Essays in Science*, Philosophical Library, New York, 1934, p.30
5. *betterexplained*, op.cit.
6. Salmon. op.cit.
7. ibid.
8. L. Susskind, G. Hrabovsky, *The Theoretical Minimum: What You Need to Know to Start Doing Physics*, Basic Books, 2014, p.47
9. *betterexplained*, op.cit.
10. *https://en.wikipedia.org/wiki/Derivative*
11. Susskind, Hrabovsky, op.cit. p.40
12. N. Yanofsky, *The Outer Limits of Reason: What Science, Mathematics, and Logic Cannot Tell Us*, The MIT Press, Cambridge, Massachusetts, 2013, p.45
13. Alexander, op.cit. p.12
14. betterexplained, op. cit.

15. R. Sheldrake, *Morphic Resonance, The Nature of Formative Causation*, Park Street Press, Rochester, Vermont, 2009, p.50
16. L. Smolen, *Time Reborn: From the Crisis in Physics to the Future of the Universe*, Mariner Books, 2014, p.42
17. A. Einstein, L. Infield, *The Evolution of Physics*, Simon & Schuster, New York, 1966, p.30
18. Susskind, Hrabovsky, op.cit. p.114
19. Einstein, Infield, op.cit. p.5
20. C. Lineweaver, P. Davies, M. Ruse eds., *Complexity and the Arrow of Time*, Cambridge University Press, New York, 2013, p.163
21. Smolin, op.cit., intro.
22. Einstein, Infield, op.cit., p.5
23. Susskind, Hrabovsky, op.cit., p.65
24. ibid. p.93
25. https://en.wikipedia.org/wiki/Laws_of_motion
26. Susskind, Hrabovsky, op. cit. p.93
27. ibid. p.86
28. Smolin, op. cit. p.27
29. ibid. p.70
30. Susskind, Hrabovsky, op. cit. p.139
31. W. Loewenstein, *Physics in Mind: A Quantum View of the Brain*, Basic Books, New York, 2013, p.13
32. M. Lange, *An Introduction to the Philosophy of Physics: Locality, Fields, Energy, and Mass*, Blackwell Publishing, 2002, p.40
33. M. Strevens, "Looking into the Black Box," *New York Times*, November 24, 2013.
34. Lineweaver, Davies, Ruse, op. cit. p.163
35. Lowenstein, op. cit. p.9

## C. CLASSICAL PHYSICS

1. L. Nottale, *Fractal Space-Time and Microphysics: Towards a Theory of Scale Relativity*, World Scientific Publishing, London, 1993, p.11
2. R.Unger and L.Smolin, *The Singular Universe and the Reality of Time: A Proposal in Natural Philosophy*, University Printing House, Cambridge, 2015, p.162
3. J. Gleick, *Chaos: Making a New Science*, Penguin Books, New York, 2008, p.15
4. L. Susskind, G. Hrabovsky, *The Theoretical Minimum: What You Need to Know to Start Doing Physics*, Basic Books, 2014, p.2
5. B. Cox and J. Forshaw, *Why Does E=mc2?: (and Why Should We Care?)*, Da Capo, 1994
6. S. Carroll, "When Nature Looks Unnatural," *The New York Times, Opinionator*, March 23, 2014
7. A. Einstein, *Essays in Science*, Philosophical Library, New York, 1934, p.114
8. http://en.wikipedia.org/wiki/Quantity
9. Cox and Forshaw op. cit. p.21
10. T. Maudlin, *Quantum Non-Locality and Relativity: Metaphysical Intimations of Modern Physics*, Wiley-Blackwell, 2011, p.132
11. L. Smolen, *Time Reborn: From the Crisis in Physics to the Future of the Universe*, Mariner Books, 2014, p.146
12. P. Hoffmann, *Life's Ratchet: How Molecular Machines Extract Order from Chaos*, Basic Books, Philadelphia, 2012, p.74
13. C. Seife, *Decoding the Universe: How the New Science of Information is Explaining Everything in the Cosmos, from Our Brains to Black Holes,* Penguin Books, 2006, p.41
14. Einstein, op. cit. p.3
15. M. Strevens, "Looking into the Black Box," *New York Times*, November 24, 2013
16. M. Lange, *An Introduction to the Philosophy of Physics: Locality, Fields, Energy, and Mass*, Blackwell Publishing, 2002, p.201
17. ibid. p.80
18. ibid. p 90
19. R. Sheldrake, *Morphic Resonance, The Nature of Formative Causation*, Park Street Press, Rochester, Vermont, 2009, p.193
20. G. Gutting, "Modern Cosmology Versus God's Creation," *Opinionator, New York Times*, June 15, 2014

21.Sheldrake, op. cit. p.264 (quoting David Bohm)
22. Smolen, op. cit. p.40
23. W. Loewenstein, *Physics in Mind: A Quantum View of the Brain*, Basic Books, New York , 2013, p.4
24. Smolen, op. cit. p.53
25. A. Einstein, L. Infield, *The Evolution of Physics*, Simon & Schuster, New York, 1966, p.180
26. M. Tegmark,. *Our Mathematical Universe: My Quest for the Ultimate Nature of Reality*, Vintage Books, 2015, p. 82
27. R. Lanza, B. Berman, *Biocentrism: How Life and Consciousness Are the Keys to Understanding the True Nature of the Universe*, BenBella Books, Dallas, 2010, p.206

## D. RELATIVITY

1. W. Loewenstein, *Physics in Mind: A Quantum View of the Brain*, Basic Books, New York, 2013, p.13
2. L. Susskind, G. Hrabovsky, *The Theoretical Minimum: What You Need to Know to Start Doing Physics*, Basic Books, 2014, p.96
3. M. Lange, *An Introduction to the Philosophy of Physics: Locality, Fields, Energy, and Mass*, Blackwell Publishing, 2002, p.128
4. ibid. p.126
5. ibid. p.114
6. ibid. p.123
7. http://teacher.nsrl.rochester.edu/phy122/Lecture_Notes/Chapter22/Chapter22.html
   Coulomb's law: "The magnitude of the electric force that a particle exerts on another particle is directly proportional to the product of their charges and inversely proportional to the square of the distance between them. The direction of the force is along the line joining the particles."
8. R. Lanza, B. Berman, *Biocentrism: How Life and Consciousness Are the Keys to Understanding the True Nature of the Universe*, BenBella Books, Dallas, 2010, p.206
9. A. Einstein, Albert, *Out of My Later Years*, Philosophical Library, New York, 1950, p.110
10. Lange, op. cit. p.231
11. Lange, op. cit. p.240
12. B. Cox and J. Forshaw, *Why Does E=mc2?: (and Why Should We Care?)*, Da Capo, 1994. p.45
13. B. Haisch, *The God Theory: Universes, Zero-point Fields, and What's Behind It All*, Red Wheel/Weiser, San Francisco, 2009, p.118
14. ibid. p. 93
15. L. Smolen, *Time Reborn: From the Crisis in Physics to the Future of the Universe*, Mariner Books, 2014, p.57
16. P.Russell, *From Science to God: The Mystery of Consciousness and the Meaning of Light*, New World Library, Novato, CA., 2002, p.62
17. Smolen, op. cit. p.56
18. ibid. p.58
19. ibid. p.61
20. C. Seife, *Decoding the Universe: How the New Science of Information is Explaining Everything in the Cosmos, from Our Brains to Black Holes,* Penguin Books, 2006, p.140
21. ibid. p.134
22. Cox and Forshaw, op. cit. p. 121
23. Smolen, op. cit. p.67
24. L. Nottale, *Fractal Space-Time and Microphysics: Towards a Theory of Scale Relativity*, World Scientific Publishing, London, 1993, p.11
25. Lanza and Berman, op. cit. p.110
26. R. Unger, and L. Smolin, *The Singular Universe and the Reality of Time: A Proposal in Natural Philosophy*, University Printing House, Cambridge, 2015, p.168
27. Lanza and Berman, op. cit. p.122
28. A. Frank, "Has Next Tuesday Already Happened?" http://www.npr.org/blogs/13.7/2014/08/26/343483711/has-next-tuesday-already-happened, August 26, 2014
29. U. Mohrhoff, "Why the Laws of Physics Are Just So," *This Quantum World,* http://thisquantumworld.com/wp/a-critique- of-quantum-mechanics/why-the-laws-of-physics-are-just-so/

## E. QUANTUM


1. H.Stapp, *Mindful Universe: Quantum Mechanics and the Participating Observer* (The Frontiers Collection) 2nd ed., Springer, Heidelberg, 2011, p. 91
2. ibid. p. 88
3. E. Frenkel, "The Reality of Quantum Weirdness," *Gray Matter, The New York Times*, February 20, 2015
4. S. Strogatz, Sync: *How Order Emerges From Chaos In the Universe, Nature, and Daily Life*, Hyperion, 2004, p.129
5. http://en.wikipedia.org/wiki/Quantum
6. P. Russell, *From Science to God: The Mystery of Consciousness and the Meaning of Light*, New World Library, Novato, CA., 2002, p.65
7. B. Haisch, *The God Theory: Universes, Zero-point Fields, and What's Behind It All*, Red Wheel/Weiser, San Francisco, 2009, p. 94
8. C. Seife, *Decoding the Universe: How the New Science of Information is Explaining Everything in the Cosmos, from Our Brains to Black Holes*, Penguin Books, 2006, p.203
9. A. Goswami, *The Self-Aware Universe: How Consciousness Creates the Material World*, Penguin Putnam Inc., New York, 1995, p.32
10. Seife, op.cit., p.204
11. J. Gribbin, *In the Beginning: The Birth of the Living Universe*, Little Brown & Co., 1995, p.246
12. A. Einstein, *Out of My Later Years,* Philosophical Library, New York, 1950, p.87
13. L. Smolen, *Time Reborn: From the Crisis in Physics to the Future of the Universe,* Mariner Books, 2014, p.111
14. Gribbin, op.cit., p.49
15. Smolen, op.cit., p.80
16. F. Capra, *The Tao of Physics*, Shambhala Publications, Inc., Boston, 1985, p.188
17. S. Kullander, B. Larsson, *Out of Sight!: From Quarks to Living Cells*, Cambridge University Press, 1994, p.160
18. A. Einstein, L. Infeld, *The Evolution of Physics*, Simon & Schuster, New York, 1966, p. 275
19. E. Schrodinger, *What Is Life? & Other Scientific Essays*, Doubleday Anchor, 1956, p.49
20. M. Byrne, "How Big Can Quantum Weirdness Get?" *Motherboard*, February 15, 2014
21. B. Rosenblum and F, Kuttner, *Quantum Enigma: Physics Encounters Consciousness*, Oxford University Press, New York, 2011, p.73
22. J. Edwards, *How Many People are There in My Head? And in Hers?: An Exploration of Single Cell Consciousness*, Imprint Academic, Exeter, UK, 2006, p.165
23. T. Hartsfield, "Does the Mind Affect Quantum Mechanics?" *RealClear Science*, July 10, 2014
24. D. Freeman, "Quantum Teleportation Feat Brings Ultrafast Computer Networks Step Closer to Reality," *The Huffington Post*, May 29, 2014
25. Capra, op. cit. p.69
26. Lange, Marc. *An Introduction to the Philosophy of Physics: Locality, Fields, Energy, and Mass*, Blackwell Publishing, 2002, p.45
27. Einstein, op.cit. p.80
28. Einstein, op.cit. p. 84
29. Lange, op.cit. p.152
30. Hartsfield, op.cit.
31. R. Sheldrake, *Morphic Resonance, The Nature of Formative Causation*, Park Street Press, Rochester, Vermont, 2009, p.257
32. Hartsfield, op.cit.
33. Smolen, op.cit., p. 292
34. R. Lanza and B. Berman, *Biocentrism: How Life and Consciousness Are the Keys to Understanding the True Nature of the Universe*, BenBella Books, Dallas, 2010, p.59
35. Rosenblum and Kuttler, op.cit. p.10
36. ibid. p.119
37. E. Schneider, and D. Sagan, *Into the Cool: Energy Flow, Thermodynamics, and Life*, University of Chicago Press, 2006, p.51
38. Hartsfield, op.cit.
39. M.Ho, *The Rainbow and Worm: The Physics of Organisms*, World Scientific Publishing, Singapore, 2008, p.272

40. Hartsfield, op.cit.
41. P. Ball, "How Big Can Schrodinger's Kittens Get?" *Nautilus*, October 8, 2015
42. Smolen, op.cit. p.143
43. Stapp, op.cit. p.104
44. T. Maudlin, *Quantum Non-Locality and Relativity: Metaphysical Intimations of Modern Physics*, Wiley-Blackwell, 2011, p.133
45. Lanza and Berman, op.cit. p.55
46. Edwards, op.cit. p.114
47. Capra, op.cit. p.122
48. Sheldrake, op.cit. p.257
49. Goswami, op.cit. p.97
50. H. Berg, *Random Walks in Biology*, Princeton University Press, 1993. p.11
51. Rosenblum and Kuttler, op. cit. p.120
52. V. Richter, "What Does an Electron Cloud Really Look Like?" *Cosmos Magazine*, September 21, 2015
53. C. Lineweaver, P. Davies, M Ruse, eds., *Complexity and the Arrow of Time*, Cambridge University Press, New York, 2013, p.225
54. ibid., p.226
55. L. Margulis, C. Asikaninen, W. Krumbein, eds., *Chimeras and Consciousness: Evolution of the Sensory Self*, MIT Press, 2011, p.12
56. Smolen, op.cit., p.113
57. University of Oregon, *http://abyss.uoregon.edu/~js/21st_century_science/lectures/lec14.html*
58. J. Lundeen, "Viewpoint: What Can We Say About a Photon's Past?" *APS Physics*, December 9, 2013
59. D. Albert, *After Physics*, Harvard University Press, 2015, p. 29
60. Ho, op.cit., p.332
61. P. Stark, SticiGui, Chapter 13, "Probability: Philosophy and Mathematical Background"
62. Linweaver, op.cit. p.102
63. Hartsfield, op.cit.
64. M. Strevens, "Looking into the Black Box," *New York Times*, November 24, 2013
65. V. Vedral, "Living in a Quantum World," *National Taiwan Normal University*, November 6, 2003
66. Maudlin, op.cit. p.233
67. W. Loewenstein, *The Touchstone of Life: Molecular Information, Cell Communication and the Foundation of Life*, Oxford University Press, New York, 1998, p.100
68. Russell, op.cit. p.60
69. Vedral, op.cit.
70. ibid.
71. The Physics arXiv Blog. "How Quantum Mechanics Forbids the Accelerating Expansion of Spacetime," January 2, 2014
72. Ball, op.cit.
73. J. Mitsch, "Pitt Researcher Lands the Cover of *Development Cell* by Uncovering an Evolutionary Secret," *University of Pittsburgh News Service,* Septermber 3, 2015
74. W. Loewenstein, *Physics in Mind: A Quantum View of the Brain*, Basic Books, New York, 2013, p.2
75. Edwards, op.cit. p.108
76. L.Susskind, G. Hrabovsky, *The Theoretical Minimum: What You Need to Know to Start Doing Physics*, Basic Books, 2014, p.190
77. J.Gorman, "Learning How Little We Know About the Brain," *New York Times*, November, 10, 2014, (comment)
78. Susskind, Hrabovsky, p.126
79. Sheldrake, op. cit. p.255
80. ibid, p.259
81. M. Kaku, *The Future of the Mind: The Scientific Quest to Understand, Enhance and Empower the Mind*, Anchor Books, New York, 2014, p.288
82. ibid, p.337

## F. THERMODYNAMICS

1. E. Schneider, D.Sagan, *Into the Cool: Energy Flow, Thermodynamics, and Life*, University of Chicago Press, 2006, P.49
2. B. Cox, Brian, J. Forshaw, *Why Does E=mc2?: (and Why Should We Care?)*, Da Capo, 1994, p.182

3. L. Smolen, *Time Reborn: From the Crisis in Physics to the Future of the Universe,* Mariner Books, 2014, p.162
4. ibid. p.87
5. ibid. p.204
6. J. van Brakel, *Philosophy of Chemistry: Between the Manifest and the Scientific Image*, Leuven University Press, Leuven, Belgium, 2000, p.2
7. ibid., p.5
8. R.Unger, L. Smolin, *The Singular Universe and the Reality of Time: A Proposal in Natural Philosophy*, University Printing House, Cambridge, 2015, p.257
9. L. Margulis, C. Asikaninen, W. Krumbein, eds., *Chimeras and Consciousness: Evolution of the Sensory Self*, MIT Press, 2011, p.12
10. S. Strogatz, Sync: *How Order Emerges From Chaos In the Universe, Nature, and Daily Life*, Hyperion, 2004, p.158
11. W. Loewenstein, *Physics in Mind: A Quantum View of the Brain*, Basic Books, New York, 2013, p.9
12. Smolen, op.cit. p.168
13. J. Gleick, *Chaos: Making a New Science*, Penguin Books, New York, 2008, p.98
14. ibid., p.99
15. L.Nottale, *Fractal Space-Time and Microphysics: Towards a Theory of Scale Relativity*, World Scientific Publishing, London, 1993. p.40
16. Gleick, op.cit. p.98
17. Nottale, op.cit. p.47
18. Boeyens, Jan C. A. "A Molecular-Structure Hypothesis," *International Journal of Molecular Sciences*, November 1, 2010
19. R. Lanza, B. Berman, *Biocentrism: How Life and Consciousness Are the Keys to Understanding the True Nature of the Universe*, BenBella Books, Dallas, 2010, p.97
20. ibid., p.98
21. Smolen, op.cit., p.198
22. C. Lineweaver, P. Davies, M. Ruse, Michael. eds., *Complexity and the Arrow of Time*, Cambridge University Press, New York, 2013. p.128
23. F. Lambert, "Entropy and the Second Law of Thermodynamics," *Occidental College*, Los Angeles, September, 2012
24. ibid.
25. D. Albert, *After Physics*, Harvard University Press, 2015, p.72
26. Lineweaver, op.cit. p.19
27. A. Pross, *What is Life? How Chemistry Becomes Biology*, Oxford University Press, Oxford, UK, 2012, p.59
28. Schneider, Sagan, op.cit., p.76
29. Strogatz, op.cit., p.198
30. Susskind, Leonard, Hrabovsky, George. *The Theoretical Minimum: What You Need to Know to Start Doing Physics*, Basic Books, 2014, p.196
31. Schneider, Sagan, op.cit., p.324
32. *https://en.wikipedia.org/wiki/Second_law_of_thermodynamics*
33. ibid.
34. Lineweaver, op.cit., p.99
35. ibid., p.22
36. ibid., p.114
36. E. Siegel, "All About Cosmic Inflation," https://medium.com/starts-with-a-bang/all-about-cosmic- in flation-211a64441ddd
37. J. Gribbin, *In the Beginning: The Birth of the Living Universe*, Little Brown & Co., 1995, p.227
38. Lineweaver, op.cit., p.51
39. ibid., p.114
40. T. Deacon, Incomplete Nature: How Mind Emerged from Matter, W.W. Norton & Co., New York, 2012, p.344
41. Strogatz, op.cit., p.
42. Lineweaver, op.cit., p.23
43. ibid., p.59
44. Schnieder, Sagan, op. cit., preface

45. M. Ho, *The Rainbow and Worm: The Physics of Organisms*, World Scientific Publishing, Singapore, 2008, p.154
46. Strogatz, op.cit., p.287
47. Gleick, op.cit. p.198
48. ibid., p.204
49. R. Cuthbertson, M. Holcombe, R. Paton, *Computation in Cellular and Molecular Biological Systems*, World Scientific, 1996, p.251
50. Gleick, op.cit., p.195
51. A. Huber, "In 125 Years, Millions of People Have Looked at this Painting. No One Really Saw it until Recently," *Upworthy*, November 14, 2014
52. Strogatz, op.cit., p.183
53. Ghose, Tia. "New Math Could Reveal Hidden Sources of Chaos," *LiveScience*, August 27, 2015
54. Smolen, op.cit., p.275
55. Strogatz, op.cit., p.192
56. Smolen, op.cit. p.278
57. Susskind, Hrabovsky, op.cit., p.162
58. ibid., p.165
59. Lineweaver, op.cit., p.13
60. ibid, p.11
61. *https://en.wikipedia.org/wiki/Thermodynamics*
62. Lineweaver, op. cit., p.338
63. Hoffmann, Peter M. *Life's Ratchet: How Molecular Machines Extract Order from Chaos*, Basic Books, Philadelphia, 2012, p.244
64. Strogatz, op. cit., p.184
65. ibid., p.54
66. ibid., p.194
67. ibid., p.196
68. ibid., p.196
69. *https://en.wikipedia.org/wiki/Self-organization*
70. Lineweaver, op.cit., p.116
71. J. Matson, "Quantum Entanglement Links 2 Diamonds," *Scientific American*, December 1, 2011
72. Strogatz, op.cit., p.231
73. Lineweaver, op.cit., p.200
74. Gleick, op.cit., p.162
75. Lineweaver, op.cit., p.203
76. Strogatz, op.cit., p.322
77. ibid., p.256
78. M. Ho, "Bioenergetics and the Coherence of Organisms," *https://ratical.org/co-globalize/MaeWanHo/prague.html*
79. Strogatz, op.cit., p.322
80. P. Ball,"The Strange Inevitability of Evolution," *Nautilus*, January 8, 2015, *http://nautil.us/issue/20/creativity/the-strange-inevitability-of-evolution*
81. C. Zimmer, Carl, *Microcosm: E. Coli and the New Science of Life*, First Vintage Books, New York, 2009, p.42
82. Loewenstein, op.cit,. p.21
83. Hoffman, op.cit., p.240
84. Smolen, op.cit., p.197
85. R. Sheldrake, *Morphic Resonance, The Nature of Formative Causation*, Park Street Press, Rochester, Vermont, 2009, p.75
86. Hoffmann, op.cit., p.76
87. ibid., p.53
88. Deacon, op.cit., p.30
89. Ibid., p.213
90. Lineweaver, op.cit., p.92
91. *https://en.wikipedia.org/wiki/Entropy_(arrow_of_time)#cite_note-5*
92. Loewenstein, op.cit., p.5
93. ibid., p.6
94. A. Friedman, "Are the Quantum World and The Real Word the Same Thing?" *The Nature of Reality*, May 7, 2015

95. ibid.
96. G. Chiribella, "Is Your Theory of Everything Pure Enough?" *Nautilus*, February 19, 2015
97. ibid.
98. N. Wolchover, "New Quantum Theory Could Explain the Flow of Time," *Wired.com*, April, 25, 2014
99. J. Ouellette, "How Quantum Pairs Stitch Space-Time," *Quantamagazine.org*, April 28, 2015
100. Loewenstein, op.cit., p.8
101. Wolchover, op.cit.
102. Lineweaver, op.cit., p.98
103. Wolchover, op.cit.
104. Ball, op.cit.
105. Ouellette, op.cit.
106. Ouellette, op.cit.
107. Ouellette, op.cit
108. H. Ooguri, "How Spacetime Is Built by Quantum Entanglement: New Insight into Unification of General Relativity and Quantum Mechanics," *Kavli Institute for the Physics and Mathematics of the Universe*, May 27, 2015
109. Chiribella, op.cit.
110. Loewenstein, op.cit., p.20
111. C. Seife, *Decoding the Universe: How the New Science of Information is Explaining Everything in the Cosmos, from Our Brains to Black Holes,* Penguin Books, 2006. p.9
112. Chiribella, op.cit.
113. Seife, op.cit. p.203
114. Center for Quantum Technologies, "Two Quantum Mysteries Merged into One," December 20, 2014
115. Wolchover, op.cit.

## II. FORM

### A. EMERGENCE

1. W. Loewenstein, *Physics in Mind: A Quantum View of the Brain*, Basic Books, New York, 2013, p.8
2. K. Taylor, "A New Physics Theory of Life," *Quanta Magazine*, January 22, 2014
3. C. Lineweaver, P. Davies, M. Ruse, eds., *Complexity and the Arrow of Time*, Cambridge University Press, New York, 2013, p.59
4. Hoffmann, Peter M. *Life's Ratchet: How Molecular Machines Extract Order from Chaos*, Basic Books, Philadelphia, 2012, p.98
5. E. Thompson, *Mind in Life: Biology, Phenomenology and the Sciences of Mind*, Harvard University Press, Cambridge, Ma., 2007, p.60
6. J. Gleick, *Chaos: Making a New Science*, Penguin Books, New York, 2008, p.103
7. ibid., p.73
8. ibid., p.260
9. Lineweaver, op.cit., p.95
10. H. Berg, *Random Walks in Biology*, Princeton University Press, 1993, p.121
11. M. Ho, *The Rainbow and Worm: The Physics of Organisms*, World Scientific Publishing, Singapore, 2008, p.87
12. Lineweaver, op.cit., p.344
13. H. Stapp, *Mindful Universe: Quantum Mechanics and the Participating Observer* (The Frontiers Collection) 2nd ed., Springer, Heidelberg, 2011, p.155
14. Ho, op.cit., p.92
15. Hoffman, op.cit., p.86
16. Lineweaver, op.cit., p.207
17. E.Schneider, D. Sagan, *Into the Cool: Energy Flow, Thermodynamics, and Life*, University of Chi cago Press, 2006, p.16
18. ibid., p.87
19. F. Lambert, "Entropy and the Second Law of Thermodynamics," *Occidental College*, Los Angeles, September, 2012, *http://secondlaw.oxy.edu/*
20. Gleick, op.cit., p.67, 114

21. Schneider, op.cit., p.198
22. L.Margulis, C. Asikaninen, W. Krumbein, eds., *Chimeras and Consciousness: Evolution of the Sensory Self*, MIT Press, 2011, p.245
23. Hoffman, op.cit., p.78
24. Ho, op.cit., p.62
25. Schneider, op.cit., p.317
26. R.Doyle, "The Two-Stage Model to the Problem of Free Will," *Free Will: The Scandal in Philosophy*, June 19, 2011
27. Schneider. op.cit., p.257
28. ibid., p.323
29. ibid., p.44
30. Hoffman, op.cit., p.81
31. Schneider, op.cit., p.32
32. Hoffman, op.cit., p.131

## B. FORM

1. M.Talbot, *The Holographic Universe: The Revolutionary Theory of Reality*, HarperCollins, New York, 2011, p.50
2. H. Jonas, *The Phenomenon of Life: Toward a Philosophical Biology*, Dell Publishing, New York, 1966, p.80
3. A. Pross, *What is Life? How Chemistry Becomes Biology*, Oxford University Press, Oxford, UK, 2012, p.23
4. M. Ho, *The Rainbow and Worm: The Physics of Organisms*, World Scientific Publishing, Singapore, 2008, p.6
5. T. Deacon, *Incomplete Nature: How Mind Emerged from Matter*, W.W. Norton & Co., New York, 2012, 270
6. Jonas, op.cit., p.84
7. U. Alon, *An Introduction to Systems Biology: Design Principles of Biological Circuits*, CRC Press, Boca Raton, FL, 2006. p.1
8. ibid., p.1
9. E.Schneider, D. Sagan, *Into the Cool: Energy Flow, Thermodynamics, and Life*, University of Chicago P ress, 2006, p.31
10. E. Schrodinger, *What Is Life? & Other Scientific Essays*, Doubleday Anchor, 1956, p.67
11. C. Lineweaver, P.Davies, M. Ruse, Michael. eds., *Complexity and the Arrow of Time*, Cambridge University Press, New York, 2013, p.166
12. L. Margulis, D. Sagan, *What Is Life?* University of California Press, Berkeley, CA., 1995, p.227
13. R. Sheldrake, *Morphic Resonance, The Nature of Formative Causation*, Park Street Press, Rochester, Vermont, 2009, p.52
14. W. Loewenstein, *The Touchstone of Life: Molecular Information, Cell Communication and the Foundation of Life*, Oxford University Press, New York, 1998, intro.
15. Margulis, op.cit., p.43
16. Jonas, op.cit., p.86
17. Schneider, op.cit., p.91
19 . Jonas, op.cit., p.58
20. F. Lambert, "Entropy and the Second Law of Thermodynamics," *Occidental College*, Los Angeles, September, 2012, *http://secondlaw.oxy.edu/*
21. N. Humphrey, *A History of the Mind: Evolution and the Birth of Consciousness*, Springer-Verlag, New York, 1999, p.187
22. Jonas, op.cit., p.76
23. Lineweaver, op.cit., p.35
24. Jonas, op.cit., p.4
25. Deacon, op.cit., p.309
26. Humphrey, op.cit., p.40
27. Pross, op.cit.,p.176
28. L. Margulis, C. Asikaninen, W, Krumbein, eds., *Chimeras and Consciousness: Evolution of the Sen sory Self*, MIT Press, 2011, p.249

29. A. Rosenberg, D. McShea, *Philosophy of Biology: A Contemporary Introduction*, Routledge, New York, 2008, p.13
30. ibid., p.13
31. B. Lipton, *The Biology of Belief: Unleashing the Power of Consciousness, Matter & Miracles*, Hay House, 2008, prologue
32. Deacon, op.cit., p.542
33. E. Thompson, *Mind in Life: Biology, Phenomenology and the Sciences of Mind*, Harvard University Press, Cambridge, Ma.,2007, p.213
34. Lineweaver, op.cit., p.168
35. E. Schrodinger, *What Is Life? & Other Scientific Essays*, Doubleday Anchor, 1956, p.274
36. Lineweaver, op.cit., p.207
37. Pross, op.cit., p.3
38. F. Boogerd, F. Bruggeman, J. Hofmeyr, H.Westerhoff, eds. *Systems Biology: Philosophical Foundations*, Elsevier, Amsterdam, 2007, p.249
39. Lineweaver, op.cit., p.207
40. G. Bateson, A. Montuori, *Mind and Nature: A Necessary Unity*, Hampton Press, 2002, p.162
41. Alon, op.cit., p.2

## C. INFORMATION

1. K. Hartnett, "The Information Theory of Life," *Quanta Magazine*, November 19, 2015
2. C. Seife, *Decoding the Universe: How the New Science of Information is Explaining Everything in the Cosmos, from Our Brains to Black Holes,* Penguin Books, 2006, p.89
3. P. Hoffmann, *Life's Ratchet: How Molecular Machines Extract Order from Chaos*, Basic Books, Philadelphia, 2012, p.223
4. R. Garrett, C.Grisham, *Biochemistry*, Cengage, Boston, 2017, p.9
5. Hoffman, op.cit., p.137
6. W. Loewenstein, *Physics in Mind: A Quantum View of the Brain*, Basic Books, New York, 2013, p.25
7. M. Strevens, "Looking into the Black Box," *New York Times*, November 24, 2013
8. C. Lineweaver, P.Davies, M. Ruse, eds., *Complexity and the Arrow of Time*, Cambridge University Press, New York, 2013, p.34
9. ibid., p.33
10. H. Jonas, *The Phenomenon of Life: Toward a Philosophical Biology*, Dell Publishing, New York, 1966, p.85
11. ibid., p.105
12. M. Ho, "The Biology of Free Will," *Journal of Consciousness Studies*, 1996
13. Lineweaver, op.cit., p.258
14. ibid. p.237
15. Seife, op.cit., p.87
16. Loewenstein, op.cit., p.63
17. ibid., p.176
18. www.truth-revelations.com
19. Loewenstein, op.cit., p.141
20. E. Thompson, *Mind in Life: Biology, Phenomenology and the Sciences of Mind*, Harvard University Press, Cambridge, Ma.,2007, p.125

## D. FREEDOM

1. H. Jonas, *The Phenomenon of Life: Toward a Philosophical Biology*, Dell Publishing, New York, 1966, p.84
2. R. Maturana, F.Varela, Francisco J., *The Tree of Knowledge: The Biological Roots of Human Understanding*, Shambhala Publications, Boston, 1992, p.124
3. M. Ho, *The Rainbow and Worm: The Physics of Organisms*, World Scientific Publishing, Singapore, 2008, p.279
4. L. Margulis, D. Sagan, *What Is Life?* University of California Press, Berkeley, CA., 1995. p.14

5. C. Lineweaver, P. Davies, M. Ruse, eds., *Complexity and the Arrow of Time*, Cambridge University Press, New York, 2013, p.342
6. P. Hoffmann, *Life's Ratchet: How Molecular Machines Extract Order from Chaos*, Basic Books, Philadelphia, 2012, p.153
7. H. Jonas, *The Phenomenon of Life: Toward a Philosophical Biology*, Dell Publishing, New York, 1966. p.108
8. U. Alon, *An Introduction to Systems Biology: Design Principles of Biological Circuits*, CRC Press, Boca Raton, FL, 2006, p.35
9. D. Hofstadter, *Godel, Escher, Bach: An Eternal Golden Braid*, Basic Books, New York, 1999 edition, p.544
10. D. Noble, "A Theory of Biological Relativity: No Privileged Level of Causation," *The Royal Society Publishing*, November 9, 2011
11. E. Thompson, *Mind in Life: Biology, Phenomenology and the Sciences of Mind*, Harvard University Press, Cambridge, Ma., 2007. p.369
12. Lineweaver, op.cit., p.183
13. http://en.wikipedia.org/wiki/Catalysis
14. A. Pross, *What is Life? How Chemistry Becomes Biology*, Oxford University Press, Oxford, UK, 2012. p.63
15. ibid., p.64
16. Lineweaver, op.cit., p.170
17. W. Loewenstein, *The Touchstone of Life: Molecular Information, Cell Communication and the Foundation of Life*, Oxford University Press, New York, 1998, intro.
18. T. Deacon, *Incomplete Nature: How Mind Emerged from Matter*, W.W. Norton & Co., New York, 2012, p.293
19. Ho, op.cit., p.29
20. Chesher, *This Magic Sea,* "The Error of Expectations," Chesher, 2011
21. Ho, op.cit., p.166
22. Hoffman, op.cit., p.232
23. Noble, op.cit.
24. Chesher, op.cit., *"Conscious"*
25. Lineweaver, op.cit. p.207
26. ibid., p.344
27. ibid., p.206
28. B. Lipton, *The Biology of Belief: Unleashing the Power of Consciousness, Matter & Miracles*, Hay House, 2008, p.72
29. Chesher, op.cit.
30. G. Lawton, "You Are Your Brain," *Slate.com*, December 8, 2013
31. Chesher, op.cit.
32. Thompson, op.cit., p.185
33. D. Noble, "Claude Bernard, the First Systems Biologist, and the Future of Physiology," *The Physiological Society*, December 18, 2007
34. Lineweaver, op.cit., p.342
35. Maturana, op.cit., p.47
36. C. Zimmer, *Microcosm: E. Coli and the New Science of Life*, First Vintage Books, New York, 2009, p.42
37. Alon, op.cit., p.41
38. ibid., p.41
39. F. Boogerd, F.Bruggeman, J. Hofmeyr, H.Westerhoff, Hans, eds. *Systems Biology: Philosophical Foundations*, Elsevier, Amsterdam, 2007, p.246
40. ibid., p.238
41. D. Bray, *Wetware: A Computer in Every Living Cell*, Yale University Press, 2009, p.120
42. Maturana, op.cit., p.75
43. Ho, op.cit., p.40
44. Boogerd, op.cit., p.294
45. Hoffman, op.cit., p.238
46. Maturana, op.cit., p.52
47. Ho, op.cit., p.327
48. Thompson, op.cit., p.426

49. http://en.wikipedia.org/wiki/Autopoiesis
50. R. Sheldrake, *Morphic Resonance, The Nature of Formative Causation*, Park Street Press, Rochester, Vermont, 2009, p.10
51. E. Schrodinger, Erwin. *What Is Life? & Other Scientific Essays*, Doubleday Anchor, 1956, p.78
52. Hofstadter, op.cit., p.503

## E. STRUCTURE

1. A. Pross, *What is Life? How Chemistry Becomes Biology*, Oxford University Press, Oxford, UK, 2012, prologue
2. J. Gribbin, *In the Beginning: The Birth of the Living Universe*, Little Brown & Co., 1995, p.187
3. E. Schneider, D. Sagan, *Into the Cool: Energy Flow, Thermodynamics, and Life*, University of Chicago Press, 2006, p.107
4. http://en.wikipedia.org/wiki/Gravitational_biology
5. L. Rothschild, A. Lister, eds. *Evolution on Planet Earth: The Impact of the Physical Environment*," Academic Press, London, 2003
6. P. Hoffmann, *Life's Ratchet: How Molecular Machines Extract Order from Chaos*, Basic Books, Philadelphia, 2012, p.68
7. R.Garrett, C. Grisham, *Biochemistry*, Cengage, Boston, 2017, p.6
8. C. Lineweaver, P. Davies, M. Ruse, eds., *Complexity and the Arrow of Time*, Cambridge University Press, New York, 2013, p.76
9. M. Ho, *The Rainbow and Worm: The Physics of Organisms*, World Scientific Publishing, Singapore, 2008, p.253
10. ibid., p.247
11. Hoffman, op.cit., p.120
12. ibid., p.123
13. ibid., p.108
14. M. Ho, "The Biology of Free Will," *Journal of Consciousness Studies*, 1996
15. D. Newman, *Bioquestions and the Mechanical Answer*, CreateSpace Independent Publishing Platform, 2013, p.14
16. W. Loewenstein, *Physics in Mind: A Quantum View of the Brain*, Basic Books, New York, 2013, p.53
17. J. Shapiro, *Evolution: A View from the 21st Century*, FT Press Science, Upper Saddle River, NJ, 2011, p.137
18. ibid., p.137
19. B. Lipton, *The Biology of Belief: Unleashing the Power of Consciousness, Matter & Miracles*, Hay House, 2008, p.7
20. ibid., prologue
21. https://en.wikipedia.org/wiki/Power_(physics)
22. R. Pico, *Consciousness in Four Dimensions: Biological Relativity and the Origins of Thought*, McGraw-Hill, 2001. p.291
23. D.Bray, *Wetware: A Computer in Every Living Cell*, Yale University Press, 2009, p.7
24. ibid., p.10
25. ibid., p.18
26. ibid., p.237
27. ibid, p.236
28. Hoffman, op.cit., p.113
29. Lipton, op.cit., p.26
30. Lineweaver, op.cit., p.167
31. Loewenstein, op.cit., p.190
32. Bray, op.cit., p.126
33. Garrett, op.cit., p.10
34. Loewenstein, op.cit., p.64
35. ibid., p.37
36. Garrett, op.cit., p.10
37. ibid., p.30
38. Ho, Biology, op.cit.
39. L. Margulis, C, Asikaninen, W. Krumbein, eds., *Chimeras and Consciousness: Evolution of the Sensory Self*, MIT Press, 2011, p.

40. Hoffman, op.cit., p.161
41. ibid., p.167
42. Bray, op.cit., p.94
43. Loewenstein, op.cit., p.147
44. Shapiro, op.cit., p.11
45. ibid., p.97
46. Ho, Rainbow, op.cit., p.137
47. B. Alberts, *Molecular Biology of the Cell*, Garland Science, 2014, p.6
48. Hoffman, op.cit., p.223
49. Shapiro, op.cit., p.57
50. C. Zimmer, *Microcosm: E. Coli and the New Science of Life*, First Vintage Books, New York, 2009, p.34
51. Garrett, op.cit., p.12
52. N. Lane, *Life Ascending: The Ten Great Inventions of Evolution*, Norton, 2010. p.7
53. Zimmer, op.cit., p.32
54. U. Alon, *An Introduction to Systems Biology: Design Principles of Biological Circuits*, CRC Press, Boca Raton, FL, 2006, p.18
55. Loewenstein, op.cit., p.277
56. Lipton, op.cit., p.161
57. Alberts, op.cit., p.626
58. ibid., p.626
59. Garrett, op.cit., p.20
60. Loewenstein, op.cit., p.38
61. ibid., p.38
62. Margulis, op.cit., p.50
63. Zimmer, op.cit., p.20
64. ibid., p.27
65. Lipton, op.cit., p.58
66. Loewenstein, op.cit., p.170
67. Ho, Rainbow, op.cit., p.142
68. Lane, op.cit., p.65
69. ibid., p.168
70. Lipton, op.cit., p.168
71. Zimmer, op.cit., p.26
72. J. Minkel, "Focus: Ripples on a Cell," *Physics Review*, November, 13, 2009
73. Garrett, op.cit., p.46
74. Bray, op.cit., p.87
75. Lane, op.cit., p.31
76. Ho, Rainbow, op.cit., p.54
77. Z. Merali, "Solving Biology's Mysteries Using Quantum Mechanics," *Discover Magazine*, October 30, 2014
78. Pross, op.cit., p.170
79. O. Froy, R. Miskin, Ruth, "Effect of feeding regimens on circadian rhythms: Implications for Aging and longevity," *Impact Journals: Open-Access Impact Journal on Aging*, January 11. 2010
80. S.Strogatz, *Sync: How Order Emerges From Chaos In the Universe, Nature, and Daily Life*, Hyperion, 2004, p.100
81. D. Noble, "Claude Bernard, the First Systems Biologist, and the Future of Physiology," *The Physiological Society*, December 18, 2007
82. A. Huffington, P.Fuller, "How Neuroscience Can Unlock Sleep's Mysteries," *Huffington Post*, March 26, 2015
83. Schneider, op.cit., p.26
84. Ho, Rainbow, op.cit., p.77
85. Pross, op.cit., p.5
86. Ho, Biology, op.cit.
87. ibid.
88. ibid.
89. Ho, Rainbow, op.cit. p.94

90. A. Scott, *Stairway to the Mind: The Controversial New Science of Consciousness*, Springer-Verlag, New York, 1995, p.170
91. J. Gleick, *Chaos: Making a New Science*, Penguin Books, New York, 2008, p.278

## F. GROUPS

1. Pross, Addy. *What is Life? How Chemistry Becomes Biology*, Oxford University Press, Oxford, UK, 2012, prologue
2. L. Margulis, C. Asikaninen, W. Krumbein, eds., *Chimeras and Consciousness: Evolution of the Sensory Self*, MIT Press, 2011, p.4
3. P. Ball, "The Strange Inevitability of Evolution," *Nautilus*, January 8, 2015
4. D. Bray, *Wetware: A Computer in Every Living Cell*, Yale University Press, 2009, p.105
5. Margulis, op.cit., p.3
6. B. Lipton, *The Biology of Belief: Unleashing the Power of Consciousness, Matter & Miracles*, Hay House, 2008, p.159
7. C. Willyard, "Bonnie Bassler, Silent Treatment," *Discover Magazine*, June 2014
8. C. Zimmer, *Microcosm: E. Coli and the New Science of Life*, First Vintage Books, New York, 2009, p.55
9. Bray, op.cit., p.169
10. Zimmer, op.cit., p.57
11. E. Conover, "Go with the Flow," *Discover Magazine*, May, 2015
12. S. Strogatz, *Sync: How Order Emerges From Chaos In the Universe, Nature, and Daily Life*, Hyperi on, 2004, p.13
13. L.Margulis, D.Sagan, *What Is Life?* University of California Press, Berkeley, CA., 1995, p.23
14. Lipton, op.cit., p.100
15. Zimmer, op.cit., p.49
16. Lipton, op.cit., p.9
17. Zimmer, op.cit., p.52
18. R. Garrett, C. Grisham, *Biochemistry*, Cengage, Boston, 2017, p.54
19. Margulis, Sagan, op.cit., p.141
20. E. Chaisson, *Epic of Evolution, Seven Ages of the Cosmos*, Columbia University Press, 2006, p.358
21. J. Shapiro, "What is the Key to a Realistic Theory of Evolution?" *Huffington Post*, February, 16, 2012
22. W. Loewenstein, *The Touchstone of Life: Molecular Information, Cell Communication and the Foundation of Life*, Oxford University Press, New York, 1998, p.253
23. B. Alberts, *Molecular Biology of the Cell*, Garland Science, 2014. p.626
24. Margulis, Asikaninen, op.cit. p.47
25. Alberts, op.cit., p.1110
26. Loewenstein, op.cit., p.288
27. ibid., p.286
28. Alberts, op.cit., p.626
29. H. Maturana, F. Varela, *The Tree of Knowledge: The Biological Roots of Human Understanding*, Shambhala Publications, Boston, 1992, p.83
30. Alberts, op.cit., p.626
31. B.Alberts, D, Bray, K, Hopkin, et.al., *Essential Cell Biology, Fourth Edition*, Garland Science, 2013, p.384
32. C. Lineweaver, P. Davies, M. Ruse, eds., *Complexity and the Arrow of Time*, Cambridge University Press, New York, 2013, p.152
33. Lowenstein, op.cit., p.277
34. A. Rosenberg, D. McShea, *Philosophy of Biology: A Contemporary Introduction*, Routledge, New York, 2008, p.145
35. Lowenstein, op.cit., p.177
36. Margulis, Asikaninen, op.cit. p.55
37. ibid., p.58
38. ibid., p.55
39. Loewenstein, op.cit., p.285
40. Chaisson, op.cit. p.358
41. Margulis, Asikaninen, op.cit. p.56

42. ibid., p.59
43. Ibid., p.61
44. Lowenstein, op.cit., p.198
45. Pross, op.cit., p.187
46. Lipton, op.cit., p.101
47. M. Ho, *The Rainbow and Worm: The Physics of Organisms*, World Scientific Publishing, Singapore, 2008, p.237
48. W. Loewenstein, *Physics in Mind: A Quantum View of the Brain*, Basic Books, New York, 2013, p.35
49. ibid., p.127
50. Ho, op.cit.,p.236
51. N. Lane, *Life Ascending: The Ten Great Inventions of Evolution*, Norton, 2010, p.214
52. Strogatz, op.cit., p.72
53. J. Gleick, *Chaos: Making a New Science*, Penguin Books, New York, 2008, p.283
54. D. Hofstadter, *Godel, Escher, Bach: An Eternal Golden Braid*, Basic Books, New York,1999 edition, p.167
55. D. Noble, "Claude Bernard, the First Systems Biologist, and the Future of Physiology," *The Physio logical Society*, December 18, 2007
56. Gleick, op.cit., p.288
57. Ho, op.cit., p.159
58. Strogatz, op.cit., p210
59. M. Pollan, "Some of My Best Friends Are Germs," *New York Times,* May 15, 2013
60. Loewenstein, *Touchstone*, op.cit., p.211
61. Bray, op.cit., p.164
62. Doyle, Robert. *Free Will: The Scandal in Philosophy,* Information Philosopher, Cambridge, MA, 2011, p.237
63. Lane, op.cit., p.246
64. J. Edwards, *How Many People are There in My Head? And in Hers?: An Exploration of Single Cell Consciousness*, Imprint Academic, Exeter, UK, 2006, p.180
65. Pollan, op.cit.

## G. EVOLUTION

1. W. Loewenstein, *The Touchstone of Life: Molecular Information, Cell Communication and the Founda tion of Life*, Oxford University Press, New York, 1998, p.329
2. S. Watts, "Assessing the Impact of Venter's 'Synthetic Life,'" *BBC News*, May 20, 2010
3. C. Lineweaver, P. Davies, M. Ruse, eds., *Complexity and the Arrow of Time*, Cambridge University Press, New York, 2013, p.344
4. A. Pross, *What is Life? How Chemistry Becomes Biology*, Oxford University Press, Oxford, UK, 2012, p.142
5. Lineweaver, op.cit., p.280
6. ibid., p.7
7. ibid., p.247
8. ibid., p.242
9. ibid., p.226
10. ibid., p.154
11. ibid., p.226
12. ibid., p.145
13. ibid., p.287
14. ibid., p.210
15. ibid., p.6
16. L. Smolen, *Time Reborn: From the Crisis in Physics to the Future of the Universe,* Mariner Books, 2014, preface
17. J. Edwards, *How Many People are There in My Head? And in Hers?: An Exploration of Single Cell Consciousness*, Imprint Academic, Exeter, UK, 2006, p.201
18. U.Alon, *An Introduction to Systems Biology: Design Principles of Biological Circuits*, CRC Press, Boca Raton, FL, 2006, p.41

19. Lineweaver, op.cit., p.247
20. A. Scott, *Stairway to the Mind: The Controversial New Science of Consciousness*, Springer-Verlag, New York, 1995, p.114
21. D. Bray, *Wetware: A Computer in Every Living Cell*, Yale University Press, 2009, p.135
22. N. Humphrey, *A History of the Mind: Evolution and the Birth of Consciousness*, Springer-Verlag, New York, 1999, p.38
23. G. Bateson, A., Montuori, Alfonso, *Mind and Nature: A Necessary Unity*, Hampton Press, 2002, f orward
24. E. Schneider, D. Sagan, Dorian, *Into the Cool: Energy Flow, Thermodynamics, and Life*, University of Chicago Press, 2006, p.46
25. L. Margulis, D. Sagan, *What Is Life?* University of California Press, Berkeley, CA., 1995, p.51
26. The Astronomist."The Universe and Life is Asymmetric: Chirality," *fieldofscience.com*, January 21, 2011
27. ibid.
28. M.Klussmann, H. Iwamura, et.al., "Thermodynamic Control of Asymmetric Amplification in Amino Acid Catalysis," *Nature*, June 1, 2006
29. Loewenstein, op.cit., p.29
30. ibid., p.39
31. Scheider, op.cit., preface
32. S. Chatterjee, "How Did Life Begin? Texas Paleontologist Says He's Found the Biological 'Holy Grail,'" interviewed in *Huffington Post*, November 1, 2013
33. N. Lane, *Life Ascending: The Ten Great Inventions of Evolution*, Norton, 2010, p.17
34. ibid., p.32
35. W. Loewenstein, *Physics in Mind: A Quantum View of the Brain*, Basic Books, New York, 2013, p.131
36. Loewenstein, *Touchstone,* op.cit., p.102
37. Lineweaver, op.cit., p.82
38. Pross, op.cit., p.136
39. Scheider, op.cit., p.255
40. Margulis, op.cit., p.229
41. Smolen, op.cit., p.219
42. Loewenstein, *Touchstone,* op.cit., p.33
43. F. Boogerd, F. Bruggeman, J. Hofmeyr, H. Westerhoff, eds. *Systems Biology: Philosophical Foundations*, Elsevier, Amsterdam, 2007, p.256
44. L. Margulis, C. Asikaninen, W. Krumbein, eds., *Chimeras and Consciousness: Evolution of the Sensory Self*, MIT Press, 2011, p.31
45. Lineweaver, op.cit., p.205
46. R. Service, "Origin of Life in RNA? Study Suggests Genetic Molecules May Have Self-Assembled," *Huffington Post*, February 12, 2013
47. Pross, op.cit. p.133
48. Loewenstein, *Touchstone,* op.cit., p.96
49. Lineweaver, op.cit., p.206
50. J. Silva, "Darwin's Pharmacy: Sex, Plants and the Evolution of the Noosphere," *Huffington Post*, June 29, 2011
51. Loewenstein, *Touchstone*, op.cit., p.81
52. Lineweaver, op.cit., p.169
53. ibid, p.169
54. Margulis, *What is Life,* op.cit., p.85
55. F. Horkey, P. Basser, A. Hecht, E. Geissler, Erik, "Effect of calcium/sodium io exchange on the osmotic properties and structure of polyelectrolyte gels," *Journal of Engineering in Medicine*, July 20, 2015
56. Lane, op.cit., p.33
57. Lineweaver, op.cit., p.325
58. C. Zimmer, *Microcosm: E. Coli and the New Science of Life*, First Vintage Books, New York, 2009, p.155
59. ibid., p.156
60. C. Zimmer, "Vitamins' Old, Old Edge," *New York Times*, December 9, 2013
61. Lineweaver, op.cit., p.214

62. Zimmer, *Microcosm*, op.cit., p.132
63. J. Drake, "A constant rate of spontaneous mutation in DNA-based microbes," *National Academy of Sciences*, August 15, 1991
64. J. Shapiro, *Evolution: A View from the 21st Century*, FT Press Science, Upper Saddle River, NJ, 2011, p.1
65. G. Bateson, A. Montuori, *Mind and Nature: A Necessary Unity*, Hampton Press, 2002, p.41
66. P. Hoffmann, *Life's Ratchet: How Molecular Machines Extract Order from Chaos*, Basic Books, Philadelphia, 2012, p.217
67. A. Rosenberg, D. McShea, *Philosophy of Biology: A Contemporary Introduction*, Routledge, New York, 2008, p.75
68. Lineweaver, op.cit., p.61
69. Rosenberg, op.cit., p.18
70. Lineweaver, op.cit., p.231
71. ibid., p.236
72. T. Deacon, *Incomplete Nature: How Mind Emerged from Matter*, W.W. Norton & Co., New York, 2012, p.423
73. Lineweaver, op.cit., p.208
74. ibid., p.162
75. Pross, op.cit., p.23
76. D. Newman, *Bioquestions and the Mechanical Answer*, CreateSpace Independent Publishing Platform, 2013, p.61
77. Lineweaver, op.cit., p.175
78. ibid., p.180
79. ibid., p.162
80. Bray, op.cit., p.97
81. Lineweaver, op.cit., p.181
82. ibid., p.176
83. ibid., p.8
84. B. Brembs, "Towards a scientific concept of free will as a biological trait: spontaneous actions and decision- making in invertebrates," *The Royal Society Publishing*, December 15, 2010
85. Rosenberg, op.cit., p.138
86. Bateson, op.cit., p.151
87. Rosenberg, op.cit., p.209
88. M. Kaku, *The Future of the Mind: The Scientific Quest to Understand, Enhance and Empower the Mind*, Anchor Books, New York, 2014, p.190
89. Scheider, op.cit., p.255
90. Hoffman, op.cit., p.61
91. P. Ball, "The Strange Inevitability of Evolution," *Nautilus*, January 8, 2015
92. Lineweaver, op.cit., p.5
93. Zimmer, *Microcosm*, op.cit., p.117
94. E. Thompson, *Mind in Life: Biology, Phenomenology and the Sciences of Mind*, Harvard University Press, Cambridge, Ma., 2007.p.198
95. Hoffman, op.cit., p.218
96. Lineweaver, op.cit., p.309
97. Thompson, op.cit. p.199
98. Margulis, *Chimeras*, op.cit., p.51
99. Lineweaver, op.cit., p.60
100. Margulis, *Chimeras*, op.cit., p.149
101. Hoffman, op.cit., p.244
102. Brembs, op.cit.
103. Lineweaver, op.cit., p.321
104. J. Gribbin, *In the Beginning: The Birth of the Living Universe*, Little Brown & Co., 1995, p.94
105. Deacon, op.cit., p.322
106. Lineweaver, op.cit., p.231
107. ibid., p.230
108. ibid., p.174
109. ibid., p.174
110. ibid., p.313

111. Lane, op.cit., p.266
112. Lineweaver, op.cit. p.139
113. ibid., p.312
114. ibid., p.143
115. ibid., p.144
116. T. Ghose, "Cambrian Explosion: Evolutionary Big Bang Was Sparked by Multiple Events, Scientists Say," *Huffington Post*, September 20, 2013
117. C. Zimmer, "Fast-Paced Evolution in the Andes," *New York Times*, November 7, 2013
118. Schneider, op.cit., p.246
119. Smolen, op.cit., p.129
120. S. Pierce, "Non-Equilibrium Thermodynamics: An Alternate Evolutionary Hypothesis," *Semantic Scholar*, 2002
121. Lineweaver, op.cit., p.295
122. Deacon, op.cit., p.473
123. Lineweaver, op.cit., p.242
124. J. Ferris, "An Orangutan Learns to Fish," *The NewYorker*, September 17, 2014
125. M. Pollan, "The Intelligent Plant," *The New Yorker*, December 23, 2013
126. M. Bennett-Smith, "How Did Early Birds Get Their Wings? New Study Suggests Surprising Answer," *Huffington Post*, September 19, 2013
127. Margulis, *What is Life,* op.cit., p.180
128. Lane, op.cit., p.147
129. E. Chaisson, *Epic of Evolution, Seven Ages of the Cosmos*, Columbia University Press, 2006, p.359
130. E. Chaisson, *Cosmic Evolution: An Inderdisciplinary Approach*, cfa.harvard.edu, "Brain Evolution"
131. Chaisson, *Epic*. op.cit., p.295
132. E. Anthes, "Coldblooded Does Not Mean Stupid," *New York Times*, November 18, 2013
133. ibid.
134. E. Wilson, "The Riddle of the Human Species," *New York Times*, February 24, 2013
135. J. Welsh, "Like Humans, Dolphins Have Brainy Genes," *NBCNews.com*, June 26, 2012
136. Margulis, *Chimeras*, op.cit., p.149
137. Lineweaver, op.cit., p.209
138. ibid., p.211
139. Rosenberg, op.cit., p.161
140. Deacon, op.cit., p.163
141. J. Von Medding, "Agent Orange, exposed: How U.S, chemical warfare in Vietnam unleashed a slow-moving disaster," Salon, October 9, 2017
142. Shapiro, op.cit., p.28
143. J. Shapiro, "What is the Key to a Realistic Theory of Evolution?" *Huffington Post*, February, 16, 2012
144. Lineweaver, op.cit., p.237
145. Loewenstein, *Touchstone*, op.cit., p.151
146. N. Comfort, "Single Gene May Hold Key to Life Itself," Huffington Post, September 23, 2012
147. B. Lipton, *The Biology of Belief: Unleashing the Power of Consciousness, Matter & Miracles*, Hay House, 2008, p.38
148. Shapiro, *What is the Key*, op.cit.
149. Shapiro, *Evolution*, op.cit., p.42
150. G. Lawton, "You Are Your Brain," *Slate.com*, December 8, 2013
151. Shapiro, *Evolution*, op.cit., p.43
152. ibid., p.68
153. ibid., p.33
154. Ibid., p.114
155. M. Ho, "The Biology of Free Will," *Journal of Consciousness Studies*, 1996
156. Z. Merali, "Solving Biology's Mysteries Using Quantum Mechanics," *Discover Magazine*, October 30, 2014
157. Hoffman, op.cit., p.62
158. Merali, op.cit.
159. Ho, op.cit.
160. Lipton, op.cit., prologue
161. Lane, op.cit., p.109

162. Shapiro, *Evolution*, p.3
163. Shapiro, *What is the Key*, op.cit.
164. Lane, op.cit., p.166
165. R. Sheldrake, *Morphic Resonance, The Nature of Formative Causation*, Park Street Press, Rochester, Vermont, 2009, p.110
166. Shapiro, James, A. "Why the 'Gene' Concept Holds Back Evolutionary Thinking," *Huffington Post*, November 30. 2012
167. D. Noble, "A Theory of Biological Relativity: No Privileged Level of Causation," *The Royal Society Publishing*, November 9, 2011
168. Shapiro, *Evolution*, p.87
169. University of Chicago, "Epigenetics: 100 Reasons to Change the Way We Think about Genetics," *ScienceDaily.com,* May 20,2009
170. Gribbin, op.cit., p.110
171. Lineweaver, op.cit., p.318
172. Lipton, op.cit., prologue
173. Margulis, *What is Life*, op.cit., p.164
174. Margulis, *Chimeras*, op.cit., p.257
175. Margulis, *What is Life*, op.cit., p.156
176. Zimmer, *Microcosm*, op.cit., p.105
177. Lineweaver, op.cit., p.236
178. Sheldrake, op.cit., p.12
179. Rosenberg, op.cit., p.93
180. Lineweaver, op.cit., p.243
181. Shapiro, *Evolution*, op.cit., p.2
182. T. Deacon, "Language and Complexity: Evolution Inside Out," *youtube,com*, August 25, 2010
183. Lineweaver, op.cit., p.243
184. University of Chicago, op.cit.
185. Lineweaver, op.cit., p.240
186. ibid., p.155
187. C. Zimmer, "In the Human Brain, Size Really Isn't Everything," *New York Times*, December 26, 2013
188. N. Pitney, "Scientists Find Alarming Deterioration in DNA of the Urban Poor," *Huffington Post*, May 8, 2015
189. C. Simon, "Deriving Common Interests from Animal Origins: The Generative Constraints of Global Polity," *http://docshare04.docshare.tips/files/4467/44670838.pdf*, September, 2008
190. Lineweaver, op.cit., p.150
191. C. Zimmer, "DNA Double Take," *New York Times*, September 16, 2013
192. Shapiro, *Evolution*, op.cit., p.93
193. N. Humphrey, *Seeing Red: A Study in Consciousness*, Harvard University Press, 2006, p.160
194. N. Rich, "Can a Jellyfish Unlock the Secret of Immortality?" *New York Times*, November 28, 2012
195. Lineweaver, op.cit., p.308
196. Thompson, op.cit., p.203
197. Lineweaver, op.cit., p.318
198. Sheldrake, op. cit., p.26
199. Ho, *Biology*, op.cit.
200. Loewenstein, *Touchstone,* op.cit., p.90
201. Lineweaver, op.cit., p.247
202. M. Rebeiz, cited in "Pitt Researcher Lands the Cover of *Development Cell*, by Uncovering an Evolutionary Secret," *University of Pittsburgh News Services*, September 3, 2015
203. E. Goldbaum, "Master orchestrator of the genome is discovered, UB stem cell scientists report," *University of Buffalo News Center*, May 7, 2015
204. Shapiro, *Evolution*, op.cit., p.35
205. N. Rich, "Can a Jellyfish Unlock the Secret of Immortality?" *New York Times*, November 28, 2012
206. Shapiro, *Evolution*, op.cit., p.117
207. Ball, op.cit.
208. Margulis, *What is Life*, p.154
209. Zimmer, Carl. *In the Human Brain*, op.cit.
210. Shapiro, *Evolution*, op.cit. p.21
211. Pierce, op.cit.

212. Smolen, op.cit. p.220
213. Chaisson, *Cosmic Evolution,* op.cit.
214. Sheldrake, op.cit. p.103
215. Lipton, op.cit. p.58
216. Kaku, op.cit., p.156
217. Sheldrake, op.cit., p.8
218. N. Subbaraman, "It's not dead! How some worms can regrow whole heads," NBCnews.com, July 24, 2013
219. Bray, op.cit., p.173
220. Subbaraman, op.cit.
221. Sheldrake, op.cit. p.36
222. Subbaraman, op.cit.
223. D. Goodsell, *The Machinery of Life*, Copernicus, 2010, p.118
224. Sheldrake, op.cit. p.8
225. Hoffman, op.cit., p233
226. R. Garrett, C. Grisham, *Biochemistry*, Cengage, Boston, 2017, p.226
227. Thompson, op.cit., p.191
228. Bateson, op.cit., p.154
229. Gribbin, op.cit., p.80
230. Pross, op.cit., p.77

## III. AWARENESS IN FORM

### A. INFORMATION

1. L. Margulis, D. Sagan, *What Is Life?* University of California Press, Berkeley, CA., 1995, p.170
2. R.Unger, L.Smolin, Lee. *The Singular Universe and the Reality of Time: A Proposal in Natural Philosophy*, University Printing House, Cambridge, 2015, p.222
3. G. Bateson, A. Montuori, *Mind and Nature: A Necessary Unity*, Hampton Press, 2002, p.96
4. Chesher, *This Magic Sea,* "The Error of Expectations," Chesher, 2011, *http://www.this-magic-sea.com/conscius.htm*
5. T. Deacon, *Incomplete Nature: How Mind Emerged from Matter*, W.W. Norton & Co., New York, 2012, p.378
6. B. Lipton, "The New Biology: Where Mind And Matter Meet," *MutualResponsibility.org*
7. J. Edwards, *How Many People are There in My Head? And in Hers?: An Exploration of Single Cell Consciousness*, Imprint Academic, Exeter, UK, 2006, p.151
8. L. Smolen, Lee. *Time Reborn: From the Crisis in Physics to the Future of the Universe* Mariner Books, 2014, \intro.
9. H. Maturana, F. Varela, *The Tree of Knowledge: The Biological Roots of Human Understanding*, Shambhala Publications, Boston, 1992, p.40
10. R. Pico, *Consciousness in Four Dimensions: Biological Relativity and the Origins of Thought*, McGraw-Hill, 2001, p.317
11. Bateson, op.cit., p.101
12. J. Shapiro, *Evolution: A View from the 21st Century*, FT Press Science, Upper Saddle River, NJ, 2011, p.10
13. B. Haisch, *The God Theory: Universes, Zero-point Fields, and What's Behind It All*, Red Wheel/Weiser, San Francisco, 2009, p.29
14. Deacon, op.cit., p.418
15. ibid., p.377
16. Pico, op.cit., p.284
17. Chesher, op.cit.
18. E. Thompson, *Mind in Life: Biology, Phenomenology and the Sciences of Mind*, Harvard University Press, Cambridge, Ma., 2007, p.156
19. Bateson, op.cit., p.160

20. N. Humphrey, *A History of the Mind: Evolution and the Birth of Consciousness*, Springer-Verlag, New York, 1999, p.40
21. ibid., p.44
22. H. Bergson, *Time and Free Will: As Essay on the Immediate Data of Consciousness*, Harper Torchbook, New York, 1960, p.13
23. Humphrey, op.cit., p.123
24. ibid., p.63
25. U. Mohrhoff, "Why the Laws of Physics Are Just So," *This Quantum World,* http://thisquantumworld.com/wp/a-critique-of-quantum-mechanics/why-the-laws-of-physics-are-just-so/
26. Deacon, op.cit., p.353
27. H. Stapp, *Mindful Universe: Quantum Mechanics and the Participating Observer* (The Frontiers Collection) 2nd ed., Springer, Heidelberg, 2011, p.137
28. D. Hofstadter, *Godel, Escher, Bach: An Eternal Golden Braid*, Basic Books, New York, 1999 edition, preface

## B. MEANING

1. E. Jaques, *The Life and Behavior of Living Organisms: A General Theory*, Praeger Publishers, Westport, CT, 2002, p.63
2. Margulis, Lynn and Sagan, Dorian, *What Is Life?* University of California Press, Berkeley, CA., 1995, p.218
3. ibid., p.226
4. H. Jonas, *The Phenomenon of Life: Toward a Philosophical Biology*, Dell Publishing, New York, 1966, p.159
5. G. Bateson, A. Montuori, *Mind and Nature: A Necessary Unity*, Hampton Press, 2002, p.85
6. T. Deacon, *Incomplete Nature: How Mind Emerged from Matter*, W.W. Norton & Co., New York, 2012, p.9
7. J. Edwards, *How Many People are There in My Head? And in Hers?: An Exploration of Single Cell Consciousness*, Imprint Academic, Exeter, UK, 2006, p.14
8. Jonas, op.cit., p.160
9. D. Hofstadter, *Godel, Escher, Bach: An Eternal Golden Braid*, Basic Books, New York,1999 edition, p.50
10. Bateson, op.cit., p.43
11. ibid., p. 92
12. Jonas, op.cit., p.85
13. Jaques, op.cit., p.63
14. A. Rosenberg, D. McShea, *Philosophy of Biology: A Contemporary Introduction*, Routledge, New York, 2008, p.180
15. N. Lane, *Life Ascending: The Ten Great Inventions of Evolution*, Norton, 2010, p.246
16. M. Ho, *The Rainbow and Worm: The Physics of Organisms*, World Scientific Publishing, Singapore, 2008, p.241
17. Deacon, op.cit., p.519
18. ibid., p.393
19. Jaques, op.cit., p.63
20. Deacon, op.cit., p.482
21. A. Noe, *Out of Our Heads: Why You Are Not Your Brain, and Other Lessons from the Biology of Consciousness*, Hill and Wang, New York, 2009, p.8
22. J. Shapiro, *Evolution: A View from the 21st Century*, FT Press Science, Upper Saddle River, NJ, 2011, p.9
23. E. Thompson, *Mind in Life: Biology, Phenomenology and the Sciences of Mind*, Harvard University Press, Cambridge, Ma., 2007, p.147
24. T. Nagel, *Mind & Cosmos: Why the Materialist Neo-Darwinian Conception of Nature is Almost Certainly False*, Oxford University Press, Oxford, 2012, p.117
25. C. Lineweaver, P. Davies, M. Ruse, eds., *Complexity and the Arrow of Time*, Cambridge University Press, New York, 2013, p.169
26. A. Damasio, *Self Comes to Mind: Constructing the Conscious Brain*, Vintage, 2012, p.27
27. Thompson, op.cit., p.146

28. ibid., p.132
29. Noe, op.cit., p.83
30. ibid., p.39
31. Bateson, op.cit., p.125
32. Noe, op.cit., p.92
33. S. Allport, *Explorers of the Black Box: The Search for the Cellular Basis of Memory*, W.W. Norton, New York, 1986, p.108
34. Edwards, op.cit., p.61
35. W. Loewenstein, *Physics in Mind: A Quantum View of the Brain*, Basic Books, New York, 2013, p.204
36. Edwards, op.cit., p.176
37. G. Marcus, *The Birth of Mind: How a Tiny Number of Genes Creates The Complexities of Human Thought*, Basic Books, New York, 2004, p.118
38. ibid., p.104
39. Allport, op.cit., p.82
40. K. Kull, "Organism as a self-reading text: anticipation and semiosis, "*Journal of Computing Anticipatory Systems,* vol. 1. 1998
41. Chesher, *This Magic Sea,* "The Error of Expectations," Chesher, 2011, *http://www.this-magic-sea.com/conscius.htm*
42. Allport, op.cit., p.257
43. R. Pico, *Consciousness in Four Dimensions: Biological Relativity and the Origins of Thought*, McGraw-Hill, 2001, p.369
44. Allport, op.cit., p.104
45. Allport, op.cit., p.102
46. Rosenberg, op.cit., p.179
47. Deacon, op.cit., p.398
48. Ho, op.cit., p.226

## C. INTERSUBJECTIVITY

1. K. Kull, "Biosemiotics in the twentieth century: A view from biology," *http://www.zbi.ee/~kalevi/semi.1999.127.385.pdf*
2. D. Noble, "A Theory of Biological Relativity: No Privileged Level of Causation," *The Royal Society Publishing*, November 9, 2011
3. E. Thompson, Evan. *Mind in Life: Biology, Phenomenology and the Sciences of Mind*, Harvard University Press, Cambridge, Ma., 2007, p.251
4. G. Gutting, "Sam Harris's Vanishing Self," *New York Times*, September, 7, 2014
5. C. Willyard,."Bonnie Bassler, Silent Treatment," *Discover Magazine*, June 2014
6. J. Edwards, *How Many People are There in My Head? And in Hers?: An Exploration of Single Cell Consciousness*, Imprint Academic, Exeter, UK, 2006, p.135
7. G. Greenberg, "The Delusions We Deserve," New York Times, August 28, 2014
8. A. Noe, *Out of Our Heads: Why You Are Not Your Brain, and Other Lessons from the Biology of Consciousness*, Hill and Wang, New York, 2009, p.33
9. S. Scheffler, "The Importance of the Afterlife. Seriously," *New York Times*, September 21, 2013
10. H. Jonas, *The Phenomenon of Life: Toward a Philosophical Biology*, Dell Publishing, New York, 1966, p.698
11. Thompson, op.cit., p.399
12. H. Maturana, F. Varela, *The Tree of Knowledge: The Biological Roots of Human Understanding*, Shambhala Publications, Boston, 1992, p.245
13. Thompson, op.cit., p.397
14. Maturana, op.cit. p.193
15. S. Bhanoo, "With Incentive, Monkeys Can Coordinate," New York Times, April 8,2013
16. Greenberg, op.cit.
17. A. Pross, *What is Life? How Chemistry Becomes Biology*, Oxford University Press, Oxford, UK, 2012, p.11
18. Noe, op.cit., p.108
19. Thompson, op.cit., p.383

20. Slobodchikoff, Con. *Chasing Doctor Dolittle: Learning the Language of Animals*, St. Martin's Press, New York, 2012, p.18
21. G. Bateson, A. Montuori, *Mind and Nature: A Necessary Unity*, Hampton Press, 2002, p.124
22. N. Fine, Editor. "Rise of the Robots, " *Time Magazine*, 2013 (single issue)
23. S. Lewis, "Of Two Minds, 'Powers of Two' by Joshua Wolf Shenk," *New York Times Book Review*, August 22, 2014
24. A. Goswami, *The Self-Aware Universe: How Consciousness Creates the Material World*, Penguin Putnam Inc., New York, 1995, p.23
25. Lewis, op.cit.
26. A. Rosenberg, D. McShea, *Philosophy of Biology: A Contemporary Introduction*, Routledge, New York, 2008. p.202
27. J. Shenk, "The End of 'Genius,'" *New York Times*, July 19, 2014
28. T. Metzinger, *The Ego Tunnel: The Science of the Mind and the Myth of the Self*, Basic Books, New York, 2009, p.177
29. Shenk, op.cit.
30. Chesher, *This Magic Sea*, "The Error of Expectations," Chesher, 2011, *http://www.this-magic -sea.com/conscius.htm*
31. L. Margulis, C. Asikaninen, W. Krumbein, eds., *Chimeras and Consciousness: Evolution of the Sensory Self*, MIT Press, 2011, p.57
32. C. Simon, "Deriving Common Interests from Animal Origins: The Generative Constraints of Global Polity," *http://docshare04.docshare.tips/files/4467/44670838.pdf*, September, 2008
33. J. Sargent, Jennifer. "Why fish don't need to be schooled in the art of swimming," *NBCNews.com*, September 17, 2013
34. M. Ho, *The Rainbow and Worm: The Physics of Organisms*, World Scientific Publishing, Singapore, 2008, p.284
35. B. Frederickson, "Your Phone vs. Your Heart," *New York Times*, March 23, 2013
36. Simon, op.cit.
37. K. Kull, "Organism as a self-reading text: anticipation and semiosis, " *Journal of Computing Anticipatory Systems*, vol. 1. 1998
38. J. Hoffmeyer, Quoted in "What is Biosemiotics," *http://www.biosemiotics.org/biosemiotics- introduc tion/*
39. P. Pericles, ed., *International Handbook of Semiotics*, Springer, 2015, p.1095
40. Kull, *Organism*, op.cit.
41. J. Deely, *Four Ages of Understanding, The First Posmodern Survey of Philosophy from Ancient Times to the Turn of the Twenty-First Century*, University of Toronto Press, Toronto, 2001, p.433
42. H. Bergson, *Time and Free Will: As Essay on the Immediate Data of Consciousness*, Harper Torchbook, New York, 1960. P.128
43. Slobodchikoff, op.cit., p.21
44. ibid., p.17
45. M. Hay, "Chemistry as the Language of Life," *New York Times*, October 14, 2011
46. Slobodchikoff, op.cit., p.40
47. E. Jaques, *The Life and Behavior of Living Organisms: A General Theory*, Praeger Publishers, Westport, CT, 2002
48. D. Hofstadter, *Godel, Escher, Bach: An Eternal Golden Braid*, Basic Books, New York,1999 edition, p.322
49. Slobodchikoff, op.cit., p.21
50. ibid., p.23
51. Duke University Medical Center. "Brain-to-brain interface allows transmission of tactile and motor information between rats," *Medical Press*, February 28, 2013
52. S. Strogatz. *Sync: How Order Emerges From Chaos In the Universe, Nature, and Daily Life*, Hyperion, 2004, p.46
53. Hofstadter, op.cit., p.371
54. Bateson, op.cit., p.174
55. R. Rosen, *Life Itself: A Comprehensive Inquiry into the Nature, Origin, and Fabrication of Life*, Columbia University Press, New York, 1991, p.7
56. Margulis, op.cit., p.62
57. Thompson, op.cit., p.403
58. J. Friedland, "Philosophy Is Not Science," New York Times, April 5, 2012

59. McWhorter, John. "Why Save a Language?" *New York Times*, December 5, 2014
60. L. Margulis, D. Sagan, *What Is Life?* University of California Press, Berkeley, CA., 1995, p.17
61. E. Schneider, D. Sagan, *Into the Cool: Energy Flow, Thermodynamics, and Life*, University of Chicago Press, 2006, p.301
62. M. Kaku, *The Future of the Mind: The Scientific Quest to Understand, Enhance and Empower the Mind*, Anchor Books, New York, 2014, p.158
63. Slobodchikoff, op.cit., p.5
64. ibid., p.44
65. ibid., p.3
66. T. Deacon, "Language and Complexity: Evolution Inside Out," *youtube,com*, August 25, 2010, *https://www.youtube.com/watch?v=OT-zZ0PMqgI*
67. Kull, *Organism*, op.cit.
68. J. Lipman, "Is Music the Key to Success?" *New York Times*, October, 12, 2013

## D. THINKING

1.H. Maturana, F. Varela, *The Tree of Knowledge: The Biological Roots of Human Understanding*, Shambhala Publications, Boston, 1992, p.34
2.J. Gorman, "Learning How Little We Know About the Brain," *New York Times*, November 10, 2014
3.H. Jonas, *The Phenomenon of Life: Toward a Philosophical Biology*, Dell Publishing, New York, 1966, p.167
4.B. Rosenblum, F. Kuttner, *Quantum Enigma: Physics Encounters Consciousness*, Oxford University Press, New York, 2011, p.193
5.D. Hofstadter, *Godel, Escher, Bach: An Eternal Golden Braid*, Basic Books, New York,1999 edition , p.673
6. ibid., p.710
7. Jonas, op.cit., p.171
8. J. Edwards, *How Many People are There in My Head? And in Hers?: An Exploration of Single Cell Consciousness*, Imprint Academic, Exeter, UK, 2006, p.21
9. R. Lloyd, "Decartes on Time," *Purdue University, http://docs.lib.purdue.edu/dissertations/AAI3403121/*
10. L. Smolen, *Time Reborn: From the Crisis in Physics to the Future of the Universe* Mariner Books, 2014, preface
11. Hoffstadter, op.cit., p.559
12. ibid., p.338
13. ibid., p.669
14. D. Gelernter, "Encounters with the Arch-genius, David Gelernter,"Time.com, http://time.com/4236974/encounters-with-the-archgenius/
15. A.. Rosenberg, D. McShea, *Philosophy of Biology: A Contemporary Introduction*, Routledge, New York, 2008, p.29
16. Hofstadter, op.cit., p.398
17. Maturana, p.174
18. A. Damasio, *Self Comes to Mind: Constructing the Conscious Brain*, Vintage, 2012, p.199
19. P. Stark, Philip, SticiGui, Chapter 13, "Probability: Philosophy and Mathematical Background" *https://www.stat.berkeley.edu/~stark/SticiGui/Text/probabilityPhilosophy.htm*
20. Smolen, op.cit. p.11
21. T. Nagel, *Mind & Cosmos: Why the Materialist Neo-Darwinian Conception of Nature is Almost Certainly False*, Oxford University Press, Oxford, 2012, p.80
22. Jonas, op.cit., p.175

## E. OBJECTIVITY

1. A. Rosenberg, D. McShea, *Philosophy of Biology: A Contemporary Introduction*, Routledge, New York, 2008, p.89
2. T. Deacon, *Incomplete Nature: How Mind Emerged from Matter*, W.W. Norton & Co., New York, 2012, p.62
3. E. Schrodinger, *What Is Life? & Other Scientific Essays*, Doubleday Anchor, 1956. p.85

4. A. Damasio, *Self Comes to Mind: Constructing the Conscious Brain*, Vintage, 2012, p.76
5. G. Bateson, A. Montuori, *Mind and Nature: A Necessary Unity*, Hampton Press, 2002, p.44
6. https://stackoverflow.com/questions/19623479/recursion-switch-return-string
7. W. Loewenstein, *The Touchstone of Life: Molecular Information, Cell Communication and the Foundation of Life*, Oxford University Press, New York, 1998, p.315
8. D. Hofstadter, *Godel, Escher, Bach: An Eternal Golden Braid*, Basic Books, New York,1999 edition, p.192
9. W. Loewenstein, *Physics in Mind: A Quantum View of the Brain*, Basic Books, New York, 2013, p. 206
10. ibid., p.207
11. L. Smolen, *Time Reborn: From the Crisis in Physics to the Future of the Universe* Mariner Books, 2014, p.7
12. R. Unger, L. Smolin, *The Singular Universe and the Reality of Time: A Proposal in Natural Philosophy*, University Printing House, Cambridge, 2015, p.321
13. C. Lineweaver, P. Davies, M. Ruse, eds., *Complexity and the Arrow of Time*, Cambridge University Press, New York, 2013 , p.86
14. E. Frenkel, "Is the Universe a Simulation?" New York Times, February 14, 2014
15. F. Capra, *The Tao of Physics*, Shambhala Publications, Inc., Boston, 1985, p.20
16. Hofstadter, op.cit., preface
17. M. Tegmark, "Why a Larger Multiverse Shouldn't Make You Feel Small," *Huffington Post*, January 7, 2014
18. ibid.
19. Loewenstein, *Physics*, op.cit., p.156
20. ibid., p159
21. Smolen, op.cit., p.294
22. Lineweaver, op.cit., p.82
23. N. Yanofsky, *The Outer Limits of Reason: What Science, Mathematics and Logic Cannot Tell Us.*, The MIT Press, Cambridge, Massachusetts, 2013, p.2
24. E.Thompson, Evan. *Mind in Life: Biology, Phenomenology and the Sciences of Mind*, Harvard University Press, Cambridge, Ma., 2007, p.42
25. *https://en.wikipedia.org/wiki/Photon*
26. Tegmark, op.cit.
27. Smolen, op.cit., p.247
28. Tegmark, op.cit.
29. Capra, op.cit., p.149
30. Loewenstein, *Physics*, op.cit., p.163
31. ibid., p.163
32. R. Rosen, *Life Itself: A Comprehensive Inquiry into the Nature, Origin, and Fabrication of Life*, Columbia University Press, New York, 1991, p.47
33. Hofstadter, op.cit. p.41
34. Tegmark, op.cit.
35. Hofstadter, op.cit. p.17
36. ibid., op.cit., p.21
37. ibid., op.cit., p.60
38. M. Suri, "How to Fall in Love with Math," *New York Times*, Sept. 15, 2013
39. N. Tyson, "What Science Is - and How and Why It Works," *Huffington Post*, November 18, 2015
40. T. Nagel, *Mind & Cosmos: Why the Materialist Neo-Darwinian Conception of Nature is Almost Certainly False*, Oxford University Press, Oxford, 2012, p.16
41. A. Einstein, *Out of My Later Years*, Philosophical Library, New York, 1950, p.98
42. Nagel, op.cit., p.16
43. Capra, op.cit., p.288
44. R. Lanza, B. Berman, *Biocentrism: How Life and Consciousness Are the Keys to Understanding the True Nature of the Universe*, BenBella Books, Dallas, 2010, p.207
45. Bateson, op.cit., p.54
46. H. Jonas, *The Phenomenon of Life: Toward a Philosophical Biology*, Dell Publishing, New York, 1966, p.132
47. Rosen, op.cit., p.57
48. Unger, op.cit., p.293
49. Bateson, op.cit., p.6
50. H. Stapp, *Mindful Universe: Quantum Mechanics and the Participating Observer* (The Frontiers

Collection) 2nd ed., Springer, Heidelberg, 2011, p.170

51. G. Gutting. "Mary and the Zombies: Consciousness Revisited," New York Times, March 22, 2013
52. A. Lee, "Untimely Metaphysics," *New York Times*, May 7, 2013
53. Smolen, op.cit., p.121
54. Unger, op.cit., p.287
55. Smolen, op.cit., p.35
56. ibid., p.275
57. ibid., p.92
58. Gutting, op.cit.
59. Smolen, op.cit., p.164
60. C. Seife, *Decoding the Universe: How the New Science of Information is Explaining Everything in the Cosmos,from Our Brains to Black Holes,* Penguin Books, 2006, p.212
61. Rosen, op.cit., p.59
62. J. Edwards, *How Many People are There in My Head? And in Hers?: An Exploration of Single Cell Consciousness*, Imprint Academic, Exeter, UK, 2006, p.27
63. Stapp, op.cit., p.28
64. Smolen, op.cit., p.153
65. Stapp, op.cit., p.124
66. Lanza, op.cit., p.51
67. Stapp, op.cit., p.126
68. Nagel, op.cit., p.45
69. T. Metzinger, *The Ego Tunnel: The Science of the Mind and the Myth of the Self*, Basic Books, New York, 2009, p.216
70. Lanza, op.cit., p.49
71. ibid., p.138
72. S. Carroll, "No Physicist Is an Island," *New York Times*, October 8,2013
73. Ibid.
74. M. Lange, *An Introduction to the Philosophy of Physics: Locality, Fields, Energy, and Mass*, Blackwell Publishing, 2002, p.128
75. Stapp, op.cit., p.38
76. Smolen, op.cit., p.17
77. B. Rosenblum, F. Kuttner, *Quantum Enigma: Physics Encounters Consciousness*, Oxford University Press, New York, 2011, p.157
78. Smolen, op.cit., p.108
79. Nagel, op.cit., p.52
80. ibid., p.56
81. J. Gleick, *Chaos: Making a New Science*, Penguin Books, New York, 2008, p.,268
82. Loewenstein, *Touchstone*, p.332
83. Smolen, op.cit., p.294
84. ibid., p.38
85. Nagel, op.cit., p.20
86. ibid., p.19
87. ibid., p.79
88. A. Pross, *What is Life? How Chemistry Becomes Biology*, Oxford University Press, Oxford, UK, 2012, p.47
89. Nagel, op.cit., p.76
90. ibid., op.cit., p.38
91. F. Ayala, R. Arp, eds., *Contemporary Debates in Philosophy of Biology*, Wiley-Blackwell, Oxford, England, 2009, p.6
92. Nagel, op.cit., p.76
93. Tegmark, op.cit.
94. R. Pico, *Consciousness in Four Dimensions: Biological Relativity and the Origins of Thought*, McGraw-Hill, 2001, p.345
95. A. Noe, *Out of Our Heads: Why You Are Not Your Brain, and Other Lessons from the Biology of Consciousness*, Hill and Wang, New York, 2009, p.39
96. G. Makari, George. "Psychiatry's Mind-Brain Problem," New York Times, November 11, 2015
97. G. Lawton, "You Are Your Brain," *Slate.com*, December 8, 2013

98. E. Kandel, "The New Science of Mind," New York Times, September 6, 2013
99. J. Shapiro,"What is the Key to a Realistic Theory of Evolution?" *Huffington Post*, February, 16, 2012
100. Gutting, op.cit.
101. R. Douthat, "Sam Harris and Scientism," *New York Times*, September 5, 2013
102. Edwards, op.cit., p.153

## F. BIOCENTRISM

1. R. Rosen, *Life Itself: A Comprehensive Inquiry into the Nature, Origin, and Fabrication of Life*, Columbia University Press, New York, 1991, preface
2. D. Mikulecky, "Robert Rosen: The Well Posed Question and Its Answer - Why Are Organisms Different from Machines?" *Virginia Commonwealth University, http://www.people.vcu.edu/~mikuleck/PPRISS3.html,* 1.1.1.
3. Rosen, op.cit., p.32
4. ibid., p.60
5. ibid., p.301
6. A. Pross, *What is Life? How Chemistry Becomes Biology*, Oxford University Press, Oxford, UK, 2012, p.49
7. T. Nagel, *Mind & Cosmos: Why the Materialist Neo-Darwinian Conception of Nature is Almost Certainly False*, Oxford University Press, Oxford, 2012. p.62
8. E. Thompson, *Mind in Life: Biology, Phenomenology and the Sciences of Mind*, Harvard University Press, Cambridge, Ma., 2007, p.334
9. R. Pico,*Consciousness in Four Dimensions: Biological Relativity and the Origins of Thought*, McGraw-Hill, 2001, p. 292
10. Rosen, op.cit., p.28
11. T. Nagel, *Mind & Cosmos: Why the Materialist Neo-Darwinian Conception of Nature is Almost Certainly False*, Oxford University Press, Oxford, 2012, p.119
12. A. Rosenberg, D. McShea, *Philosophy of Biology: A Contemporary Introduction*, Routledge, New York, 2008, p.178
13. E. Thompson, *Mind in Life: Biology, Phenomenology and the Sciences of Mind*, Harvard University Press, Cambridge, Ma., 2007, p.149
14. W. Loewenstein, Werner. *Physics in Mind: A Quantum View of the Brain*, Basic Books, New York, 2013, p.117
15. A. Pross, op.cit., p.56
16. ibid., p.53
17. Nagel, op.cit., p.57
18. J. Edwards, *How Many People are There in My Head? And in Hers?: An Exploration of Single Cell Consciousness*, Imprint Academic, Exeter, UK, 2006, p.153
19. T. Metzinger, *The Ego Tunnel: The Science of the Mind and the Myth of the Self*, Basic Books, New York, 2009, p.216
20. H. Maturana, F. Varela, *The Tree of Knowledge: The Biological Roots of Human Understanding*, Shambhala Publications, Boston, 1992, p.34
21. Nagel, op.cit. p.85
22. Edwards, op.cit., p.24
23. Mikulecky, op.cit., 2.2.4
24. Rosen, op.cit., p.7
25. ibid., p.8
26. ibid., p.176
27. Mikulecky, op.cit., 1.1.5
28. Rosen, op.cit., p.68
29. ibid., p.300
30. G. Bateson, A. Montuori, *Mind and Nature: A Necessary Unity*, Hampton Press, 2002, p.81
31. J. Shapiro, *Evolution: A View from the 21st Century*, FT Press Science, Upper Saddle River, NJ, 2011, p.4
32. Rosen, op.cit., p.120
33. ibid., p.246
34. ibid., p.275

35. ibid., p.268
36. ibid., p.76
37. ibid., p.78
38. ibid., p.76
39. ibid., p.49
40. ibid., p.54
41. ibid., p.280
42. ibid., p.54
43. ibid., p.173
44. ibid., p.236
45. Thompson, op.cit., p.69
46. Mikulecky, op.cit., 2.1.3.
47. ibid., p.123
48. Jonas, op.cit., p.251
49. Mikulecky, op.cit., 2.2.3.
50. ibid., 2.2.4.

## G. CONSCIOUSNESS

1. A. Noe, *Out of Our Heads: Why You Are Not Your Brain, and Other Lessons from the Biology of Consciousness*, Hill and Wang, New York, 2009, p.45
2. T. Metzinger, *The Ego Tunnel: The Science of the Mind and the Myth of the Self*, Basic Books, New York, 2009, p.170
3. A. Regalado, "What Will It Take for Computers to Be Conscious," *MIT Technology Review*, October 2, 2014
4. R. Lanza, B. Berman, *Biocentrism: How Life and Consciousness Are the Keys to Understanding the True Nature of the Universe*, BenBella Books, Dallas, 2010, p.36
5. Regalado, op.cit.
6. J. Edwards, *How Many People are There in My Head? And in Hers?: An Exploration of Single Cell Consciousness*, Imprint Academic, Exeter, UK, 2006, p.135
7. N. Humphrey, *A History of the Mind: Evolution and the Birth of Consciousness*, Springer-Verlag, New York, 1999, p.204
8. M. Kaku, *The Future of the Mind: The Scientific Quest to Understand, Enhance and Empower the Mind*, Anchor Books, New York, 2014, p. 268
9. R. Pico, *Consciousness in Four Dimensions: Biological Relativity and the Origins of Thought*, McGraw-Hill, 2001, p.345
10. Humphrey, op.cit., p.235
11. M. Ho, *The Rainbow and Worm: The Physics of Organisms*, World Scientific Publishing, Singapore, 2008, p.238
12. Metzinger, op.cit., p.35
13. Pico, op.cit., p.221
14. N. Papageorgiou, "How the Brain Produces Consciousness in 'Time Slices,'" *News Mediacom, https://actu.epfl.ch/news/how- the-brain-produces-consciousness-in-time-slice/*
15. A. Scott, *Stairway to the Mind: The Controversial New Science of Consciousness*, Springer-Verlag, New York, 1995
16. F. Boogerd, F. Bruggeman, J. Hofmeyr, H, Westerhoff, eds. *Systems Biology: Philosophical Foundations*, Elsevier, Amsterdam, 2007, p.272
17. G. Bateson, A. Montuori, Alfonso, *Mind and Nature: A Necessary Unity*, Hampton Press, 2002, p.118
18. J. Gleick, *Chaos: Making a New Science*, Penguin Books, New York, 2008, p.223
19. R. Unger, L. Smolin, *The Singular Universe and the Reality of Time: A Proposal in Natural Philosophy*, University Printing House, Cambridge, 2015, p.273
20. Pico, op.cit., p.303
21. Metzinger, op.cit., p.179
22. Edwards, op.cit., p.170
23. D. Hofstadter, Douglas, R. *Godel, Escher, Bach: An Eternal Golden Braid*, Basic Books, New York,1999 edition, preface
24. Lanza, op.cit., p.170

25. A. Damasio, *Self Comes to Mind: Constructing the Conscious Brain*, Vintage, 2012, p.67
26. S. Blackmore, *Conversations on Consciousness: What the Best Minds Think about the Brain, Free Will, and What It Means to Be Human*," Oxford University Press, Oxford, 2006, p.59
27. Ho, op.cit., p.239
28. A. Rosenberg, D. McShea, *Philosophy of Biology: A Contemporary Introduction*, Routledge, New York, 2008, p.167
29. H. Bergson, *Time and Free Will: As Essay on the Immediate Data of Consciousness*, Harper Torchbook, New York, 1960, \p.207
30. Metzinger, op.cit., p.192
31. L. Margulis, D.Sagan, *What Is Life?* University of California Press, Berkeley, CA., 1995, p.169
32. Lanza, op.cit., p.37
33. N. Lane, *Life Ascending: The Ten Great Inventions of Evolution*, Norton, 2010, p.100
34. Metzinger, op.cit., p.216
35. Regalado, op.cit.
36. H. Stapp, *Mindful Universe: Quantum Mechanics and the Participating Observer* (The Frontiers Collection) 2nd ed., Springer, Heidelberg, 2011, p.106
37. Regalado, op.cit.
38. Metzinger, op.cit., p.50
39. E. Thompson, *Mind in Life: Biology, Phenomenology and the Sciences of Mind*, Harvard University Press, Cambridge, Ma., 2007, p.12
40. Stapp, op.cit., p,8
41. A. Goswami, *The Self-Aware Universe: How Consciousness Creates the Material World*, Penguin Put nam Inc., New York, 1995, p.20
42. Noe, op.cit., p.59
43. Stapp, op.cit., p.105
44. Thompson, op.cit., p.313
45. ibid., p.256
46. Pico, op.cit, p.295
47. Metzinger, op.cit., p.11
48. ibid., p.208
49. Lane, op.cit. p.245
50. T. Deacon, *Incomplete Nature: How Mind Emerged from Matter*, W.W. Norton & Co., New York, 2012, p.527
51. T. Stokes, "Clownfish 'Talk' To Assert & Reinforce Social Status, Scientists Say," Huffington Post, November 7, 2012
52. R. Rosen, *Life Itself: A Comprehensive Inquiry into the Nature, Origin, and Fabrication of Life*, Columbia University Press, New York, 1991, p.69
53. W. Loewenstein, *Physics in Mind: A Quantum View of the Brain*, Basic Books, New York, 2013, p.9
54. Hofstadter, op.cit., p.478
55. J. Gorman, "Considering the Humanity of Nonhumans," *New York Times*, December, 9, 2013
56. Humphrey, op.cit., p.228
57. Metzinger, op.cit., p.8
59. K. Nelson, "The Spiritual Doorway in the Brain: A Neurologist's Search for the God Experience," Plume, 2012, p.233
60. H. Jonas, *The Phenomenon of Life: Toward a Philosophical Biology*, Dell Publishing, New York, 1966, p.709
61. Damasio, op.cit., p.243
62. D. Chalmers, *The Conscious Mind: In Search of a Fundamental Theory*, Oxford University Press, 1997, p.15
63. L. Margulis, C. Asikaninen, W. Krumbein, eds., *Chimeras and Consciousness: Evolution of the Sensory Self*, MIT Press, 2011, p.228
64. Edwards, op.cit., p.59
65. Noe, op.cit., p.24
66. Metzinger, op.cit., p.194
67. Chalmers, op.cit., p.12
68. ibid., p.11
69. S.Allport, *Explorers of the Black Box: The Search for the Cellular Basis of Memory*, W.W. Norton, New York, 1986, p.95

70. Chalmers, op.cit., p.15
71. Allport, op.cit., p.83
72. T. Lewis, "Ants Use 'Math' To Find Fastest Route, Study Shows," *Huffington Post*, April 17, 2013
73. T. Nagel, *Mind & Cosmos: Why the Materialist Neo-Darwinian Conception of Nature is Almost Certainly False*, Oxford University Press, Oxford, 2012, p.68
74. Goswami, op.cit., p.87
75. Thompson, op.cit., p.147
76. ibid., p.283
77. Damasio, p.178
78. C. Slobodchikoff, *Chasing Doctor Dolittle: Learning the Language of Animals*, St. Martin's Press, New York, 2012, p.4
79. Chesher, *This Magic Sea,* "The Error of Expectations," *Chesher*, 2011, *http://www.this-magic -sea.com/conscius.htm*
80. Thompson, op.cit., p.301
81. Kaku, op.cit., p.50
82. C. Simon, "Deriving Common Interests from Animal Origins: The Generative Constraints of Global Polity," *http://docshare04.docshare.tips/files/4467/44670838.pdf*, September, 2008

## H. FREE WILL

1. S. Harris, *Free Will,* Free Press, 2012, p.38
2. E. Parens, "The Benefits of 'Binocularity,'" *New York Times*, September 28, 2014
3. C. Simon, "Deriving Common Interests from Animal Origins: The Generative Constraints of Global Polity," *http://docshare04.docshare.tips/files/4467/44670838.pdf*, September, 2008
4. E. Nahmias, "Is Neuroscience the Death of Free Will?" *New York Times*, November 13, 2011
5. B. Brembs, "Towards a scientific concept of free will as a biological trait: spontaneous actions and decision- making in invertebrates," *The Royal Society Publishing*, December 15, 2010
6. C. Zimmer, *Microcosm: E. Coli and the New Science of Life*, First Vintage Books, New York, 2009, p.187
7. R. Pico, *Consciousness in Four Dimensions: Biological Relativity and the Origins of Thought*, McGraw-Hill, 2001, p.292
8. P. Tompkins, C. Bird, *The Secret Life of Plants*, Harper & Row, 1989, p.213
9. H. Jonas, *The Phenomenon of Life: Toward a Philosophical Biology*, Dell Publishing, New York, 1966, p.23
10. M. Ho, "The Biology of Free Will," *Journal of Consciousness Studies*, 1996
11. Zimmer, op.cit., p.187
12. A. Damasio, *Self Comes to Mind: Constructing the Conscious Brain*, Vintage, 2012, p.315
13. Brembs, op.cit.
14. Harris, op.cit., p.5
15. Simon, op.cit.
16. D. Newman, *Bioquestions and the Mechanical Answer*, CreateSpace Independent Publishing Platform, 2013, p.112
17. A. Goswami, Amit. *The Self-Aware Universe: How Consciousness Creates the Material World*, Penguin Putnam Inc., New York, 1995, p.173
18. R. Doyle, Robert O. "The Two-Stage Model to the Problem of Free Will," *Free Will: The Scandal in Philosophy, informationphilosopher.com,* June 19,2011
19. H. Bergson, *Time and Free Will: As Essay on the Immediate Data of Consciousness*, Harper Torchbook, New York, 1960, p.221
20. ibid., p.239
21. Simon, op.cit.
22. D. Hofstadter, *Godel, Escher, Bach: An Eternal Golden Braid*, Basic Books, New York,1999 edition, p.37
23. B. Lipton, "The New Biology: Where Mind And Matter Meet," *MutualResponsibility.org*
24. Bergson, op.cit., p.231
25. Damasio, op.cit., p.56
26. J. Corrigan, *An Introduction to Awareness*, BookSurge Publishing, 2006, p.149

27. J. Edwards, *How Many People are There in My Head? And in Hers?: An Exploration of Single Cell Consciousness*, Imprint Academic, Exeter, UK, 2006, p.147
28. Harris, op.cit., p.32
29. Damasio, op.cit., p.297
30. Lipton, op.cit., p.138
31. Brembs, op.cit.
32. Simon, op.cit.
33. C. Aschwanden, "Where Is Thought," *Discover Magazine*, June, 2013
34. Harris, op.cit. p.37
35. T. Metzinger, *The Ego Tunnel: The Science of the Mind and the Myth of the Self*, Basic Books, New York, 2009, p.121
36. Harris, op.cit., p.8
37. E. Nahmias, "Is Neuroscience the Death of Free Will?" *New York Times*, November 13, 2011
38. Brembs, op.cit.
39. A. Noe, "Art and the Limits of Neuroscience," *New York Times*, December 4, 2011
40. Brembs, op.cit.
41. Metzinger, op.cit., p.122
42. M. Tegmark, "Why a Larger Multiverse Shouldn't Make You Feel Small," *Huffington Post*, January 7, 2014
43. Bergson, op.cit., p.219
44. T. Deacon, *Incomplete Nature: How Mind Emerged from Matter*, W.W. Norton & Co., New York, 2012, p.360
45. Brembs, op.cit.
46. G. Bateson, A. Montuori, *Mind and Nature: A Necessary Unity*, Hampton Press, 2002, p.168
47. Nahmias, op.cit.
48. Goswami, op.cit., p.201
49. E. Thompson, *Mind in Life: Biology, Phenomenology and the Sciences of Mind*, Harvard University Press, Cambridge, Ma., 2007, p.313
50. Aschwanden, op.cit.
51. U. Mohrhoff, "Why the Laws of Physics Are Just So," *This Quantum World*, http://thisquantumworld.com
52. Jonas, op.cit., p.119
53. Metzinger, op.cit., p.118
54. T. Nagel, *Mind & Cosmos: Why the Materialist Neo-Darwinian Conception of Nature is Almost Certainly False*, Oxford University Press, Oxford, 2012, p.113
55. Bateson, op.cit., p.34
56. M. Ho, *The Rainbow and Worm: The Physics of Organisms*, World Scientific Publishing, Singapore, 2008, p.335
57. Hofstadter, op.cit., preface
58. M. Pollan, "The Intelligent Plant," *The New Yorker*, December 23, 2013
59. E. Jaques, *The Life and Behavior of Living Organisms: A General Theory*, Praeger Publishers, Westport, CT, 2002, p.90
60. Simon, op.cit.
61. R. Sheldrake, *Morphic Resonance, The Nature of Formative Causation*, Park Street Press, Rochester, Vermont, 2009, p.197
62. Simon, op.cit.
63. Pollan, op.cit.
64. J. Gorman, "Learning How Little We Know About the Brain," *New York Times*, November, 10, 2014
65. C. Lineweaver, P.Davies, M. Ruse, Michael. eds., *Complexity and the Arrow of Time*, Cambridge University Press, New York, 2013
66. Goswami, op.cit., p.225

## IV. AWARENESS IN MATTER

### A. NERVOUS SYSTEM

1. A. Noe, *Out of Our Heads: Why You Are Not Your Brain, and Other Lessons from the Biology of Consciousness*, Hill and Wang, New York, 2009, p.169
2. D. Chalmers, *The Conscious Mind: In Search of a Fundamental Theory*, Oxford University Press, 1997, p.104
3. E. Jaques, *The Life and Behavior of Living Organisms: A General Theory*, Praeger Publishers, Westport, CT, 2002, p.61
4. Noe, op.cit., p.65
5. K. Nelson, *The Spiritual Doorway in the Brain: A Neurologist's Search for the God Experience*, Plume, 2012, p.205
6. G. Vince, "Hacking the Nervous System," *Huffington Post*, May 29, 2015
7. N. Lane, *Life Ascending: The Ten Great Inventions of Evolution*, Norton, 2010, p.110
8. M. Ho, Mai-Wan. *The Rainbow and Worm: The Physics of Organisms*, World Scientific Publishing, S Singapore, 2008, p.176
9. Orange County. "Introduction to Learning Disabilities," *Orange County Learning Disabilities Association, http://www.oclda.org/understanding.html*
10. Noe, op.cit., p.77
11. A. Damasio, *Self Comes to Mind: Constructing the Conscious Brain*, Vintage, 2012, p.76
12. H. Jonas, *The Phenomenon of Life: Toward a Philosophical Biology*, Dell Publishing, New York, 1966, p.85
13. H. Bergson, *Time and Free Will: As Essay on the Immediate Data of Consciousness*, Harper Torchbook, New York, 1960, p.33
14. P. Gibbs, "Can a Human See a Single Photon?" University of California Riverside, 1996 http://math.ucr.edu/home/baez/physics/Quantum/see_a_photon.html
15. W. Loewenstein, Werner. *Physics in Mind: A Quantum View of the Brain*, Basic Books, New York, 2013, p.242
16. L. Margulis, C. Asikaninen, W. Krumbein, eds., *Chimeras and Consciousness: Evolution of the Sensory Self*, MIT Press, 2011, p.147
17. Loewenstein, op.cit., p.71
18. R. Lanza, B. Berman, *Biocentrism: How Life and Consciousness Are the Keys to Understanding the True Nature of the Universe*, BenBella Books, Dallas, 2010, p.20
19. Damasio, op.cit., p.275
20. Lowenstein, op.cit., p.246
21. ibid., p.48
22. Margulis, op.cit., p.146
23. N. Humphrey, *A History of the Mind: Evolution and the Birth of Consciousness*, Springer-Verlag, New York, 1999, p.189
24. T. Deacon, *Incomplete Nature: How Mind Emerged from Matter*, W.W. Norton & Co., New York, 2012, p.205
25. Loewenstein, op.cit., p.98
26. Margulis, op.cit., p.148
27. S. Blackmore, *Conversations on Consciousness: What the Best Minds Think about the Brain, Free Will, and What It Means to Be Human*," Oxford University Press, Oxford, 2006, p.59
28. T. Metzinger, *The Ego Tunnel: The Science of the Mind and the Myth of the Self*, Basic Books, New York, 2009, p.60
29. V.S.Ramachandran, *The Tell-Tale Brain: A Neuroscientist's Quest for What Makes Us Human*, W.W. Norton, 2012, p.57
30. Humphrey, op.cit., p.82
31. Bergson, op.cit., p.84
32. H. Maturana, F. Varela, *The Tree of Knowledge: The Biological Roots of Human Understanding*, Shambhala Publications, Boston, 1992, p.23
33. Humphrey, op.cit., p.159
34. Loewenstein, op.cit., p.3

35. Deacon, op.cit., p.509
36. Damasio, op.cit., p.39
37. Margulils, op.cit., p.144
38. Ramachandran, op.cit., p.58
39. Loewenstein, op.cit., p.201
40. Damasio, op.cit., p.335
41. ibid., p.273
42. ibid., p.109
43. ibid., p.56
44. Nelson, op.cit., p.24
45. Humphrey, op.cit., p.105
46. G. Bateson, A. Montuori, *Mind and Nature: A Necessary Unity*, Hampton Press, 2002, p.66
47. H. Stapp, *Mindful Universe: Quantum Mechanics and the Participating Observer* (The Frontiers Collection) 2nd ed., Springer, Heidelberg, 2011, p.24
48. D. Ropeik, "Inside the Mind of Worry," *New York Times*, September 28, 2012
49. Nelson, op.cit., p.127
50. Damasio, op.cit., p.257
51. Lane, op.cit., p.259
52. Deacon, op.cit., p.274
53. D. Hofstadter, *Godel, Escher, Bach: An Eternal Golden Braid*, Basic Books, New York,1999 edition, p.338
54. Damasio, op.cit. p.335

## B. BRAIN

1. P. Tse, *The Neural Basis of Free Will: Criterial Causation*, MIT Press, 2013, p.3
2. D. Bray, *Wetware: A Computer in Every Living Cell*, Yale University Press, 2009, p.211
3. K. Nelson, *The Spiritual Doorway in the Brain: A Neurologist's Search for the God Experience*, Plume, 2012. p.131
4. A. Scott, *Stairway to the Mind: The Controversial New Science of Consciousness*, Springer-Verlag, New York, 1995, p.71
5. T. Metzinger, *The Ego Tunnel: The Science of the Mind and the Myth of the Self*, Basic Books, New York, 2009, p.41
6. S. Allport, *Explorers of the Black Box: The Search for the Cellular Basis of Memory*, W.W. Norton, New York, 1986, p.157
7. J. Edwards, *How Many People are There in My Head? And in Hers?: An Exploration of Single Cell Consciousness*, Imprint Academic, Exeter, UK, 2006, p.38
8. Tse, op.cit., p.34
9. W. Loewenstein, *Physics in Mind: A Quantum View of the Brain*, Basic Books, New York, 2013, p.128
10. Scott, op.cit., p.49
11. Loewenstein, op.cit., p.145
12. Bray, op.cit., p.218
13. Loewenstein, op. cit., p.445
14. Tse, op.cit., p.9
15. Loewenstein, op.cit., p.229
16. Allport, op.cit., p.42
17. ibid., p.251
18. Bray, op.cit., p.224
19. V. Ramachandran, *The Tell-Tale Brain: A Neuroscientist's Quest for What Makes Us Human*, W.W. Norton, 2012, p.31
20. Tse, op.cit., p.167
21. Allport, op.cit., p.47
22. ibid., p.124
23. Scott, op.cit., p.76
24. M. Talbot, *The Holographic Universe: The Revolutionary Theory of Reality*, HarperCollins, New York, 2011,, p.31
25. N. Lane, *Life Ascending: The Ten Great Inventions of Evolution*, Norton, 2010, p.247

26. Ramachandran, op.cit., p.126
27. J. Gorman, "A Mouse Switch Turns Off Appetite," *New York Times*, August 11, 2014
28. M. Kaku, *The Future of the Mind: The Scientific Quest to Understand, Enhance and Empower the Mind*, Anchor Books, New York, 2014, p.112
29. C. Zimmer, "Phantom Melodies Yield Real Clues to Brain's Workings," New York Times, February 13, 2014
30. E. Reas, "Your Brain Has 2 Clocks," *Scientific American*, November 26, 2013
31. Loewenstein, op.cit., p.227
32. E. Kandel, "The New Science of Mind," New York Times, September 6, 2013
33. M. Cantor, "Scientists Discover 'New' Area of Brain," *www.Newser.com*, January 29, 2014
34. Kandel, op.cit.
35. V.Ramachandran, *A Brief Tour of Human Consciousness: From Imposter Poodles to Purple Numbers*, Plume, 2005, p.4
36. Lowenstein, op.cit., p.195
37. Ramachandran, *Tell-Tale,* op.cit., p.64
38. Edwards, op.cit., p.41
39. T. Deacon, *Incomplete Nature: How Mind Emerged from Matter*, W.W. Norton & Co., New York, 2012, p.510
40. R. Pico, *Consciousness in Four Dimensions: Biological Relativity and the Origins of Thought*, McGraw-Hill, 2001, p.172
41. Ibid., p.181
42. Nelson, op.cit., p.81
43. Kaku, op.cit., p.36
44. E. Sternberg, "When People With Schizophrenia Hear Vocies, They're Really Hearing Their Own Subvocal Speech," *Slate,* March 2, 2016
45. Kaku, op.cit., p.58
46. S. Horowitz, "The Science and Art of Listening," *New York Times*, November 9, 2012
47. Kaku, op.cit., p.138
48. Nelson, op.cit., p.259
49. G. Stuart, N. Spruston, "Dendritic Integration: 60 Years of Progress," *Nature Neuroscience*, November 25, 2015
50. Edwards, op.cit., p.99
51. D. Hofstadter, *Godel, Escher, Bach: An Eternal Golden Braid*, Basic Books, New York,1999 edition, p preface
52. Edwards, op.cit., p.68
53. A. Damasio, *Self Comes to Mind: Constructing the Conscious Brain*, Vintage, 2012, p.70
54. Tse, op.cit, p.119
55. Scott, op.cit., p.81
56. J. Gorman, "Navigational Cells Located in Human Brains," *New York Times*, August 6, 2013
57. S. Satel, S. Lilienfeld, *Brainwashed: The Seductive Appeal of Mindless Neuroscience,* Basic Books, 2013, p.12
58. Edwards, op.cit.,p.120
59. Tse, op.cit., p.7
60. Scott, op.cit., p.81
61. J. Gorman, "Learning How Little We Know About the Brain," *New York Times*, November, 10, 2014
62. G. Lawton, "You Are Your Brain," *Slate.com*, December 8, 2013
63. W. Loewenstein, *The Touchstone of Life: Molecular Information, Cell Communication and the Foundation of Life*, Oxford University Press, New York, 1998, p.309
64. Edwards, op.cit., p.151
65. Ho, Mai-Wan. *The Rainbow and Worm: The Physics of Organisms*, World Scientific Publishing, Singapore, 2008, p.242
66. S. Strogatz, *Sync: How Order Emerges From Chaos In the Universe, Nature, and Daily Life*, Hyperion, 2004, p.237
67. J. Gleick, *Chaos: Making a New Science*, Penguin Books, New York, 2008, p.299
68. Tse, op.cit., p.5
69. Allport, op.cit., p.69
70. Strogatz, op.cit., p.271

71. Satel, op.cit., p.16
72. Ho, op.cit., p.241
73. Scott, op.cit., p.84
74. S. Fan, "Naughty or Nice? One Brain Scan Is Now All It Takes to Find Out," *SingularityHub*, October 11, 2015
75. S. Hameroff, "The 'Conscious Pilot' - Dendritic Synchrony Moves Through the Brain to Mediate Consciousness," *https://www.ncbi.nlm.nih.gov/pmc/articles/PMC2791805/*, April 2,2009
76. Metzinger, op.cit., p.32
77. Loewenstein, *Physics*, op.cit., p.134
78. Kaku, op.cit., p.205
79. Damasio, op.cit., p.263
80. Kaku, op.cit., p.48
81. Loewenstein, Physics, op.cit., p.167
82. ibid., p.228
83. KU Leuven, "'Brain Waves' Challenge Area-Specific View of Brain Activity," *ScienceDaily.com*, March 20, 2013
84. U. Mohrhoff, "Why the Laws of Physics Are Just So," *This Quantum World*, http://thisquantumworld.com/wp/a-critique-of-quantum-mechanics/why-the-laws-of-physics-are-just-so/
85. Strogatz, op.cit., p.282
86. Damasio, op.cit., p.268
87. L. Margulis, C. Asikaninen, W. Krumbein, eds., *Chimeras and Consciousness: Evolution of the Sensory Self*, MIT Press, 2011, p.227
88. G. Cook, "Mind Games," *NY Times Magazine*, January 11, 2015
89. Tse, op.cit., p.75
90. B. Yirka, "Researchers Image Most of Vertebrae Brain at Single Cell Level," *https://medicalxpress.com/news/2013-03-image-vertebrae-brain-cell-video.html*, March 20, 2013
91. P. Catapano, E. Gehr, "Can We See Philosophy? A Dialogue With Ernie Gehr," *New York Times*, October 3, 2013
92. Lane, op.cit., p.244
93. https://en.wikipedia.org/wiki/Gamma_wave#cite_note-pollack-4
94. E. Thompson, *Mind in Life: Biology, Phenomenology and the Sciences of Mind*, Harvard University Press, Cambridge, Ma., 2007, p.332
95. Hameroff, op.cit.
96. Lane, op.cit., p.249
97. G. Hickok, "It's Not a 'Stream' of Consciousness," *New York Times*, May 8, 2015
98. R. Zatorre, V. Salimpoor, "Why Music Makes Our Brain Sing," *New York Times*, June 7, 2013
99. Damasio, op.cit., p.99
100. Ho, op.cit., p.320
101. Reas, op.cit.
102. Pico, op.cit., p.209
103. N. Humphrey, *Seeing Red: A Study in Consciousness*, Harvard University Press, 2006, p.119
104. T. Lewis, "New Record for Human Brain: Fastest Time to See an Image," *LiveScience.com*, January 17, 2014
105. A. Goswami, *The Self-Aware Universe: How Consciousness Creates the Material World*, Penguin Putnam I Inc., New York, 1995, p.194
106. Thompson, op.cit., p.332
107. Pico, op.cit., p.95
108. ibid., p.199
109. ibid., p.197
110. Strogratz, op.cit., p.76
111. ibid., p.87
112. Ibid., p.69
113. Nelson, op.cit., p.157
114. Strogratz, op.cit., p.69
115. Nelson, op.cit., p.157
116. Tse, op.cit., p.120
117. ibid., p.152

118. ibid., p.121
119. ibid., p.32
120. ibid., p.130
121. ibid., p.15
122. ibid., p.74
123. ibid., P.123
124. ibid., p.164
125. ibid., p.161
126. ibid., p.124
127. ibid., p.126
128. ibid., p.13
129. ibid., p.27
130. ibid., p.70
131. ibid., p.238
132. ibid., p.2
133. ibid., p.12
134. ibid., p.119
135. ibid., p.24
136. ibid., p.266
137. Nelson, op.cit., p.172
138. Tse, op.cit., p.129
139. ibid., p.22
140. B. Brembs, "Towards a scientific concept of free will as a biological trait: spontaneous actions and decision- making in invertebrates," *The Royal Society Publishing*, December 15, 2010
141. Tse, op.cit., p.14
142. ibid., p.77
143. Brembs, op.cit.
144. Tse, op.cit., p.175
145. ibid., p.223
146. ibid., p.211
147. Damasio, op.cit., p.268
148. Scott, op.cit., p.172
149. S. Blackmore, *Conversations on Consciousness: What the Best Minds Think about the Brain, Free Will, and What It Means to Be Human*," Oxford University Press, Oxford, 2006, p.13
150. Kaku, op.cit., p.230
151. Nelson, op.cit., p.52
152. Hameroff, op.cit.
153. L. Smolen, *Time Reborn: From the Crisis in Physics to the Future of the Universe*, Mariner Books, 2014, p.268
154. Tse, op.cit., p.197
155. Horowitz, op.cit.
156. Tse, op.cit., p.237
157. ibid., p.202
158. ibid., p.211
159. A. Noe, Alva. *Out of Our Heads: Why You Are Not Your Brain, and Other Lessons from the Biology of Consciousness*, Hill and Wang, New York, 2009, p.154
160. Tse, op.cit., p.193
161. Zimmer, op.cit.
162. Kaku, op.cit., p.113
163. Tse, op.cit., p.86
164. ibid., p.189
165. ibid., p.204
166. Metzinger, op.cit., p.38
167. Tse, op.cit., p.188
168. D. Chalmers, *The Conscious Mind: In Search of a Fundamental Theory*, Oxford University Press, 1997, p.115

169. S. Larson, Stephen. *The Healing Power of Neurofeedback: The Revolutionary LENS Technique for Restoring Optimal Brain Function*, Healing Arts, 2006, p.377
170. Tse, op.cit., p.48
171. Damasio, op.cit., p.264
172. Loewenstein, *Physics*, op.cit., p.217
173. Metzinger, op.cit., p.127
174. S. Fan, Shelly. "Think Your Conscious Brain Directs Your Actions? Think Again," *Huffington Post,* August 3, 2015
175. Tse, op.cit., p.195
176. Kaku, op.cit., p.338
177. Metzinger, op.cit., p.35
178. Ibid., p.127
179. Tse, op.cit., p.134
180. J. Rosen, "What Free Will Looks Like in the Brain," *Johns Hopkins University*, July 13,2016, *http://releases.jhu.edu/2016/07/13/what-free-will-looks-like-in-the-brain/*
181. Tse, op.cit., p.20
182. Ibid., p.206
183. Scott, op.cit., p.135
184. Tse, op.cit., p.199

## C. OTHER ANIMALS

1. M. Bekoff, "Scientists Conclude Nonhuman Animals Are Conscious Beings," *Psychology Today*, August 10,,2012
2. A. Halberstadt, "Zoo Animals and Their Discontents," *New York Times*, July 3, 2014
3. F. de Waal, "What I Learned From Tickling Apes," *New York Times*, April 8, 2016
4. J. Kluger, *The Animal Mind*, Time Special Edition, 2017. p.10
5. M. Kaku, *The Future of the Mind: The Scientific Quest to Understand, Enhance and Empower the Mind*, Anchor Books, New York, 2014, p.57
6. Kluger, op.cit.,p.18
7. R. Pico, *Consciousness in Four Dimensions: Biological Relativity and the Origins of Thought*, McGraw-Hill, 2001, p.248
8. Bekoff, op.cit.
9. G. Dvorksy, "Prominent Scientists Sign Declaration That Animals Have Conscious Awareness, Just Like Us," *IO9giszmodo.com*, August 23, 2012
10. Halberstadt, op.cit.
11. B. Bosker, "Cyborg Neil Harbisson On Life With Extra Senses," *Huffington Post*, February 23, 2013
12. Kaku, op.cit., p.74
13. S. Allport, *Explorers of the Black Box: The Search for the Cellular Basis of Memory*, W.W. Norton, New York, 1986, p.60
14. S. Kullander, B. Larsson, *Out of Sight!: From Quarks to Living Cells*, Cambridge University Press, 1994, p.6
15. V. Vedral,"Living in a Quantum World," *National Taiwan Normal University*, November 6, 2003, *http://phy.ntnu.edu.tw/~chchang/Notes10b/0611038.pdf*
16. J. Howard, "Here's How Birds Make Do Without External Ears," *Huffington Post*, December 14, 2014
17. A. Zolli, "Seeing the Whole: A Reflection for Earth Day," *www.garrisoninstitute.org*, April 22, 2016
18. Kaku, op.cit., p.304
19. D. Overbye, "Stars, Gold, Dung Beetles and Us," *New York Times*, August 6, 2013
20. G. Marcus, *The Birth of Mind: How a Tiny Number of Genes Creates The Complexities of Human Thought*, Basic Books, New York, 2004, p.80
21. Halberstadt, op.cit.
22. E. Chaisson, "Cosmic Evolution: An Inderdisciplinary Approach", *cfa.harvard.edu*, "Brain Evolution"
23. J. Cambell, *The Way of the Animal Powers, Volume 1, Historical Atlas of World Mythology*, Alfred Van Der Marck Editions; 1st edition 1983
24. T. Matsuzawa, Interviewed in "Chimpazees Beat Humans In Memory," *Huffington Post*, September 30, 2013
25. P. Churchland, *Touching a Nerve: The Self as Brain*, W.W. Norton, 2013. p.21

26. J. Balcombe, *What a Fish Knows: The Inner Lives of Our Underwater Cousins*, Scientific American / Farrar, Straus and Giroux; Reprint edition, 2017, p.107
27. R. Pico, *Consciousness in Four Dimensions: Biological Relativity and the Origins of Thought*, McGraw-Hill, 2001, p.207
28. K. Harmon, "Octopuses Gain Consciousness (According to Scientists' Declaration)," *Scientific American*, August 21, 2012
29. N. Wade, "In Tiny Worm, Unlocking Secrets of the Brain," *New York Times*, June 20, 2011
30. G. Berns, "Dogs Are People, Too," *New York Times*, October 5, 2013
31. Kluger, op.cit., p.14
32. Kaku, op.cit., p.305
33. C. Ray, "Plants in Peril," *New York Times*, December 23, 2013
34. G. Marcus, Gary. *The Birth of Mind: How a Tiny Number of Genes Creates The Complexities of Human Thought*, Basic Books, New York, 2004, p.114
35. J. Gorman, "Learning How Little We Know About the Brain," *New York Times*, November, 10, 2014
36. B. Brembs, "Towards a scientific concept of free will as a biological trait: spontaneous actions and decision- making in invertebrates," *The Royal Society Publishing*, December 15, 2010
37. Gorman, op.cit.
38. Harmon, op.cit.
39. Brembs, op.cit.
40. Marcus, op.cit., p.95
41. Wade, op.cit.
42. Brembs, op.cit.
43. Kluger, op.cit., p.72
44. E. Howard, "Sea Lion Is First Non-human Mammal Shown to Keep a Beat," *EarthSky.org*, April 5, 2013
45. Marcus, op.cit., p.104
46. Kaku, op.cit., p.159
47. Kluger, op.cit. p.111
48. ibid., p.75
49. ibid., p.106
50. ibid., p.78
51. E. Mazza, "Chimps Can Learn to Communicate with Each Other, New Study Shows," *Huffington Post*, February 10, 2015
52. Kluger, op.cit., p.26
53. ibid., p.23
54. L. Margulis, C. Asikaninen, W. Krumbein, eds., *Chimeras and Consciousness: Evolution of the Sensory Self*, MIT Press, 2011, p.65
55. Kaku, op.cit., p.307
56. Kluger, op.cit., p.31
57. J. Welsh, "Ants Vaccinate One Another To Prevent Epidemics, Colony Study Suggests," *Huffington Post*, April 3, 2012
58. J. Gorman, "No Glass Ceiling for Worker Bees," *New York Times*, September 9, 2014
59. R. Dicker, "Cockroaches Shown to Have Distinct Personalities," *Huffington Post*, February 9, 2015
60. Kluger, op.cit., p.16
61. Mazza, op.cit.
62. Kluger, op.cit. p.42
63. Balcombe, op.cit.
64. Kluger, op.cit. p.53
65. Halberstadt, op.cit.
66. Kluger, op.cit., p.60
67. ibid., p.87
68. Balcombe, op.cit.
69. Kluger, op.cit., p.93
70. Margulis, op.cit. p.211.
71. V. Virga, *The Soul of All Living Creatures: What Animals Can Teach Us About Being Human*, Broadway Books, 2014, p.207
72. Kluger, op.cit., p.10

73. Chaisson, op.cit.
74. Halberstadt, op.cit.
75. P. Tse, *The Neural Basis of Free Will: Criterial Causation*, MIT Press, 2013, p.211
76. Balcombe, op.cit.
77. J. Howard, "Cockatoos 'Pick Locks' For Food In Bird Intelligence Study," *Huffington Post,* July 26, 2013
78. Harmon, op.cit.
79. Kluger, op.cit., p.13
80. S. Montgomery, "Deep Intellect," *Orion Magazine, https://orionmagazine.org/article/deep- intellect/*
81. Kluger, op.cit., p.17
82. Mazza, op.cit.
83. F. Jabr, "Brainless Slime Molds Redefine Intelligence," Scientific American, November 7, 2012

## D. PLANTS

1. B. Ford, *The Secret Language of Life: How Animals and Plants Feel and Communicate*, Fromm International, New York, 1999, p.185
2. Tompkins, Peter & Bird, Christopher. *The Secret Life of Plants*, Harper & Row, 1989, p.xi
3. Ford, op.cit., p.195
4. J. Stansbury, "Plant Intelligence - A Philosophical Discourse on Cognitive Awareness in Plants," *Battlegroundhealingarts.com,* May 4, 2016
5. S. Hanel, "Plant Intelligence - Our 5 Senses + 15 More," *Kosmos Journal*, May 2, 2017
6. D. Chamovitz, *What a Plant Knows*, Scientific American 2014, p.170
7. Ford, op.cit., p.185
8. G. Witzany, "Plant Communication from Biosemiotic Perspective," *Plant Signaling & Behavior, https://www.ncbi.nlm.nih.gov/pmc/articles/PMC2634023/*
9. Ford, op.cit., p.185
10. ibid., p.222
11. ibid., p.239
12. Chamovitz, op.cit., p.26
13. D, Chamovitz, "Plants Exhibit The Same Senses As Humans And See, Touch, Smell, Hear and Even Taste," *preventdisease.com,* November 7, 2013
14. G. Cook,. "Do Plants Think?" *Scientific American*, June 5, 2012
15. Chamovitz, Plants, op.cit.
16. Witzany, op.cit.
17. Ford, op.cit., p.239
18. F. Cyrckova, Fatima. "Plant Intelligence: Why, why not or where?" *Plant Signaling & Behavior*, May 5, 2009 *http://www.whataplantknows.com/home/plant-senses*
19. Ford, op.cit., p.197
20. Stansbury, op.cit.
21. P. Lenda, "Plant Consciousness," *https://www.shift.is/plant-consciousness*
22. ibid.
23. Stansbury, op.cit.
24. Ford, op.cit., p.197
25. M. Pollan, "The Intelligent Plant," *The New Yorker*, December 23, 2013
26. P. Wohlleben, Peter. *The Hidden Life of Trees*, Greystone Books, 2016, p.47
27. Witzany, op.cit.
28. Wohlleben, op.cit., p.148
29. Ford, op.cit., p.213
30. Stansbury, op.cit.
31. Ford, op.cit., p.215
32. A. Scialdone, H, Martin, Howard. "How plants manage food reserves at night: Quantitative models and open questions," *Frontiers in Plant Science*, March 31, 2015
33. Pollan, op.cit.
34. ibid.
35. Stansbury, op.cit.
36. Ford, op.cit., p.188
37. Chamovitz, What a Plant, op.cit.

38. Ford, op.cit., p.185
39. Pollan, op.cit.
40. Witzany, op.cit.
41. Pollan, op.cit.
42. Witzany, op.cit.
43. Pollan, op.cit.
44. Witzany, op.cit.
45. A. Trewavas. Anthony. "Intelligence, Cognition, and Language of Green Plants," *Frontiers in Psychology*, April 26, 2016
46. Wohlleben, op.cit., p.23
47. Lenda, op.cit.
48. Wohlleben, op.cit., p.86
49. Ford, op.cit., p.185
50. Lenda, op.cit.
51. Witzany, op.cit.
52. Cook, op.cit.
53. Wohlleben, op.cit., p.15
54. J. Van Mater, "Our Intelligent Companions, the Plants," *http://www.theosophy-nw.org/theosnw/brother/br-jvmj2.htm*
55. Witzany, op.cit.
56. Lenda, op.cit.
57. Ford, op.cit., p.185
58. Carnegie Science. "The Heart of the Plant," *Department of Plant Biology*, December 8, 2011
59. Cook, op.cit.
60. Trewavas, op.cit.
61. Pollan, op.cit.
62. Lenda, op.cit.
63. Scialdone, op.cit.
64. Trewavas, op.cit.
65. Stansbury, op.cit.
66. Scialdone, op.cit.
67. Witzany, op.cit.
68. ibid.
69. ibid.
70. Ford, op.cit., p.239
71. Witzany, op.cit.
72. F. Baluska, S. Mancuso, et.al, "The 'root-brain' hypothesis of Charles and Francis Darwin," *Plant Signaling & Behavior*, December 1, 2009
73. Wohlleben, op.cit., p.83
74. Stansbury, op.cit.
75. Wohlleben. op.cit. p.171
76. Witzany. op.cit.
77. Stansbury, op.cit.
78. F. Baluska, S. Mancuso, et.al., *Root apices as Plant Command Centres: The Unique 'brain-like' Status of the Root Apex Transition Zone*, https://www.uv.mx/personal/tcarmona/files/2016/08/Baluzka-et-al-2004.pdf, 2004, p.3
79. Baluska, Darwin, op.cit.
80. Baluska, Ibid.
81. Baluska, Root, op.cit., p.3
82. ibid., p.6
83. ibid., p.9
84. ibid., p.7
85. Cook, op.cit.
86. Lenda, op.cit.
87. Baluska. Root, op.cit., p.2
88. ibid., p.6
89. Baluska, Darwin, op.cit.

90. Trewavas, op.cit.
91. Lenda, op.cit.
92. Ford, op.cit., p.185
93. Witzany, op.cit.
94. Ford, op.cit., p.239
95. R. Hardwick, "The Growth and Form of Modular Organisms - Physiological Consequences of Modular Growth in Plants," *The Royal Society*, August 14, 1986
96. Trewavas, op.cit.
97. Harkwick, op.cit.
98. ibid.
99. ibid.
100. Baluska, Root, op.cit., p.3
101. Khan. "Light and Photosynthetic Pigment," *Khan Academy, https://www.khanacademy.org/science/biology/photosynthesis-in- plants/the-light-dependent-reactions-of-photosynthesis/a/light-and-photosynthetic-pigments*
102. Trewavas. op.cit.
103. Ford, op.cit., p.185
104. A. Trewavas, "Aspects of Plant Intelligence," *Annals of Botany*, July, 2003
105. M. Marder, "Plant Intelligence and Attention," *Plant Signaling & Behavior*, February 20, 2013
106. ibid.
107. Pollan, op.cit.
108. Trewavas Intelligence, op.cit.
109. Pollan, op.cit.
110. Baluska, root, op.cit.
111. Pollan, op.cit.
112. ibid.

## E. FUNGI

1. P. Stamets, *Mycelium Running: How Mushrooms Can Help Save the World,* Ten Speed Press, First Edition, 2005
2. http://www.answers.com/Q/What_makes_up_a_cell_wall?#slide=1
3. D. Love, "Are Mushrooms Intellligent? *Backtotheroots.com*, October 13, 2016
4. Stamets, op.cit., p.1
5. Stamets, op.cit., p.304
6. Stamets, op.cit., p.1
7. S. Worrall,"There Is Such a Thing as Plant Intelligence," *National Geographic*, February 21, 2016
8. S. Blackstone, "Gift of the Fungi: Mushrooms - Yes, Mushrooms - Could Help Save the World," *Salon*, December 25, 2016
9. Worrall, op.cit.
10. Stamets, op.cit., p.2
11. J. Sayer, "Are Fungi the Earth's Natural Internet?" *GreenMedInfo.com*, February 20, 2012
12. Lumen, "Characteristics of Fungi," *Courses.lumenlearning.com/boundless-biology/chapter/characteristics-of- fungi/*
13. P. Stamets, "Fungal Intelligence and the 21st Psychedelic Journey - How Mushrooms Can Help Save the World," *Podcasts, personallifemedia.com*
14. Stamets, Mycelium, op.cit., p.2
15. ibid., p.7
16. A. Adamatzky, A, Ilachinski, "The Wisdom of Slime," *New York Times*, May 12, 2012
17. Stamets, Fungal, op.cit.
18. Stamets, Mycelium, op.cit., p.7
19. ibid., p.4
20. ibid., p.4
21. ibid., p.2
22. Stamets, Fungal, op.cit.
23. Stamets, Mycelium, op.cit. p.7
24. ibid., p.10
25. https://worldbuilding.stackexchange.com/questions/48528/super-intelligent-fungus

26. D. Jenson, "Nature's Internet: The Vast, Intelligent Network Beneath Our Feet," *Peakenergy.blogspot,com*, July 21, 2008
27. Worldbuilding, op.cit.
28. Stamets, Mycelium, op.cit., p.9
29. ibid., p.11

## F. ARTIFICIAL INTELLIGENCE

1. S. Hawking, M. Tegmark, S. Russell, F. Wilczek, "Transcending Complacency on Superintelligent Machines," *Huffington Post*, April 19, 2014
2. D. Chopra, "Good News: You Are Not Your Brain," *Huffington Post*, March 27, 2012
3. S. Blackmore, *Conversations on Consciousness: What the Best Minds Think about the Brain, Free Will, and What It Means to Be Human*," Oxford University Press, Oxford, 2006, p.5
4. N. Carr, "Why Robots Will Always Need Us," *New York Times*, May 20, 2015
5. J. Lee, "6 Human Jobs That Computers Will Never Replace," *Makeuseof.com*, July 21, 2014
6. J. Renstrom, "If I Only Had a Brain: How AI 'Thinks,'" *Daily Beast*, February 19, 2017
7. B. Gardner, "How Humans Can Keep Superintelligent Robots From Murdering Us All," *Mother Jones*, May 2, 2015
8. B. Bosker, "Cyborg Neil Harbisson On Life With Extra Senses," *Huffington Post*, February 23, 2013
9. D. Chopra, "Can Your Brain Fall in Love," *Huffington Post*, April 5, 2012
10. A. Regalado, "What Will It Take for Computers to Be Conscious," *MIT Technology Review*, October 2, 2014
11. J. Markkoff, "Brainlike Computers, Learning From Experience," *New York Times*, December 28. 2013
12. ibid.
13. ibid.
14. ibid.
15. ibid.
16. M. Strevens, "Looking into the Black Box," *New York Times*, November 24, 2013
17. R. Rosenkranz, "Artificial Intelligence: Stop or Go?" *Huffington Post*, February 1, 2016
18. J.Tankersley, "Science Fiction Come True: Moving a Paralyzed Hand with the Power of Thought," *Washington Post*, June 24, 2014
19. M. Kaku, *The Future of the Mind: The Scientific Quest to Understand, Enhance and Empower the Mind*, Anchor Books, New York, 2014, p.91
20. ibid., p.67
21. ibid., p.71
22. ibid., p.7
23. R. Kuhn, "Is Our Universe a Fake?" *Yahoo.com*, August 3, 2015
24. Kaku, op.cit., p.63
25. D. Freeman, "Quantum Teleportation Feat Brings Ultrafast Computer Networks Step Closer to Reality," *The Huffington Post*, May 29, 2014
26. R. Brooks, "What Robots Can Tell Us," *Time Magazine - Rise of the Robots*, 2013
27. Kaku, op.cit., p.107
28. Brooks, op.cit.
29. K. Miles,"Ray Kurzweil: In The 2030s, Nanobots In Our Brains Will Make Us 'Godlike,'" *Huffington Post*, October 1, 2015
30. C. Chocano, Carina. "The Last Medium; Virtual Reality Is Here," *California Sunday Magazine*, October 5, 2014
31. Kaku, op.cit., p.243
32. I. Nourbakhsh, "Complexity, Intelligence and How to Ignore the Meaning of Life," *Huffington Post*, June 14, 2013
33. Kaku, op.cit., p.239
34. Regalado, op.cit.
35. H. Price, "Cambridge, Cabs and Copehagen: My Route to Existential Risk," *New York Times*, January 27, 2013
36. ibid.
37. G. Gutting, "Modern Cosmology Versus God's Creation," *Opinionator, New York Times*, June 15, 2014

38. Kaku, op.cit., p.216
39. Price, op.cit.
40. Brooks, op.cit.
41. Gutting, op.cit.

# V. LIVING AWARENESS

## A. TIME

1. R. Lanza, B. Berman, *Biocentrism: How Life and Consciousness Are the Keys to Understanding the True Nature of the Universe*, BenBella Books, Dallas, 2010, p.101
2. W. Egginton, "On Borges, Particles and the Paradox of the Perceived," *New York Times*, April 28, 2013
3. ibid.
4. H. Macdonald, "A Bestiary of the Mind," *New York Times Magazine*, May 21,2017
5. U. Mohrhoff, "Why the Laws of Physics Are Just So," *This Quantum World*
6. M. Kaku, *The Future of the Mind: The Scientific Quest to Understand, Enhance and Empower the Mind*, Anchor Books, New York, 2014, p.325
7. ibid., p.324
8. M. Talbot, *The Holographic Universe: The Revolutionary Theory of Reality*, HarperCollins, New York, 2011, (quoting David Bohm), p.41
9. Kaku, op.cit., p.325
10. M. Humphrey, *A History of the Mind: Evolution and the Birth of Consciousness*, Springer-Verlag, New York, 1999, p.183
11. M. Ho, "The Biology of Free Will," *Journal of Consciousness Studies*, 1996
12. R. Unger, L. Smolin, *The Singular Universe and the Reality of Time: A Proposal in Natural Philosophy*, University Printing House, Cambridge, 2015, p.241
13. H.Bergson, *Time and Free Will: As Essay on the Immediate Data of Consciousness*, Harper Torchbook, New York, 1960, p. 102
14. Strevens, "Looking into the Black Box," *New York Times*, November 24, 2013
15. L. Smolen, *Time Reborn: From the Crisis in Physics to the Future of the Universe* Mariner Books, 2014, p.182
16. D. Hofstadter, *Godel, Escher, Bach: An Eternal Golden Braid*, Basic Books, New York,1999 edition, p.661
17. Smolen, op.cit., p.215
18. Lanza, op.cit., p.46
19. Bergson, op.cit., p.
20. E. Eckhart, *A New Earth: Awakening to Your Life's Purpose*, Penguin,2008, p.201
21. Kaku, op.cit., p.340
22. J. Brogan, "Our Virtual Future," *Slate*, September 1, 2017
23. Lanza, op.cit., p.110
24. D. Albert, David Z. *After Physics*, Harvard University Press, 2015, p.70
25. E. Thompson, *Mind in Life: Biology, Phenomenology and the Sciences of Mind*, Harvard University Press, Cambridge, Ma., 2007, p.318
26. J. Corrigan, M. *An Introduction to Awareness*, BookSurge Publishing, 2006, p.95
27. Ho, op.cit.
28. ibid.
29. ibid.
30. C. Zimmer, "Seeking the Gears of Our Inner Clock," *New York Times*, December 28, 2015
31. D. Quenqua. "Circadian Clocks May Be Tuned to Vibrations, Too," *New York Times*, February 4, 2014
32. J. Gleick, *Chaos: Making a New Science*, Penguin Books, New York, 2008, p.285
33. L. Margulis, C. Asikaninen, W. Krumbein, eds., *Chimeras and Consciousness: Evolution of the Sensory Self*, MIT Press, 2011, p.111
34. Quenqua, op.cit.
35. F.S. Brainard, *Reality's Fugue: Reconciling Worldviews in Philosophy, Religion, and Science,* The Pennsylvania State University Press, University Park, PA, 2017, p.184
36. https://en.wikipedia.org/wiki/Relativity_of_simultaneity

## B. AGENCY

1. G. Bateson, A. Montuori, *Mind and Nature: A Necessary Unity*, Hampton Press, 2002, p.6
2. U. Mohrhoff, "Why the Laws of Physics Are Just So," *This Quantum World*
3. A. Damasio, *Self Comes to Mind: Constructing the Conscious Brain*, Vintage, 2012, p.336
4. S. Harris, *Free Will,* Free Press, 2012, p.39
5. H. Bergson, *Time and Free Will: As Essay on the Immediate Data of Consciousness*, Harper Torchbook, New York, 1960, p.167
6. M. Ho, *The Rainbow and Worm: The Physics of Organisms*, World Scientific Publishing, Singapore, 2008, p.331
7. Bergson, op.cit., p.167

## C. SELF

1.G. Gutting, "What Would Krishna Do? Or Shiva? Or Vishnu?" *New York Times,* August 3, 2014
2. M. Ho, *The Rainbow and Worm: The Physics of Organisms*, World Scientific Publishing, Singapore, 2008, p.319
3. E. Thompson, *Mind in Life: Biology, Phenomenology and the Sciences of Mind*, Harvard University Press, Cambridge, Ma., 2007, p.149 (quoting Hans Jonas)
4. Ho, op.cit., p.331
5. ibid., p.332
6. B. Haisch, *The God Theory: Universes, Zero-point Fields, and What's Behind It All*, Red Wheel/Weiser, San Francisco, 2009, p.113
7. B. Stanley, "I Am Not This Body," *New York Times*, May 6, 2013
8. H. Morch, "Is Matter Conscious?" Nautilus, April 6, 2017
9. Thompson, op.cit., p.153
10. R. Lanza, B. Berman, *Biocentrism: How Life and Consciousness Are the Keys to Understanding the True Nature of the Universe*, BenBella Books, Dallas, 2010,p.180
11. Thompson, op.cit., p.76, quoting Merleau-Ponty
12. ibid., p.163
13. K. Nelson, *The Spiritual Doorway in the Brain: A Neurologist's Search for the God Experience*, Plume, 2012, p.232
14. ibid., p.228
15. Stanley, op.cit.
16. L. Margulis, C. Asikaninen, W. Krumbein, eds., *Chimeras and Consciousness: Evolution of the Sensory Self*, MIT Press, 2011, p.209
17. L. Smolen, *Time Reborn: From the Crisis in Physics to the Future of the Universe* Mariner Books, 2014, introduction
18. T. Deacon, *Incomplete Nature: How Mind Emerged from Matter*, W.W. Norton & Co., New York, 2012, p.142
19. ibid., p.155
20. ibid., p.171
21. ibid., p.367
22. ibid., p.175
23. ibid., p.177
24. C. Lineweaver, P. Davies, M. Ruse, eds., *Complexity and the Arrow of Time*, Cambridge University Press, New York, 2013, p.71
25. M. Ho, "The Biology of Free Will," *Journal of Consciousness Studies*, 1996
26. ibid.
27. J. Cepelewicz, "Seeing Emergent Physics Behind Evolution," *Quanta Magazine*, August 31, 2017
28. ibid.
29. ibid.
30. ibid.
31. W. Loewenstein, *The Touchstone of Life: Molecular Information, Cell Communication and the Foundation of Life*, Oxford University Press, New York, 1998, p.235

32. S. Blackmore, *Conversations on Consciousness: What the Best Minds Think about the Brain, Free Will, and What It Means to Be Human*," Oxford University Press, Oxford, 2006, p.8
33. A. Noe, *Out of Our Heads: Why You Are Not Your Brain, and Other Lessons from the Biology of Consciousness*, Hill and Wang, New York, 2009, p.95

## D. AWARENESS

1. A. Noe, *Out of Our Heads: Why You Are Not Your Brain, and Other Lessons from the Biology of Consciousness*, Hill and Wang, New York, 2009, p.41
2. H. Jonas, *The Phenomenon of Life: Toward a Philosophical Biology*, Dell Publishing, New York, 1966, p.150
3. R. Lanza, B. Berman, *Biocentrism: How Life and Consciousness Are the Keys to Understanding the True Nature of the Universe*, BenBella Books, Dallas, 2010, p.96
4. ibid., p.99
5. B. Lipton, *The Biology of Belief: Unleashing the Power of Consciousness, Matter & Miracles*, Hay House, 2008, p.166
6. ibid., p.167
7. Lanza, op.cit., p.38 (quoting Einstein)
8. J.Corrigan, *An Introduction to Awareness*, BookSurge Publishing, 2006, p.41
9. M. Gleiser, Marcelo."There Is No Now," *http://www.pbs.org/wgbh/nova/blogs/physics/2014/06/there-is-no-now/*, June 30, 2014
10. J. Cambell, *The Way of the Animal Powers, Volume 1, Historical Atlas of World Mythology*, by Joseph Campbell, Alfred Van Der Marck Editions; 1st edition 1983, p.25
11. Gleiser, op.cit.
12. D. Albert, *After Physics*, Harvard University Press, 2015, p.516
13. L. Smolen, *Time Reborn: From the Crisis in Physics to the Future of the Universe* Mariner Books, 2014, preface
14. E. Parens, "The Benefits of 'Binocularity,'" *New York Times*, September 28, 2014, comment
15. Gleiser, op.cit.
16. https://en.wikipedia.org/wiki/Orders_of_magnitude_(time)
17. Gleiser, op.cit.
18. T. Lewis," New Record for Human Brain: Fastest Time to See an Image," *LiveScience.com*, January 17, 2014
19. M. Strevens, "Looking into the Black Box," *New York Times*, November 24, 2013
20. Gleiser, op.cit.
21. D. Hoffman, "Conscious Realism and the Mind-Body Problem," *Mind & Matter, Vol. 6(1),* 2008
22. L. Nottale, *Fractal Space-Time and Microphysics: Towards a Theory of Scale Relativity*, World Scientific Publishing, London, 1993, p.11 (quoting Einstein)
23. J. Lent, "The Dangerous Delusions of Richard Dawkins," *Salon,* August 7, 2017
24. D. Overbye, Dennis. "A Quantum of Solace," New York Times, July 1, 2013, quoting Einstein
25. M. Tegmark, "Life Is a Braid in Spacetime," *Nautilus*, October 1, 2015
26. Quanta, "Interactive: What Is Space?" *https://www.quantamagazine.org/interactive-what-is-space-20150430*
27. U. Mohrhoff, "Why the Laws of Physics Are Just So," *This Quantum World*
28. R. Cowen, "The Quanum Source of Space-Time," *Nature*, November 16, 2015
29. M. Tegmark, *Our Mathematical Universe: My Quest for the Ultimate Nature of Reality*, Vintage Books, 2015, p.282
30. M. Tegmark, "Life Is a Braid in Spacetime," *Nautilus*, October 1, 2015
31. Gleiser, op.cit.
32. Hoffman, op.cit.
33. ibid.
34. ibid.
35. A. Rosenberg, D. McShea, *Philosophy of Biology: A Contemporary Introduction*, Routledge, New York, 2008, p.44
36. T. Metzinger, *The Ego Tunnel: The Science of the Mind and the Myth of the Self*, Basic Books, New York, 2009, p.216
37. Lent, op.cit.

38. M. Kaku, *The Future of the Mind: The Scientific Quest to Understand, Enhance and Empower the Mind*, Anchor Books, New York, 2014, p.52
39. Gleiser, op.cit.
40. H. Bergson, *Time and Free Will: As Essay on the Immediate Data of Consciousness*, Harper Torchbook, New York, 1960, p.123
41. M. Wilson, "AI Is Inventing Languages Humans Can't Understand. Should We Stop It?" *Fastcodesign.com*, July 14, 2017
42. ibid.
43. M. Jasper, "Artificial Intelligence Invents New Language Humans Can't Understand," *Themarysue.com*, July 29, 2017
44. T. Nagel, *Mind & Cosmos: Why the Materialist Neo-Darwinian Conception of Nature is Almost Certainly False*, Oxford University Press, Oxford, 2012
45. Smolen, op.cit., preface
46. A. Goswami, *The Self-Aware Universe: How Consciousness Creates the Material World*, Penguin Putnam I Inc., New York, 1995, p.141
47. E. Tolle, *A New Earth: Awakening to Your Life's Purpose*, Penguin, 2008, p.259

## BIBLIOGRAPHY


Adamatzky, Andrew and Ilachinski, Andrew. "The Wisdom of Slime," *New York Times*, May 12, 2012

Alexander, Amir. *Infinitesimal: How a Dangerous Mathematical Theory Shaped the Modern World,* Scien tific American/Farrar, Straus and Giroux, New York, 2014

Albert, David Z. *After Physics*, Harvard University Press, 2015

Alberts, Bruce. *Molecular Biology of the Cell*, Garland Science, 2014

Alberts, Bruce, Bray, Dennis, Hopkin, Karen et.al., *Essential Cell Biology, Fourth Edition*, Garland Science, 2013

Allport, Susan. *Explorers of the Black Box: The Search for the Cellular Basis of Memory*, W.W. Norton, New York, 1986

Alon, Uri. *An Introduction to Systems Biology: Design Principles of Biological Circuits*, CRC Press, Boca Raton, FL, 2006

Anthes, Emily. "Coldblooded Does Not Mean Stupid," *New York Times*, November 18, 2013

Aschwanden, Christie. "Where Is Thought," *Discover Magazine*, June, 2013

Ayala, Francisco J. and Arp, Robert, eds., *Contemporary Debates in Philosophy of Biology*, Wiley-Blackwell, Oxford, England, 2009

Balcombe, Jonathan. *What a Fish Knows: The Inner Lives of Our Underwater Cousins*, Scientific American / Farrar, Straus and Giroux; Reprint edition, 2017

Baluska, Frantisek & Mancuso, Stefano, et.al., *Root apices as Plant Command Centres: The Unique 'brain-like' Status of the Root Apex Transition Zone*, https://www.uv.mx/personal/tcarmona/files/2016/08/Baluzka-et-al-2004.pdf, 2004

Baluska, Frantisek & Mancuso, Stefano, et.al. "The 'root-brain' hypothesis of Charles and Francis Darwin," *Plant Signaling & Behavior*, December 1, 2009

Ball, Phillip. "How Big Can Schrodinger's Kittens Get?" *Nautilus*, October 8, 2015, *http://nautil.us/issue/29/scaling/how-big-can-schr246dingers-kittens-get*

Ball, Phillip."The Strange Inevitability of Evolution," *Nautilus*, January 8, 2015, *http://nautil.us/issue/20/creativity/the-strange-inevitability-of-evolution*

Bateson, Gregory and Montuori, Alfonso. *Mind and Nature: A Necessary Unity*, Hampton Press, 2002

Bekoff, Mark. "Scientists Conclude Nonhuman Animals Are Conscious Beings," *Psychology Today*, August 10, 2012

Bennett-Smith, Meredith. "How Did Early Birds Get Their Wings? New Study Suggests Surprising Answer," *Huffington Post*, September 19, 2013

Betterexplained.com. *betterexplained.com/articles/a-calculus-analogy-integrals-as-multiplication/*

Berg, Howard C. *Random Walks in Biology*, Princeton University Press, 1993

Bergson, Henri. *Time and Free Will: As Essay on the Immediate Data of Consciousness*, Harper Torchbook, New York, 1960

Berns, Gregory. "Dogs Are People, Too," *New York Times*, October 5, 2013

Bhanoo, Sindya N. "With Incentive, Monkeys Can Coordinate," New York Times, April 8,2013

Blackmore, Susan. *Conversations on Consciousness: What the Best Minds Think about the Brain, Free Will, and What It Means to Be Human*," Oxford University Press, Oxford, 2006

Blackstone, Samuel. "Gift of the Fungi: Mushrooms - Yes, Mushrooms - Could Help Save the World," *Salon*, December 25,2016

Boeyens, Jan C. A. "A Molecular-Structure Hypothesis," *International Journal of Molecular Sciences*, November 1, 2010, *https://www.ncbi.nlm.nih.gov/pmc/articles/PMC3000081/*

Boogerd, Fred C. and Bruggeman, Frank J. and Hofmeyr, Jan-Hendrik and Westerhoff, Hans V., eds. *\Systems Biology: Philosophical Foundations*, Elsevier, Amsterdam, 2007

Bosker, Bianca. "Cyborg Neil Harbisson On Life With Extra Senses," *Huffington Post*, February 23, 2013

Brainard, F. Samuel. *Reality's Fugue: Reconciling Worldviews in Philosophy, Religion, and Science,* The Pennsylvania State University Press, University Park, PA, 2017

Bray, Dennis. *Wetware: A Computer in Every Living Cell*, Yale University Press, 2009

Brembs, Bjorn. "Towards a scientific concept of free will as a biological trait: spontaneous actions and decision- making in invertebrates," *The Royal Society Publishing*, December 15, 2010

Brogan, Jacob. "Our Virtual Future," *Slate*, September 1, 2017

Brooks, Rodney. "What Robots Can Tell Us," *Time Magazine - Rise of the Robots*, 2013

Byrne, Michael. "How Big Can Quantum Weirdness Get?" *Motherboard*, February 15, 2014 *(https://motherboard.vice.com/en_us/article/wnj7mq/how-big-can-classical-weirdness-get)*

Cambell, Joseph. *The Way of the Animal Powers, Volume 1, Historical Atlas of World Mythology* ,Alfred Van Der Marck Editions; 1st edition 1983

Cantor, Matt. "Scientists Discover 'New' Area of Brain," *www.Newser.com*, January 29, 2014

Capra, Fritjof. *The Tao of Physics*, Shambhala Publications, Inc., Boston, 1985

Carnegie Science. "The Heart of the Plant," *Department of Plant Biology*, December 8, 2011

Carr, Nicholas. "Why Robots Will Always Need Us," *New York Times*, May 20, 2015

Carroll, Sean, "No Physicist Is an Island," *New York Times*, October 8, 2013

Carroll, Sean. "When Nature Looks Unnatural," *The New York Times, Opinionator*, March 23, 2014

Catapano, Peter and Gehr, Ernie. "Can We See Philosophy? A Dialogue With Ernie Gehr," *New York Times*, October 3, 2013

Cepelewicz, Jordana. "Seeing Emergent Physics Behind Evolution," *Quanta Magazine*, August 31, 2017

Center for Quantum Technologies, "Two Quantum Mysteries Merged into One," December 20, 2014, *http://www.quantumlah.org/highlight/141220_wave_particle.php*

Chaisson, Eric J. "Cosmic Evolution: An Inderdisciplinary Approach", *cfa.harvard.edu*, "Brain Evolution"

Chaisson, Eric J. *Epic of Evolution, Seven Ages of the Cosmos*, Columbia University Press, 2006

Chalmers, David J. *The Conscious Mind: In Search of a Fundamental Theory*, Oxford University Press, 1997

Chamovitz, Daniel. "Plants Exhibit The Same Senses As Humans And See, Touch, Smell, Hear and Even Taste," *preventdisease.com,* November 7, 2013

Chamovitz, Daniel. *What a Plant Knows*, Scientific American 2014

Chatterjee, Sankar. "How Did Life Begin? Texas Paleontologist Says He's Found the Biological 'Holy Grail,'" `interviewed in *Huffington Post*, November 1, 2013

Chesher, *This Magic Sea,* "The Error of Expectations," *Chesher*, 2011, *http://www.this-magic -sea.com/error.htm*

Chiribella, Giulio. "Is Your Theory of Everything Pure Enough?" Nautilus, February 19, 2015, *http://nautil.us/issue/21/information/is-your-theory-of-everything-pure-enough*

Chocano, Carina. "The Last Medium; Virtual Reality Is Here," *California Sunday Magazine*, October 5, 2014

Chopra, Deepak. "Can Your Brain Fall in Love," *Huffington Post*, April 5, 2012

Chopra, Deepak. "Good News: You Are Not Your Brain," *Huffington Post*, March 27, 2012

Churchland, Patricia S. *Touching a Nerve: The Self as Brain*, W.W. Norton, 2013

Comfort, Nathanial. "Single Gene May Hold Key to Life Itself," Huffington Post, September 23, 2012

Conover, Emily. "Go with the Flow," *Discover Magazine*, May, 2015

Cook, Gareth. "Do Plants Think?" *Scientific American*, June 5, 2012

Cook, Gareth. "Mind Games," *NY Times Magazine*, January 11, 2015

Corrigan, James M. *An Introduction to Awareness*, BookSurge Publishing, 2006

Cowen, Ron. "The Quanum Source of Space-Time," *Nature*, November 16, 2015

Cox , Brian and Forshaw, James. *Why Does E=mc2?: (and Why Should We Care?)*, Da Capo, 1994

Cuthbertson, Roy & Holcombe, Mike & Paton, Ray. *Computation in Cellular and Molecular Biological Systems*, World Scientific, 1996

Cyrckova, Fatima. "Plant Intelligence: Why, why not or where?" *Plant Signaling & Behavior*, May 5, 2009 *http://www.whataplantknows.com/home/plant-senses*

Damasio, Antonio. *Self Comes to Mind: Constructing the Conscious Brain*, Vintage, 2012

Deacon, Terrence, W. *Incomplete Nature: How Mind Emerged from Matter*, W.W. Norton & Co., New York, 2012

Deacon, Terrence, "Language and Complexity: Evolution Inside Out," *youtube,com*, August 25, 2010, *https://www.youtube.com/watch?v=OT-zZ0PMqgI*

Deely, John. *Four Ages of Understanding, The First Posmodern Survey of Philosophy from Ancient Times to the Turn of the Twenty-First Century*, University of Toronto Press, Toronto, 2001

de Waal, Frans. "What I Learned From Tickling Apes," *New York Times*, April 8, 2016

Dijkgraaf, Robbert. "How to Solve the Biggest Mystery in Physics." *Wired.com*, September 9, 2017

Dicker, Rob. "Cockroaches Shown to Have Distinct Personalities," *Huffington Post*, February 9, 2015

Douthat, Ross. "Sam Harris and Scientism," *New York Times*, September 5, 2013

Doyle, Robert O. "The Two-Stage Model to the Problem of Free Will," *Free Will: The Scandal in Philoso phy*, June 19,2011, *informationphilosopher.com*

Doyle, Robert. *Free Will: The Scandal in Philosophy,* Information Philosopher, Cambridge, MA,2011

Drake, J.W. "A constant rate of spontaneous mutation in DNA-based microbes," *National Academy of Sciences*, August 15, 1991

Duke University Medical Center. "Brain-to-brain interface allows transmission of tactile and motor infor mation between rats," *Medical Press*, February 28, 2013, *https://medicalxpress.com/news/2013-02-brain-to-brain-interface-transmission-tactile- motor.html*

Dvorksy, George. "Prominent Scientists Sign Declaration That Animals Have Conscious Awareness, Just Like Us," *IO9giszmodo.com*, August 23, 2012

Edwards, Jonathan, C.W., *How Many People are There in My Head? And in Hers?: An Exploration of Single Cell Consciousness*, Imprint Academic, Exeter, UK, 2006

Egginton, William. "On Borges, Particles and the Paradox of the Perceived," *New York Times*, April 28, 2013

Einstein, Albert. *Essays in Science*, Philosophical Library, New York, 1934

Einstein, Albert, *Out of My Later Years*, Philosophical Library, New York, 1950

Einstein, Albert, Leopold Infeld. *The Evolution of Physics*, Simon & Schuster, New York, 1966

Fan, Shelly. "Naughty or Nice? One Brain Scan Is Now All It Takes to Find Out," *SingularityHub*, October 11, 2015

Fan, Shelly. "Think Your Conscious Brain Directs Your Actions? Think Again," *Huffington Post*, August 3, 2015

Fine, Neil. Editor. "Rise of the Robots, " *Time Magazine*, 2013 (single issue)

Ford, Brian J. *The Secret Language of Life: How Animals and Plants Feel and Communicate*, Fromm International, New York, 1999

Frank, Adam. "Has Next Tuesday Already Happened?" *http://www.npr.org/blogs/13.7/2014/08/26/343483711/has- next-tuesday-already-happened*, August 26, 2014

Frederickson, Barbara L. "Your Phone vs. Your Heart," *New York Times*, March 23, 2013

Freeman, David. "Quantum Teleportation Feat Brings Ultrafast Computer Networks Step Closer to Reality," *The Huffington Post*, May 29, 2014

Frenkel, Edward. "Is the Universe a Simulation?" New York Times, February 14, 2014

Frenkel, Edward. "The Reality of Quantum Weirdness," *Gray Matter, The New York Times*, February 20, 2015

Friedland, Julian. "Philosophy Is Not Science," New York Times, April 5, 2012

Friedman, Andres. "Are the Quantum World and The Real Word the Same Thing?" *The Nature of Reality*, May 7, 2015, *http://www.pbs.org/wgbh/nova/blogs/physics/2015/05/quantum-word-real-world-thing/*

Froy, Oren and Miskin, Ruth. "Effect of feeding regimens on circadian rhythms: Implications for Aging and longevity," *Impact Journals: Open-Access Impact Journal on Aging*, January 11, 2010, *https://www.ncbi.nlm.nih.gov/pmc/articles/PMC2837202/*

Gardner, Bill. "How Humans Can Keep Superintelligent Robots From Murdering Us All," *Mother Jones*, May 2, 2015

Garrett, Reginald H., Grisham, Charles M., *Biochemistry*, Cengage, Boston, 2017

Gelernter, David. "Encounters with the Arch-genius, David Gelernter,"Time.com, http://time.com/4236974/encounters-with- the-archgenius/

Ghose, Tia. "Cambrian Explosion: Evolutionary Big Bang Was Sparked by Multiple Events, Scientists Say," *Huffington Post*, September 20, 2013

Ghose, Tia. "New Math Could Reveal Hidden Sources of Chaos," *LiveScience*, August 27, 2015, *https://www.livescience.com/52003-new-definition-chaos.html*

Gibbs, Philip."Can a Human See a Single Photon?" University of California Riverside, 1996 http://math.ucr.edu/home/baez/physics/Quantum/see_a_photon.html

Gleick, James. *Chaos: Making a New Science*, Penguin Books, New York, 2008

Gleiser, Marcelo."There Is No Now," *http://www.pbs.org/wgbh/nova/blogs/physics/2014/06/there-is-no-now/*, June 30, 2014

Goodsell, Davis S. *The Machinery of Life*, Copernicus, 2010

Goldbaum, Ellen. "Master orchestrator of the genome is discovered, UB stem cell scientists report," *University of Buffalo News Center*, May 7, 2015

Gorman, James. "A Mouse Switch Turns Off Appetite," *New York Times*, August 11, 2014

Gorman, James. "Considering the Humanity of Nonhumans," *New York Times*, December, 9, 2013

Gorman, James. "Learning How Little We Know About the Brain," *New York Times*, November, 10, 2014

Gorman, James. "Navigational Cells Located in Human Brains," *New York Times*, August 6, 2013

Gorman, James. "No Glass Ceiling for Worker Bees," *New York Times*, September 9, 2014

Goswami, Amit. *The Self-Aware Universe: How Consciousness Creates the Material World*, Penguin Putnam Inc., New York, 1995

Greenberg, Gary. "The Delusions We Deserve," New York Times, August 28, 2014

Gribbin, John. *In the Beginning: The Birth of the Living Universe*, Little Brown & Co., 1995

Gutting, Gary. "Mary and the Zombies: Consciousness Revisited," New York Times, March 22, 2013

Gutting, Gary. "Modern Cosmology Versus God's Creation," *Opinionator, New York Times*, June 15, 2014

Gutting, Gary. "Sam Harris's Vanishing Self," *New York Times*, September, 7, 2014

Gutting, Gary. "What Would Krishna Do? Or Shiva? Or Vishnu?" *New York Times,* August 3, 2014

Haisch, Bernard. *The God Theory: Universes, Zero-point Fields, and What's Behind It All*, Red Wheel/Weiser, San Francisco, 2009

Halberstadt, Alex. "Animals and Their Discontents," *New York Times*, July 3, 2014

Hameroff, Stuart. "The 'Conscious Pilot' - Dendritic Synchrony Moves Through the Brain to Mediate Consciousness," *https://www.ncbi.nlm.nih.gov/pmc/articles/PMC2791805/*, April 2, 2009

Hanel, Stephanie. "Plant Intelligence - Our 5 Senses + 15 More," *Kosmos Journal*, May 2, 2017

Hardwick, R.C. "The Growth and Form of Modular Organisms - Physiological Consequences of Modular Growth in Plant," *The Royal Society*, August 14, 1986

Harmon, Katherine. "Octopuses Gain Consciousness (According to Scientists' Declaration)," *Scientific American*, August 21, 2012

Harris, Sam. *Free Will,* Free Press, 2012

Hartnett, Kevin. "The Information Theory of Life," *Quanta Magazine*, November 19, 2015

Hartsfield, Tom. "Does the Mind Affect Quantum Mechanics?" *RealClear Science*, July 10, 2014,http://www.realclearscience.com/blog/2014/07/does_the_mind_affect_quantum_mechanics.html

Hawking, Stephen & Tegmark, Max & Russell, Stuart & Wilczek, Frank, "Transcending Complacency on Superintelligent Machines," *Huffington Post*, April 19, 2014

Hay, Mark."Chemistry as the Language of Life," *New York Times*, October 14, 2011

Hickok, Gregory. "It's Not a 'Stream' of Consciousness," *New York Times*, May 8, 2015

Ho, Mai-Wan, "Bioenergetics and the Coherence of Organisms," *https://ratical.org/co-globalize/MaeWanHo/prague.html*

Ho, Mai-Wan."The Biology of Free Will," *Journal of Consciousness Studies*, 1996

Ho, Mai-Wan. *The Rainbow and Worm: The Physics of Organisms*, World Scientific Publishing, Singapore, 2008

Hoffman, Donald. "Conscious Realism and the Mind-Body Problem," *Mind & Matter, Vol. 6(1),* 2008

Hoffmann, Peter M. *Life's Ratchet: How Molecular Machines Extract Order from Chaos*, Basic Books, Philadelphia, 2012

Hoffmeyer, Jesper. Quoted in "What is Biosemiotics," *http://www.biosemiotics.org/biosemiotics-introduction/*

Hofstadter, Douglas, R. *Godel, Escher, Bach: An Eternal Golden Braid*, Basic Books, New York,1999 edition

Horkey, Ferenc & Basser, Peter & Hecht, Anne-Marie & Geissler, Erik. "Effect of calcium/sodium io ex change on the osmotic properties and structure of polyelectrolyte gels," *Journal of Engineering in Medicine*, July 20, 2015

Horowitz, Seth S. "The Science and Art of Listening," *New York Times*, November 9, 2012

Howard, Emily. "Sea Lion Is First Non-human Mammal Shown to Keep a Beat," *EarthSky.org*, April 5, 2013

Howard, Jaqueline, "Cockatoos 'Pick Locks' For Food In Bird Intelligence Study," *Huffington Post,* July 26, 2013

Howard, Jaqueline. "Here's How Birds Make Do Without External Ears," *Huffington Post*, December 14, 2014

Huber, Alisha. "In 125 Years, Millions of People Have Looked at this Painting. No One Really Saw it until Recently," *Upworthy*, November 14, 2014, *http://www.upworthy.com/in-125-years-millions-of- people-have-looked-at-this-painting-no-one-really-saw-it-until-recently*

Huffington, Arianna with Patrick Fuller, "How Neuroscience Can Unlock Sleep's Mysteries," *Huffington Post*, March 26, 2015

Humphrey, Nicholas. *A History of the Mind: Evolution and the Birth of Consciousness*, Springer-Verlag, New York, 1999

Humphrey, Nicholas. *Seeing Red: A Study in Consciousness*, Harvard University Press, 2006

http://en.wikipedia.org/wiki/Autopoiesis

http://en.wikipedia.org/wiki/Catalysis

https://en.wikipedia.org/wiki/Derivative

https://en.wikipedia.org/wiki/Entropy_(arrow_of_time)#cite_note-5
https://en.wikipedia.org/wiki/Gamma_wave#cite_note-pollack-4
http://en.wikipedia.org/wiki/Gravitational_biology
https://en.wikipe https://en.wikipedia.org/wiki/Laws_of_motion dia.org/wiki/Orders_of_magnitude_(time)
https://en.wikipedia.org/wiki/Photon
https://en.wikipedia.org/wiki/Power_(physics)
https://en.wikipedia.org/wiki/Quantity
https://en.wikipedia.org/wiki/Relativity_of_simultaneity
https://en.wikipedia.org/wiki/Second_law_of_thermodynamics
https://en.wikipedia.org/wiki/Self-organization
https://en.wikipedia.org/wiki/Thermodynamics
https://stackoverflow.com/questions/19623479/recursion-switch-return-string
https://worldbuilding.stackexchange.com/questions/48528/super-intelligent-fungus
http://www.answers.com/Q/What_makes_up_a_cell_wall?#slide=1

Jabr, Ferris. "An Orangutan Learns to Fish," *The NewYorker*, September 17, 2014

Jabr, Ferris. "Brainless Slime Molds Redefine Intelligence," Scientific American, November 7, 2012

Jaques, Elliott. *The Life and Behavior of Living Organisms: A General Theory*, Praeger Publishers, Westport, CT, 2002

Jasper, Marykate. "Artificial Intelligence Invents New Language Humans Can't Understand," *Themarysue.com*, July 29, 2017

Jenson, Derrick. "Nature's Internet: The Vast, Intelligent Network Beneath Our Feet," *Peakenergy.blogspot,com*, July 21, 2008

Jonas, Hans. *The Phenomenon of Life: Toward a Philosophical Biology*, Dell Publishing, New York, 1966

Kaku, Michio. *The Future of the Mind: The Scientific Quest to Understand, Enhance and Empower the Mind*, Anchor Books, New York, 2014

Kandel, Eric R. "The New Science of Mind," New York Times, September 6, 2013

Khan. "Light and Photosynthetic Pigment," *Khan Academy, https://www.khanacademy.org/science/biology/photosynthesis-in- plants/the-light-dependent-reactions-of-photosynthesis/a/light-and-photosynthetic-pigments*

Kluger, Jeffrey. *The Animal Mind*, Time Special Edition, 2017

Klussmann, Martin and Iwamura, Hiroshi, et.al., "Thermodynamic Control of Asymmetric Amplification in Amino Acid Catalysis," *Nature*, June 1, 2006

Kuhn, Robert Lawrence. "Is Our Universe a Fake?" *Yahoo.com*, August 3, 2015

Kull, Kalevi. "Biosemiotics in the twentieth century: A view from biology," *http://www.zbi.ee/~kalevi/semi.1999.127.385.pdf*

Kull, Kalevi. "Organism as a self-reading text: anticipation and semiosis, " *Journal of Computing Anticipatory Systems,* vol. 1. 1998

Kullander, Sven and Larsson, Borje, *Out of Sight!: From Quarks to Living Cells*, Cambridge University Press, 1994

Lambert, Frank L. "Entropy and the Second Law of Thermodynamics," *Occidental College*, Los Angeles, September, 2012, *http://secondlaw.oxy.edu/*

Lane, Nick. *Life Ascending: The Ten Great Inventions of Evolution*, Norton, 2010

Lange, Marc. *An Introduction to the Philosophy of Physics: Locality, Fields, Energy, and Mass*, Blackwell Publishing, 2002

Lanza, Robert and Berman, Bob. *Biocentrism: How Life and Consciousness Are the Keys to Understanding the True Nature of the Universe*, BenBella Books, Dallas, 2010

Larson, Stephen. *The Healing Power of Neurofeedback: The Revolutionary LENS Technique for Restoring Optimal Brain Function*, Healing Arts, 2006

Lawton, Graham. "You Are Your Brain," *Slate.com*, December 8, 2013

Lee, A.C., "Untimely Metaphysics," *New York Times*, May 7, 2013

Lee, Joel. "6 Human Jobs That Computers Will Never Replace," *Makeuseof.com*, July 21, 2014

Lenda. Paul. "Plant Consciousness," *https://www.shift.is/plant-consciousness/*

Lent, Jeremy. "The Dangerous Delusions of Richard Dawkins," *Salon,* August 7, 2017

Leuven, KU. "'Brain Waves' Challenge Area-Specific View of Brain Activity," *ScienceDaily.com*, March 20, 2013

Lewis, Sarah. "Of Two Minds, 'Powers of Two' by Joshua Wolf Shenk," *New York Times Book Review*, August 22, 2014

Lewis, Tanya. "Ants Use 'Math' To Find Fastest Route, Study Shows," *Huffington Post*, April 17, 2013

Lewis, Tanya. "New Record for Human Brain: Fastest Time to See an Image," *LiveScience.com*, January 17, 2014

Lineweaver, Charles H., Davies, Paul C. W., Ruse, Michael. eds., *Complexity and the Arrow of Time*, Cambridge University Press, New York, 2013

Lipman, Joanne. "Is Music the Key to Success?" *New York Times*, October, 12, 2013

Lipton, Bruce H. *The Biology of Belief: Unleashing the Power of Consciousness, Matter & Miracles*, Hay House, 2008

Lipton, Bruce. "The New Biology: Where Mind And Matter Meet," *MutualResponsibility.org*

Lloyd, Rebecca. "Decartes on Time," *Purdue University, http://docs.lib.purdue.edu/dissertations/AAI3403121/*

Loewenstein, Werner. *The Touchstone of Life: Molecular Information, Cell Communication and the Foundation of Life*, Oxford University Press, New York, 1998

Loewenstein, Werner. *Physics in Mind: A Quantum View of the Brain*, Basic Books, New York, 2013

Love, Dylan. "Are Mushrooms Intellligent? *Backtotheroots.com*, October 13, 2016

Lumen. "Characteristics of Fungi," *Courses.lumenlearning.com/boundless- biolo gy/chapter/characteristics-of- fungi/*

Lundeen, Jeff. "Viewpoint: What Can We Say About a Photon's Past?" *APS Physics*, December 9, 2013. *https://physics.aps.org/articles/v6/133*

Macdonald, Helen. "A Bestiary of the Mind," *New York Times Magazine*, May 21,2017

Makari, George. "Psychiatry's Mind-Brain Problem," New York Times, November 11, 2015

Marcus, Gary. *The Birth of Mind: How a Tiny Number of Genes Creates The Complexities of Human Thought*, Basic Books, New York, 2004

Marder, Michael. "Plant Intelligence and Attention," *Plant Signaling & Behavior*, February 20, 2013

Markkoff, John. "Brainlike Computers, Learning From Experience," *New York Times*, December 28. 2013

Margulis, Lynn and Asikaninen, Celeste A. and Krumbein, Wolfgang E., eds., *Chimeras and Consciousness: Evolution of the Sensory Self*, MIT Press, 2011

Margulis, Lynn and Sagan, Dorian, *What Is Life?* University of California Press, Berkeley, CA., 1995

Matson, John. "Quantum Entanglement Links 2 Diamonds," *Scientific American*, December 1, 2011

Matsuzawa, Tetsuro. Interviewed in "Chimpazees Beat Humans In Memory," *Huffington Post*, September 30, 2013

Maturana, Humberto R, Varela, Francisco J. *The Tree of Knowledge: The Biological Roots of Human Understanding*, Shambhala Publications, Boston, 1992

Maudlin, Tim. *Quantum Non-Locality and Relativity: Metaphysical Intimations of Modern Physics*, Wiley-Blackwell, 2011

Mazza, Ed. "Chimps Can Learn to Communicate with Each Other, New Study Shows," *Huffington Post*, February 10, 2015

McWhorter, John. "Why Save a Language?" *New York Times*, December 5, 2014

Merali, Zeeya. "Solving Biology's Mysteries Using Quantum Mechanics," *Discover Magazine*, October 30, 2014

Metzinger, Thomas. *The Ego Tunnel: The Science of the Mind and the Myth of the Self*, Basic Books, New York, 2009

Mikulecky, Donald C., "Robert Rosen: The Well Posed Question and Its Answer - Why Are Organisms Different from Machines?" *Virginia Commonwealth University, http://www.people.vcu.edu/~mikuleck/PPRISS3.html*

Miles, Kathleen. "Ray Kurzweil: In The 2030s, Nanobots In Our Brains Will Make Us 'Godlike,'" *Huffington Post*, October 1, 2015

Minkel, J.R. "Focus: Ripples on a Cell," *Physics Review*, November, 13, 2009, https://physics.aps.org/story/v24/st19

Mitsch, Joe. "Pitt Researcher Lands the Cover of *Development Cell* by Uncovering an Evolutionary Secret," *University of Pittsburgh News Service*, Septermber 3, 2015

Mohrhoff, Ulrich. "Why the Laws of Physics Are Just So," *This Quantum World*, http://thisquantumworld.com/wp/a-critique- of-quantum-mechanics/why-the-laws-of-physics-are-just-so/

Montgomery, Sy. "Deep Intellect," *Orion Magazine, https://orionmagazine.org/article/deep-intellect/*

Morch, Hedda. "Is Matter Conscious?" *Nautilus*, April 6, 2017

Nahmias, Eddy. "Is Neuroscience the Death of Free Will?" *New York Times*, November 13, 2011

Nagel, Thomas. *Mind & Cosmos: Why the Materialist Neo-Darwinian Conception of Nature is Almost Certainly False*, Oxford University Press, Oxford, 2012

Nelson, Kevin. *The Spiritual Doorway in the Brain: A Neurologist's Search for the God Experience*, Plume, 2012

Newman, Didier. *Bioquestions and the Mechanical Answer*, CreateSpace Independent Publishing Platform, 2013

Noe, Alva. "Art and the Limits of Neuroscience," *New York Times*, December 4, 2011

Noe, Alva. *Out of Our Heads: Why You Are Not Your Brain, and Other Lessons from the Biology of Consciousness*, Hill and Wang, New York, 2009

Noble, Denis. "A Theory of Biological Relativity: No Privileged Level of Causation," *The Royal Society Publishing*, November 9, 2011, *https://www.ncbi.nlm.nih.gov/pmc/articles/PMC3262309/*

Noble, Denis. "Claude Bernard, the First Systems Biologist, and the Future of Physiology," *The Physio logical Society*, December 18, 2007, *http://onlinelibrary.wiley.com/doi/10.1113/expphysiol.2007.038695/pdf*

Nottale, Laurent. *Fractal Space-Time and Microphysics: Towards a Theory of Scale Relativity*, World Scientific Publishing, London, 1993

Nourbakhsh, Illah. "Complexity, Intelligence and How to Ignore the Meaning of Life," *Huffington Post*, June 14, 2013

Ooguri, Hirosi. "How Spacetime Is Built by Quantum Entanglement: New Insight into Unification of General Relativity and Quantum Mechanics," *Kavli Institute for the Physics and Mathematics of the Universe*, May 27, 2015, *http://www.ipmu.jp/node/2174*

Orange County. "Introduction to Learning Disabilities," *Orange County Learning Disabilities Association, http://www.oclda.org/understanding.html*

Ouellette, Jennifer. "How Quantum Pairs Stitch Space-Time," *Quantamagazine.org*, April 28, 2015

Overbye, Dennis. "A Quantum of Solace," New York Times, July 1, 2013

Overbye, Dennis. "Stars, Gold, Dung Beetles and Us," *New York Times*, August 6, 2013

Papageorgiou, Nik. "How the Brain Produces Consciousness in 'Time Slices,'" *News Mediacom, https://actu.epfl.ch/news/how- the-brain-produces-consciousness-in-time-slice/*

Parens, Erik. "The Benefits of 'Binocularity,'" *New York Times*, September 28, 2014

Pericles, Peter. Editor, *International Handbook of Semiotics*, Springer, 2015, p.1095

Perkowitz, Sidney. "Light Dawns," *https://aeon.co/essays/why-is-the-speed-of-light-the-speed-of-light*, 2015

Pico, Richard M. *Consciousness in Four Dimensions: Biological Relativity and the Origins of Thought*, McGraw- Hill, 2001

Pierce, Stephanie E. "Non-Equilibrium Thermodynamics: An Alternate Evolutionary Hypothesis," *Semantic Scholar*, 2002, *https://www.semanticscholar.org*

Pitney, Nico. "Scientists Find Alarming Deterioration in DNA of the Urban Poor," *Huffington Post*, May 8, 2015

Pollan, Michael. "Some of My Best Friends Are Germs," *New York Times*, May 15, 2013

Pollan, Michael. "The Intelligent Plant," *The New Yorker*, December 23, 2013

Price, Huw. "Cambridge, Cabs and Copehagen: My Route to Existential Risk," *New York Times*, January 27, 2013

Pross, Addy. *What is Life? How Chemistry Becomes Biology*, Oxford University Press, Oxford, UK, 2012

Quanta. "Interactive: What Is Space?" *https://www.quantamagazine.org/interactive-what-is-space-20150430*

Quenqua. Douglas. "Circadian Clocks May Be Tuned to Vibrations, Too," *New York Times*, February 4, 2014

Ramachandran, V.S. *A Brief Tour of Human Consciousness: From Imposter Poodles to Purple Numbers*, Plume, 2005

Ramachandran, V.S. *The Tell-Tale Brain: A Neuroscientist's Quest for What Makes Us Human*, W.W. Norton, 2012

Ray, C. Claiborne. "Plants in Peril," *New York Times*, December 23, 2013

Reas, Emilie. "Your Brain Has 2 Clocks," *Scientific American*, November 26, 2013

Rebeiz, Mark. Cited in "Pitt Researcher Lands the Cover of *Development Cell*, by Uncovering an Evolutionary Secret, "*University of Pittsburgh News Services*, September 3, 2015

Rich, Nathanial. "Can a Jellyfish Unlock the Secret of Immortality?" *New York Times*, November 28, 2012

Richter, Viviane. "What Does an Electron Cloud Really Look Like?" *Cosmos Magazine*, September 21, 2015 *https://cosmosmagazine.com/technology/what-does-electron-cloud-really-look*

Regalado, Antonio, "What Will It Take for Computers to Be Conscious," *MIT Technology Review*, October 2, 2014

Renstrom, Joelle. "If I Only Had a Brain: How AI 'Thinks,'" *Daily Beast*, February 19, 2017

Ropeik, David. "Inside the Mind of Worry," *New York Times*, September 28, 2012

Rosen, Jill. "What Free Will Looks Like in the Brain," *Johns Hopkins University*, July 13, 2016, *http://releases.jhu.edu/2016/07/13/what-free-will-looks-like-in-the-brain/*

Rosen, Robert. *Life Itself: A Comprehensive Inquiry into the Nature, Origin, and Fabrication of Life*, Columbia University Press, New York, 1991

Rosenberg, Alex, McShea, Daniel W., *Philosophy of Biology: A Contemporary Introduction*, Routledge, New York, 2008

Rosenblum, Bruce and Kuttner, Fred, *Quantum Enigma: Physics Encounters Consciousness*, Oxford University Press, New York, 2011

Rosenkranz, Robert. "Artificial Intelligence: Stop or Go?" *Huffington Post*, February 1, 2016

Rothschild, lynn, Lister, Adrian, eds. *Evolution on Planet Earth: The Impact of the Physical Environment*," Academic Press, London, 2003

Russell, Peter. *From Science to God: The Mystery of Consciousness and the Meaning of Light*, New World Library, Novato, CA., 2002

Salmon, Wesley C. *Space, Time and Motion, A Philosophical Introduction*, Chapter Two, "A Contemporary Look at Zeno's Paradoxes, University of Minnesota Press, Minneapolis, *https://math.dartmouth.edu/~matc/Readers/HowManyAngels/SpaceTimeMotion/STM.html*

Sargent, Jennifer. "Why fish don't need to be schooled in the art of swimming," *NBCNews.com*, September 17, 2013

Satel, Sally and Lilienfeld, Scott. *Brainwashed: The Seductive Appeal of Mindless Neuroscience*, Basic Books, 2013

Sayer, Ji. "Are Fungi the Earth's Natural Internet?" *GreenMedInfo.com*, February 20, 2012

Scheffler, Samuel. "The Importance of the Afterlife. Seriously," *New York Times*, September 21, 2013

Scialdone, Antonio, Martin, Howard. "How plants manage food reserves at night: Quantitative models and open questions," *Frontiers in Plant Science*, March 31, 2015, *https://www.ncbi.nlm.nih.gov/pmc/articles/PMC4379750/*

Schneider, Eric D. and Sagan, Dorian, *Into the Cool: Energy Flow, Thermodynamics, and Life*, University of Chicago Press, 2006

Schrodinger, Erwin. *What Is Life? & Other Scientific Essays*, Doubleday Anchor, 1956

Scott, Alwyn. *Stairway to the Mind: The Controversial New Science of Consciousness*, Springer-Verlag, New York, 1995

Seife, Charles. *Decoding the Universe: How the New Science of Information is Explaining Everything in the Cosmos, from Our Brains to Black Holes*, Penguin Books, 2006

Service, Robert F., "Origin of Life in RNA? Study Suggests Genetic Molecules May Have Self-Assembled," *Huffington Post*, February 12, 2013

Sheldrake, Robert. *Morphic Resonance, The Nature of Formative Causation*, Park Street Press, Rochester, Vermont, 2009

Shapiro, James A. *Evolution: A View from the 21st Century*, FT Press Science, Upper Saddle River, NJ, 2011

Shapiro, James, A. "What is the Key to a Realistic Theory of Evolution?" *Huffington Post*, February, 16, 2012

Shapiro, James, A. "Why the 'Gene' Concept Holds Back Evolutionary Thinking," *Huffington Post*, November 30. 2012

Shenk, Joshua Wolf. "The End of 'Genius,'" *New York Times*, July 19, 2014

Siegel, Ethan. "All About Cosmic Inflation," https://medium.com/starts-with-a-bang/all-about-cosmic-inflation-211a64441ddd

Silva, Jason. "Darwin's Pharmacy: Sex, Plants and the Evolution of the Noosphere," *Huffington Post*, June 29, 2011

Simon, Craig L. "Deriving Common Interests from Animal Origins: The Generative Constraints of Global Polity," *http://docshare04.docshare.tips/files/4467/44670838.pdf*, September, 2008

Slobodchikoff, Con. *Chasing Doctor Dolittle: Learning the Language of Animals*, St. Martin's Press, New York, 2012

Smolen, Lee. *Time Reborn: From the Crisis in Physics to the Future of the Universe* Mariner Books, 2014

Stansbury, Jillian. "Plant Intelligence - A Philosophical Discourse on Cognitive Awareness in Plants," *Battlegroundhealingarts.com,* May 4, 2016

Stamets, Paul. "Fungal Intelligence and the 21st Psychedelic Journey - How Mushrooms Can Help Save the World," *Podcasts, personallifemedia.com*

Stamets, Paul. *Mycelium Running: How Mushrooms Can Help Save the World,* Ten Speed Press, First Edition, 2005

Stanley, Brian Jay. "I Am Not This Body," *New York Times*, May 6, 2013

Stapp, Henry P. *Mindful Universe: Quantum Mechanics and the Participating Observer* (The Frontiers Collection) 2nd ed., Springer, Heidelberg, 2011

Stark, Philip B. SticiGui, Chapter 13, "Probability: Philosophy and Mathematical Background" *https://www.stat.berkeley.edu/~stark/SticiGui/Text/probabilityPhilosophy.htm*

Sternberg, Eliezer. "When People With Schizophrenia Hear Vocies, They're Really Hearing Their Own Subvocal Speech," *Slate,* March 2, 2016

Stokes, Trevor. "Clownfish 'Talk' To Assert & Reinforce Social Status, Scientists Say," *Huffington Post*, November 7, 2012

Strevens, Michael. "Looking into the Black Box," *New York Times*, November 24, 2013

Strogatz. Steven H. *Sync: How Order Emerges From Chaos In the Universe, Nature, and Daily Life*, Hyperion, 2004

Stuart, Greg and Spruston, Nelson. "Dendritic Integration: 60 Years of Progress," *Nature Neuroscience*, November 25, 2015

Subbaraman, Nidhi. "It's not dead! How some worms can regrow whole heads," NBCnews.com, July 24, 2013

Susskind, Leonard, Hrabovsky, George. *The Theoretical Minimum: What You Need to Know to Start Doing Physics*, Basic Books, 2014

Suri, Manil. "How to Fall in Love with Math," *New York Times*, Sept. 15, 2013

Talbot, Michael. *The Holographic Universe: The Revolutionary Theory of Reality*, HarperCollins, New York, 2011

Tankersley, Jim. "Science Fiction Come True: Moving a Paralyzed Hand with the Power of Thought," *Washington Post*, June 24, 2014

Taylor, Katherine. "A New Physics Theory of Life, *Quanta Magazine*, January 22, 2014

Tegmark, Max. "Life Is a Braid in Spacetime," *Nautilus*, October 1, 2015

Tegmark, Max. *Our Mathematical Universe: My Quest for the Ultimate Nature of Reality*, Vintage Books, 2015

Tegmark, Max."Why a Larger Multiverse Shouldn't Make You Feel Small," *Huffington Post*, January 7, 2014

The Astronomist."The Universe and Life is Asymmetric: Chirality," *fieldofscience.com*, January 21, 2011, *http://theastronomist.fieldofscience.com/2011/01/universe-and-life-is-asymmetric.html*

The Physics arXiv Blog. "How Quantum Mechanics Forbids the Accelerating Expansion of Spacetime," January 2, 2014 *https://medium.com/the-physics-arxiv-blog/how-quantum-mechanics-forbids-the-accelerating- expansion-of-spacetime- cbcd754531dd*

Thompson, Evan. *Mind in Life: Biology, Phenomenology and the Sciences of Mind*, Harvard University Press, Cambridge, Ma., 2007

Tolle, Eckhart. *A New Earth: Awakening to Your Life's Purpose*, Penguin,2008

Tompkins, Peter & Bird, Christopher. *The Secret Life of Plants*, Harper & Row, 1989

Trewavas, Anthony. "Aspects of Plant Intelligence," *Annals of Botany*, July, 2003, *https://www.ncbi.nlm.nih.gov/pmc/articles/PMC4243628/*

Trewavas, Anthony. "Intelligence, Cognition, and Language of Green Plants," *Frontiers in Psychology*, April 26, 2016, *https://www.ncbi.nlm.nih.gov/pmc/articles/PMC4845027/*

Tse, Peter Ulric. *The Neural Basis of Free Will: Criterial Causation*, MIT Press, 2013

Tyson, Neil deGrasse. "What Science Is - and How and Why It Works," *Huffington Post*, November, 18, 2015

Unger, Roberto Mangabeira and Smolin, Lee. *The Singular Universe and the Reality of Time: A Proposal in Natural Philosophy*, University Printing House, Cambridge, 2015

University of Chicago, "Epigenetics: 100 Reasons to Change the Way We Think about Genetics," *ScienceDaily.com,* May 20,2009

University of Oregon, *http://abyss.uoregon.edu/~js/21st_century_science/lectures/lec14.html*

Van Brakel, Jaap. *Philosophy of Chemistry: Between the Manifest and the Scientific Image*, Leuven Uni versity Press, Leuven, Belgium, 2000

Van Mater, John. "Our Intelligent Companions, the Plants," *http://www.theosophy-nw.org/theosnw/brother/br- jvmj2.htm*

Vedral, Vlatco. "Living in a Quantum World," *National Taiwan Normal University*, November 6, 2003, *http://phy.ntnu.edu.tw/~chchang/Notes10b/0611038.pdf*

Vince, Gaia. "Hacking the Nervous System," *Huffington Post*, May 29, 2015

Virga, Vint. *The Soul of All Living Creatures: What Animals Can Teach Us About Being Human*, Broadway Books, 2014

Von Medding, Jason. "Agent Orange, exposed: How U.S, chemical warfare in Vietnam unleashed a slow-moving disaster," Salon, October 9, 2017

Wade, Nicholas. "In Tiny Worm, Unlocking Secrets of the Brain," *New York Times*, June 20, 2011

Watts, Susan. "Assessing the Impact of Venter's 'Synthetic Life,'" *BBC News*, May 20, 2010

Welsh, Jennifer. "Ants Vaccinate One Another To Prevent Epidemics, Colony Study Suggests," *Huffington Post*, April 3, 2012

Welsh, Jennifer."Like Humans, Dolphins Have Brainy Genes," *NBCNews.com*, June 26, 2012

Willyard, Cassandra."Bonnie Bassler, Silent Treatment," *Discover Magazine*, June 2014

Wilson, Edward O. "The Riddle of the Human Species," *New York Times*, February 24, 2013

Wilson, Mark. "AI Is Inventing Languages Humans Can't Understand. Should We Stop It?" *Fastcodesign.com*, July 14, 2017

Witzany, Gunther. "Plant Communication from Biosemiotic Perspective," *Plant Signaling & Behavior, https://www.ncbi.nlm.nih.gov/pmc/articles/PMC2634023/*

Wohlleben, Peter. *The Hidden Life of Trees*, Greystone Books, 2016

Wolchover, Natalie. "New Quantum Theory Could Explain the Flow of Time," *Wired.com*, April, 25, 2014

Worrall, Simon. "There Is Such a Thing as Plant Intelligence," *National Geographic*, February 21, 2016

Yanofsky, Noson S. *The Outer Limits of Reason: What Science, Mathematics, and Logic Cannot Tell Us*, The MIT Press, Cambridge, Massachusetts, 2013

Yirka, Bob."Researchers Image Most of Vertebrae Brain at Single Cell Level," *https://medicalxpress.com/news/2013-03-image-vertebrae-brain-cell-video.html*, March 20, 2013

Zatorre, Robert and Salimpoor, Valorie. "Why Music Makes Our Brain Sing," *New York Times*, June 7, 2013

Zimmer, Carl. "DNA Double Take," *New York Times*, September 16, 2013

Zimmer, Carl. "Fast-Paced Evolution in the Andes," *New York Times*, November 7, 2013

Zimmer, Carl. "In the Human Brain, Size Really Isn't Everything," *New York Times*, December 26, 2013

Zimmer, Carl. *Microcosm: E. Coli and the New Science of Life*, First Vintage Books, New York, 2009

Zimmer, Carl. "Phantom Melodies Yield Real Clues to Brain's Workings," *New York Times*, February 13, 2014

Zimmer, Carl. "Seeking the Gears of Our Inner Clock," *New York Times*, December 28, 2015

Zimmer, Carl. "Vitamins' Old, Old Edge," *New York Times*, December 9, 2013

Zolli, Andrew. "Seeing the Whole: A Reflection for Earth Day," *www.garrisoninstitute.org*, April 22, 2016

## IMAGES

1. Cartesian Coordinates. (https://www.khanacademy.org/math/algebra/linear-equations-and-inequalitie/more-analytic-geometry/v/distance-formula)
2. Wave Properties.
3. Photo Image of Pentacine and diagram.
http://intergalacticrobot.blogspot.com/2009/08/
https://commons.wikimedia.org/wiki/File:Perfluoropentacene.svg
4. "Fractal Tree No. 2", Robert Fathauer © 2007, www.robertfathauer.com
4a. Self-Similar Trajectory. https://commons.wikimedia.org/wiki/File:A_Trajectory_Through_Phase_Space_in_a_Lorenz_Attractor.gif
5. Universe.
6. Bio-molecule fit.
7. Carbon Molecular Structure.
8. Glucose
9. Water.
10. Protein Folding.. https://commons.wikimedia.org/wiki/File:Protein_folding_schematic.png
11. Lock and Key.
12. Catalytic Loop.
13. Peptide Synthesis. https://commons.wikimedia.org/wiki/File:Peptide_syn.png
14. Cell Wall
15. Proteins in Cell Membrane. https://en.wikipedia.org/wiki/Artificial_cell (usage rights: free to use, share or modify, even commercially)
16. Biochemical Flow Chart.
17. Simple Metabolism Flow Chart.
18. Hemoglobin and Chlorophyll.
19. Subroutine Flow Chart.
20. Self-Reference.
21. Recursion.
22. Self-Referential Awareness.
23. Infinite Regress.
24. Recursive Awareness.
25. In Time.
26, Recursive Division.
27. Geometric Proportion.
28. Golden Rectangle.. https://en.wikipedia.org/wiki/File:SimilarGoldenRectangles.png
29. Golden Triangle.
30. Golden Ratio. https://commons.wikimedia.org/wiki/File:Fibonacci_spiral_34.svg
31. Living Ratio
32. Fibonacci Spiral.
33. Non-Fibonacci Spiral
34. Parthenon. https://www.flickr.com/photos/fruey/1372547275 - cropped, b/w, and superimposed golden ratio. Usage rights: Free to use, share or modify commercially
35. Observation Circle.
36. Limit of Measurement and Awareness.
37. Component Flow Chart.
38. Whole Organism Flow Chart.
39. Circle of Living Awareness.
40. Limit of Measurement and Awareness.
41. Form and Matter Intersection.
42. Neuron. https://commons.wikimedia.org/wiki/File:Neuron.svg
43. Synapse. https://commons.wikimedia.org/wiki/File:Chem._Synapse_scheme.jpg
44. Nuclei. https://commons.wikimedia.org/wiki/File:1306_Lobes_of_Cerebral_Cortex.jpg
45. Calculating Synchrony
46. Phagocytosis. https://en.wikipedia.org/wiki/Amoeba. Google usage rights: Free to use or share com mercially
47. Root and Shoot. http://maxpixel.freegreatpicture.com/Nature-Plant-Root-Ecology-Tree-Branches-Eco-309046

48. Root Tip. https://media1.britannica.com/eb-media/42/62942-004-312C3FCA.gif
49. Modular Growth is Fractal. https://pixabay.com/en/tree-blue-sky-branch-branches-164915/ "Fractal Tree No. 2", Robert Fathauer © 2007
50. Modular Plant Growth.
51. Animal vs. Plant Flow Chart.
52. Fungi Anatomy. https://commons.wikimedia.org/wiki/File:Basidie-2.jpg
53. Mycelium.
54. Mycelium Flow Chart.
55. Magnification.
56. Can't See Edges.
57. Can See Edge.
58. Catching the Present. https://pixabay.com/en/optical-illusion-black-pattern

CPSIA information can be obtained
at www.ICGtesting.com
Printed in the USA
FSHW04n1150170418
46968FS

9 780692 081440